The near-Earth and interplanetary plasma

Volume 1

General properties and fundamental theory

The near-Earth and interplanetary plasma

Volume 1
General properties and fundamental theory

YA. L. AL'PERT

Formerly Deputy Director, Institute of Terrestrial Magnetism, Ionosphere and Propagation of Radio Waves (IZMIRAN) of the Academy of Sciences of the USSR

CAMBRIDGE UNIVERSITY PRESS

CAMBRIDGE

LONDON NEW YORK NEW ROCHELLE

MELBOURNE SYDNEY

Published by the Press Syndicate of the University of Cambridge
The Pitt Building, Trumpington Street, Cambridge CB2 1RP
32 East 57th Street, New York, NY 10022, USA
296 Beaconsfield Parade, Middle Park, Melbourne 3206, Australia

First published 1983

Printed in Great Britain at the University Press, Cambridge

Library of Congress catalogue card number: 82-12879

British Library Cataloguing in Publication Data

Al'pert, Ya L.
The near-Earth and interplanetary plasma.
Vol. 1: General properties and fundamental theory
1. Plasma (Ionised gases) 2. Solar system
I. Title
523.2 QL717.6
ISBN 0 521 24364 5

TM

Contents

Preface ix

Introduction 1

1 General remarks 5

2 General properties of the near-Earth and interplanetary plasma 10
2.1 Structure and properties of the plasma at different distances from the Earth 11
2.2 Distribution of charged-particle density inside and outside the plasmasphere 15
2.3 Structure of the magnetic field in the near-Earth plasma 24

3 Fundamental equations 28

4 Refractive indexes for a cold magnetoplasma 35
4.1 Permittivity-tensor elements for a collisional plasma 35
4.1.1 Tensor elements taking into account collisions between different kinds of charged particles 37
4.1.2 Effect of collisions between like particles (electrons, or ions of one kind), in the ELF range $0 \leqslant \omega \simeq \Omega_{Hs}$ 42
4.2 Formulas for $\tilde{n}_{12}^2$ in the presence of ions of one kind over the entire frequency range $0 \leqslant \omega \rightarrow \infty$ 47
4.3 Properties of the different branches of waves 53
4.3.1 Polarization coefficients 53
4.3.2 Refractive indexes. Resonances. Classification of the waves 56
4.4 Refractive indexes for a multicomponent plasma 65
4.4.1 Properties of n_{12}^2 in a collisionless plasma 65
4.4.2 Relationships for n_{12} and κ_{12} in a collisional plasma 70

5 Refractive indexes and attenuation factors in a warm plasma 82
5.1 Tensor elements and values of n_{12}^2 allowing for the thermal motion of the particles ($T_e \neq T_i \neq 0$) 83

5.2 High-frequency, Langmuir, and low-frequency waves ($\omega \gtrsim \omega_0, \omega_H; \omega_L \ll \omega \lesssim \omega_H$) 84
5.2.1 Refractive-index formulas 85
5.2.2 'Kinetic' attenuation of LF and HF waves 89
5.3 Very-low-frequency (VLF) and extremely-low-frequency (ELF) waves ($0 \leqslant \omega \ll \omega_H$) 93
5.4 Electron-acoustic waves 96
5.5 Ion-acoustic waves 98
5.6 Magnetoacoustic waves 101
5.7 Electron and ion cyclotron waves 102

6 Growth rates for the different oscillation branches 106
6.1 Generation of high-frequency (HF) waves ($\omega \gg \Omega_0$) in an isotropic plasma 107
6.1.1 Beam instability 107
6.1.2 'Anisotropic' instability 110
6.2 Excitation of HF and LF waves ($\omega \gg \omega_L$) in a magnetoplasma 110
6.2.1 Beam instability of longitudinal waves 111
6.2.2 Beam instability of transverse electromagnetic waves 115
6.2.3 Cyclotron instability for an anisotropic distribution of electron velocities 116
6.3 Excitation of the very-low-frequency (VLF) and extremely-low-frequency (ELF) oscillation branches ($0 < \omega \ll \omega_H$) 118
6.3.1 Beam instability of ion-acoustic waves ($\omega \ll \omega_0$) in an isotropic plasma 118
6.3.2 Instability of ion-acoustic waves caused by motion of the electrons relative to the ions in an isotropic plasma 119
6.3.3 Beam instability of longitudinal waves in a magnetoplasma 120
6.3.4 Beam instability of transverse electromagnetic waves in a magnetoplasma 123
6.3.5 Cyclotron instability for an anisotropic distribution of ion velocities 126
6.4 Some remarks about growth rates 127

7 Nonlinear effects in a plasma 130
7.1 General remarks 130
7.2 Nonlinear phenomena of the 'heating' type in a collisional plasma 136
7.2.1 Electron temperature and collision frequency 138
7.2.2 Large-scale changes in electron density 153
7.2.3 Cloud-like inhomogeneous structure of the perturbed plasma region. Field-aligned structure 160
7.2.4 Self-action and focusing of waves and other effects of changes in their field structure 184

7.3 Parametric nonlinear phenomena 192
7.3.1 Nonlinear spectra of the resonance branches of longitudinal waves 194
7.3.2 Parametric excitation of longitudinal waves 198
7.3.3 Parametric excitation of transverse waves 202
7.4 Nonlinear waves in a plasma. Changes in plasma properties in a nonuniform RF field 204
7.4.1 Nonlinear wave processes 207
7.4.2 Pressure on the electrons in a nonuniform RF electric field 213

8 Group velocity, trajectories, and trapping of electromagnetic waves in a magnetoplasma 216
8.1 Group velocity **u** in a homogeneous magnetoplasma 218
8.1.1 Angle α between the group-velocity vector **u** and the magnetic field $\mathbf{H}_0$ 221
8.1.2 Langmuir, ion-acoustic, and magnetoacoustic waves 236
8.1.3 Magnetoacoustic transverse waves in a nonisothermal plasma 237
8.2 Trajectories of electromagnetic waves in a smoothly varying magnetoplasma 238
8.2.1 Trajectories guided by the magnetic field $\mathbf{H}_0$ 239
8.2.2 Unguided ray trajectories ($\omega < \omega_L$) 241
8.3 Trapping of electromagnetic waves by the magnetic field $\mathbf{H}_0$ 243
8.3.1 Turning of the Poynting vector of plane waves on entering a plane-stratified inhomogeneous medium 247
8.3.2 Trapping of plane electromagnetic waves in a plane-stratified layer of magnetoplasma 251
8.3.3 Distribution of the field of electromagnetic waves emitted by a point source in directions close to $\mathbf{H}_0$ 260

References 271
Author index 283
Subject index 289

Contents of Volume 2

Preface

PART 1. FLOW AROUND BODIES MOVING IN A PLASMA

9 Brief description of some theoretical problems and experiments

10 Some remarks about conditions at the boundaries of bodies moving in plasmas

11 Disturbances of the plasma near fast-moving bodies ($V_0 \gg v_i$)

12 Disturbances of the plasma near quasi-stationary bodies ($V_0 \lesssim v_i$)

13 Scattering of radio waves in the wake of a fast-moving body

14 Some remarks about the excitation of waves and plasma instability near a fast-moving body

PART 2. WAVES AND OSCILLATIONS IN THE NEAR-EARTH AND INTERPLANETARY PLASMA

15 Introductory remarks

16 Results of studies of ELF waves

17 Results of studies of VLF waves

18 Results of studies of LF waves

19 Results of studies of HF waves ($\omega \gtrsim \omega_H$), ($\omega \gtrsim \omega_0$)

20 Energy densities of various types of waves

References

Author index

Subject index

Preface

The basic subject matter of this book first appeared in 1976 as part of the encyclopedia *Handbuch der Physik* (Volume XLIX/5, pp. 217–349, Editor: K. Rawer. Publisher: Springer-Verlag, Berlin). That original paper has been extensively revised and expanded, and this book now presents a much more detailed and comprehensive treatment of wave processes and of the motion of bodies through plasma. The limitations placed on the length of the paper did not then allow me to include a number of sections that I deemed important; this book is about four times longer than the original paper.

During the years which elapsed, a great many most interesting experimental results were obtained, results which are to a certain degree fundamental. They enrich and expand considerably the content of the branch of plasma physics in question. Moreover, several theoretical works which appeared during these years enabled a more profound interpretation to be made of the experimental findings.

This book is divided into two volumes. The first is concerned with the general properties of the near-Earth and interplanetary plasma, the second with the flow of plasma around moving bodies such as artificial satellites, and with natural plasma waves and oscillations.

The theoretical parts of the book (Volume 1) have been supplemented, in Chapter 5 and in later chapters, by a description of the plasma properties caused by spatial dispersion. The formulas of the kinetic theory of a plasma presented in these sections provide an orientation with regard to the topics under consideration. It is my hope that they will also be useful for analysis of the various experimental results.

The description of the properties of a plasma in the hydrodynamic approximation (Chapter 4) has also been expanded and supplemented by new results and formulas, in particular for a collisional and multicomponent plasma. Problems related to the guiding of waves in a magnetoplasma have been considered, as well (Chapter 8).

Chapter 7, which includes some results of the nonlinear theory of a plasma and the corresponding experimental data, is new. The findings of experimental studies of nonlinear effects presented in these sections are of great significance. These findings made it possible in part to verify some predictions of the theory, and also to discover a number of nonlinear plasma properties which were hitherto unknown in theory, and which were quite remarkable, perhaps even somewhat strange. This progress in the experimental study of the nonlinear properties of a plasma pertains to both the ionosphere and the magnetosphere. In several cases a good understanding of the observed effects has already been arrived at. However, on the whole their theoretical description is still far from complete.

During the above-mentioned period the results of a number of theoretical and experimental studies investigating the supersonic flow of a different form of plasma around a body also became known (Volume 2, Part 1). These findings were in part interpreted theoretically, as were some results of Earth-satellite experiments from previous years. With regard to this work, however, it should be noted that the development has been much too slow. This slowness may well have been associated with an underestimation of the heuristic nature of such studies, which should give us a deeper understanding of, and should reveal, a number of processes taking place in a plasma, the results of which should prove useful for plasma diagnostics. At the present level of development of experimental science using Earth satellites, it is quite feasible to carry out in space appropriate experiments which will be comprehensive and at the same time physically adequate. Naturally, the present lack of such results is explained partly by the difficulty of meeting the specific requirements placed on such experiments and on the methods of their theoretical interpretation. On the other hand, another important factor with respect to this is the fact that many experiments carried out on satellites and space rockets have not been sufficiently purposeful. It is my hope that we have nevertheless reached a turning point in this area of 'space' experiments.

During the last decade, experimental investigations of various wave processes in the near-Earth and interplanetary plasma *in situ* have developed at an increasing tempo, and many new data have been obtained. It has frequently been difficult to explain all these data from a single point of view, and thus to describe them. To some extent the very abundance of these data and their frequency of occurrence may complicate matters. Such difficulties were in fact experienced by me while writing the corresponding chapters of the book. Most of the findings of experiments like these have been described in Volume 2, Part 2. Just as in the paper of 1976, I have endeavoured to include those results which seem to me to be more permanent in nature, and which will thus retain their value in the future as well. The theory of wave phenomena has also developed quite comprehensively during these years, including the nonlinear description of a number of effects. However, the bulk of this experimental material still awaits a comprehensive theoretical explanation. At first it was proposed that the description of the experimental facts be accompanied by an analysis of the known results of their theoretical interpretation. However, a more thoroughgoing familiarity with a number of theoretical works revealed a need to reconsider these results and make them more precise. It was also found that the corresponding theoretical problems had often not been posed sufficiently rigorously and accurately, and in a manner suitable to the experimental conditions. In spite of the fact that great progress in the development of the theory describing many of the phenomena considered here has been achieved in recent years, the time is not yet ripe for a systematic, all-embracing theoretical interpretation of the totality of experimental data examined.

The many references cited in this book are an indication that some new, important results have inevitably been omitted or incorrectly understood. I hope that the reader will be indulgent with regard to any such defects of the book. This monograph includes many figures and diagrams, since I attach great significance to illustrative material in a work like this. Such data increase considerably the volume of information cited, and aid in its interpretation; this often helps in perceiving and understanding better the nature and essence of the experiments described. Naturally many illustrations have been borrowed from the original sources. Frequently they have been modified somewhat and put into a form more suitable for inclusion.

This book was written under conditions of isolation from my normal scientific surroundings, during years of waiting in agonizing, indefinite suspense for permission to emigrate to Israel. The comments of certain of my colleagues would undoubtedly have improved the book. Constant, invaluable aid in preparing and designing the manuscript of the book was rendered to me by my wife Svetlana, and I would like to express my cordiality and deepest gratitude for this.

I would also like to thank Dr A.V. Gurevich for his comments on the Russian manuscript, Mr R.E. Hardin of the Israel Program for Scientific Translations for translating it into English, and Dr and Mrs L.R.O. Storey and Dr K.G. Budden for representing me in my dealings with the publisher. Moreover I thank the numerous colleagues in various countries who have regularly sent me reprints and preprints. Without this material it would have been practically impossible to write this monograph in complete form. Finally I will be very grateful to any readers who will be kind enough to inform me via the publisher of any errors and shortcomings in the book.

YA. L. AL'PERT

Introduction

After the launching of artificial Earth satellites and space probes, two new directions of study in modern plasma physics appeared and are now developing at an ever-increasing pace. One of these considers the effects of interactions of bodies with the near-Earth and interplanetary plasma, which the bodies pass through during their motion. The other studies the oscillations and waves arising in a plasma, in particular as a result of its interaction with incoming particle fluxes.

The launching of satellites into near-Earth and outer space created a situation in some respects analogous to the situation in the mechanics of continuous media after the invention of the aeroplane. Just as the development of aviation required a study of the *aerodynamics of flow* of a compressible gas around a body, so did the flight of satellites in near-Earth and outer space necessitate a study of the *kinetics of flow* of a plasma around a body. The effects arising when a body moves through a plasma do not, however, determine the motion of the body, unlike, for instance, the motion of an aeroplane, since the frictional forces between a satellite or rocket and the plasma are small. But these phenomena are nevertheless important in themselves, first because their study reveals a number of features of general significance for plasma physics, and second because familiarity with them aids in the performance and interpretation of many experiments aboard spacecraft used as laboratories for investigating the properties of the surrounding medium.

On the other hand, the possibility of making direct measurements, one could say, of the laboratory type, in the near-Earth and interplanetary plasma made feasible a direct *in-situ* study of the wave processes taking place there. Under laboratory conditions it is practically impossible to create a plasma with the wide range of physical parameters observed in a natural plasma, and this is a limiting factor. Moreover, the analysis of experimental data obtained on bodies in space constitutes one of the most precise methods of plasma diagnostics. It enables a determination of the main quantities characterizing the state of a plasma to be made in cases where the experimental results can be identified with theoretical data (for instance, to determine the type of waves observed, the type of wave processes, the mechanism of their excitation, etc.). If the wave processes are investigated using instruments aboard the moving bodies themselves (satellites and rockets), then the observed waves and plasma oscillations in their vicinity may well be connected with effects arising due to the interaction between the body itself and the plasma, that is, with the influence of the inhomogeneous ionized cloud and nonuniform electric field formed around the body. Plasma oscillations can also be caused by particle fluxes emitted by the body itself, for example, electrons, or by electromagnetic radio waves transmitted from aboard it.

Consequently, the two above-mentioned directions of study in modern plasma physics are interrelated in many respects. They are also unified by sharing the theoretical methods employed to solve the various problems arising in this field. In most cases the kinetic theory of gases is used for this.

The subjects discussed in this book pertain mainly to a highly rarefied magnetized plasma. In such a medium the mean free paths of particles will be much greater than the characteristic dimensions of the body and very often are much larger than the wavelengths of the observed oscillations. The corresponding equations are written in the phase space of the particles. The problems to be solved are considerably more complicated than problems in hydrodynamics, and are of a different kind. A basic feature of these problems is the need to take into account the electric and magnetic fields. In the theory three new parameters make their appearance, each with the dimensions of length: the Debye length D, the Larmor radius of the ions ρ_{Hi}, and the Larmor radius of the electrons ρ_H. The nature of the plasma flow around the body, as well as the spectra and types of waves and oscillations to be expected in it, become even more diverse, due to the nonisothermal nature of the plasma ($T_e \neq T_i$), which is observed in a number of its regions. The diversity of the observed wave processes is also promoted by the irregularities and ionized clouds of various sizes which are always observed in the near-Earth plasma, and which in turn also often apparently arise due to wave excitation in this plasma.

The theoretical problems dealing with plasma flow around bodies, instability of a plasma, and the types of oscillations and waves excited in a plasma are frequently *intrinsically nonlinear*. This stage in the development of the indicated directions of study in the physics of the near-Earth and interplanetary plasma also governs the present state of the theory.

Most of the effects described by the linear theory have already been investigated quite thoroughly both theoretically and experimentally. This undoubtedly constitutes a major success for this new field of experimental physics, a success which was moreover achieved during a quite short period. On the other hand, a further development of the theory will require the solution of nonlinear problems.

Here each step forward will require a great deal of effort. Present-day computers are able to solve only some such problems, and so far the results of studies of nonlinear problems are still insufficiently complete. Often this is because, due to the multiplicity of factors affecting the phenomena being studied, it is difficult to choose the principal factors, that is, the ones determining the observed experimental effects in question. To sum up, a number of experimental data have accumulated for which there is as yet no clear and definite theoretical explanation. This does not contradict the fact that the overall state of the indicated fields of physics is complete in many respects. A general consideration of the set of known data gives, figuratively speaking, an orderly picture. Some of the experimental and theoretical results are quite elegant, and they illustrate the wealth of plasma physics, as well as the possibilities offered by it for studying the nature of, and analysing, the near-Earth and interplanetary plasma.

It is the purpose of this book to describe the main results that have been obtained in the indicated fields. It seemed advisable to me to unify these results, since there is often an internal relationship, albeit not yet such an obvious one, between the phenomena accompanying the flow of plasma around a body and the wave processes observed in the plasma. We present general equations describing the set of topics being considered, and in a number of cases we pose the corresponding problems in general form. Also given are fundamental formulas making it possible to calculate various effects in a plasma; the various types of phenomena are classified as well. In some cases the experimental results are compared with the theory. The great volume of the material under consideration prevented me from discussing in detail all aspects of the subjects in question. Thus interesting, important data sometimes had to be left out, or in some cases only presented schematically. However, the literature cited should enable the reader to become more acquainted with those subjects which are of interest to him, and also to get an idea of how this field of plasma physics has developed so far.

Most of the equations in this book are written in forms appropriate to the use of SI units, but the values of physical quantities are also given in CGS or other units when these are more familiar: for instance, electron densities are commonly given in cm^{-3} and altitudes in km or in Earth radii. Vector quantities are represented by bold-faced symbols, and their moduli by the corresponding italic symbols: thus $A = |\mathbf{A}|$. The terminology for the various frequency ranges differs from that used by many authors, since I prefer to classify the wave frequencies in terms of their values relative to the local characteristic frequencies of the plasma, rather than in terms of their absolute values.

1

General remarks

The phenomena discussed in this monograph were observed starting from altitudes of 300 to 400 km, where the maximum electron density N_M of the F-2 region of the ionosphere is situated, and ending at distances tens or hundreds of thousands of kilometers from the Earth, in interplanetary space. Data from observations in the solar wind at distances of a million or more kilometers were also used.

There is by now a plethora of experimental data yielded by artificial Earth satellites in the outer ionosphere (plasmasphere) as far as its upper boundary. This boundary, known as the plasmapause, is situated usually at geocentric distances R ranging from $4R_0$ to $5R_0$, where R_0 is the Earth's mean radius. Numerous observations have also been made outside the plasmasphere in the magnetosphere, in the region encompassing its upper boundary – the magnetopause – and in the magnetosheath which is the transition region from the magnetosphere to interplanetary space, these regions being situated, respectively, at distances $R \simeq (9\text{–}11)R_0$ and $R \simeq (11\text{–}15)R_0$ (see Figs. 2.1 and 2.9 and Tables 2.1 and 2.2). More rarely, direct observations have been carried out in interplanetary space at great distances from the Earth, for example in cases where space probes have plied through it on their way to planets in the Solar System. However, the amount of information about the properties of the natural plasma obtained from direct measurements carried out on bodies moving through it turned out to be very different for different regions of the plasma. For this specific reason, despite the exceptional, sometimes decisive, role played by *in-situ* studies using artificial satellites and space probes in shaping modern conceptions of the phenomena of interest to us here, results concerning the near-Earth plasma derived from ground-based observations continue to be very important. In studies of the flow of plasma around bodies, moreover, laboratory findings are of substantial significance.

Since the properties of the plasma regions under review vary in the extreme, the nature of the physical phenomena occurring in different regions varies appreciably in a number of cases. On the other hand, the same phenomena are observed both at the lower boundary of the plasma and in the magnetosphere. For this reason, when examining the totality of phenomena of interest to us, it is important first of all to divide the plasma into zones within each of which physical processes of a uniform character may take place. For different processes the boundaries of these zones naturally do not have to coincide. Next, it is advisable to break up the entire range of frequencies of the oscillation phenomena being studied into typical sectors, not according to a quantitative principle, as has often been done in the literature so far, but rather on the basis of a specific physical approach to the various phenomena. The following examples illustrate what should be the basis for such a classification.

When a plasma flows around a body, the effects arising in the vicinity of this body change radically, depending on the velocity of its motion V_0 or the velocity V_s of the external particle fluxes impinging upon it; the subscript s denotes the species of particle. It is known that the maximum velocities of bodies launched into a plasma vary from 8 to 11 $\mathrm{km\,s^{-1}}$. Near the earth $V_0 \simeq 10^4\,\mathrm{m\,s^{-1}}$, while at great distances from Earth $V_0 \simeq 2 \times 10^3$ to $5 \times 10^3\,\mathrm{m\,s^{-1}}$. In the plasma regions under discussion the average thermal velocity of electrons is

$$v_e \simeq (2kT_e/m)^{1/2} \simeq 2 \times 10^5 \text{ to } 2 \times 10^6\,\mathrm{m\,s^{-1}}, \tag{1.1}$$

where k is Boltzmann's constant, T_e the electron temperature, and m the mass of an electron (see Tables 2.1 and 2.2), i.e., $v_e \gg V_0$. Therefore, with regard to the electrons, bodies in the near-Earth plasma may always be considered quasi-stationary. However, with regard to the average thermal velocity of the ions,

$$v_i \simeq (2kT_i/M)^{1/2} \simeq 10^3 \text{ to } 5 \times 10^4\,\mathrm{m\,s^{-1}} \tag{1.2}$$

which increases with increasing distance from the Earth, artificial satellites initially move at a supersonic velocity $V_0 \gg v_i$, then in some intermediate zone $V_0 \simeq v_i$, and finally there are regions where a body may be considered quasi-stationary: $V_0 <$ or $\ll v_i$. In (1.2), T_i is the ion temperature and M the mass of an ion. We see that in studies of plasma flow around bodies it is reasonable to consider three regions:

Zone I, supersonic motion of a body, $V_0 \gg v_i$; this extends to altitudes $z \simeq 1000$ to 2000 km;

Zone II, a transition zone, where $V_0 \simeq v_i$; this encompasses the outer regions of the ionosphere: 2000 km $< z < 3R_0$ to $5R_0$ (where R_0 is the Earth's mean radius);

Zone III, a quasi-stationary body, $V_0 \ll v_i$; this constitutes mainly the outer regions of the magnetosphere and interplanetary space: $R > 10R_0$ to $15R_0$ and the solar wind.

Note that, relative to the solar-wind plasma, the velocity of which $V_s \simeq 300$ to $500\,\mathrm{km\,s^{-1}}$, the motion of artificial bodies, like that of planets, is always supersonic.

However, the nature of the plasma disturbances in the vicinity of a moving body is not just defined by the ratio V_0/v_i. Also important is the ratio of the linear size of the body ρ_b to the electronic or ionic Debye length

$$D_e = (\varepsilon_0 kT_e/Ne^2)^{1/2} \quad \text{or} \quad D_i = (\varepsilon_0 kT_i/Ne^2)^{1/2} \tag{1.3}$$

and to the Larmor radii of the electrons and the ions

$$\rho_{He} = v_e/f_H, \quad \rho_{Hi} = v_i/F_H, \tag{1.4}$$

In formula (1.3) the quantities D_e and D_i are written for a quasi-neutral plasma (i.e., one with $N_i = N_e = N$), while ε_0 is the electric permittivity of free space, e is the absolute value of the electronic charge, and N is the electron density of the plasma. Also, in (1.4),

$$f_H = \omega_H/2\pi = (1/2\pi)(\mu_0 eH_0/m), \quad F_H = \Omega_H/2\pi = (1/2\pi)(\mu_0 eH_0/M) \tag{1.5}$$

where f_H and F_H are, respectively, the electron and ion gyrofrequencies (in the strict sense of the word 'frequency'), ω_H and Ω_H are the corresponding angular frequencies or pulsatances (also often referred to loosely as 'frequencies'), μ_0 is the magnetic permeability of free space, and H_0 is the external magnetic field of the Earth. (*Note*: many authors prefer to define the Larmor radii as $\rho^*_{He} = v_e/\omega_H$ and $\rho^*_{Hi} = v_i/\Omega_H$, but here the definitions (1.4), namely $\rho_{He} = 2\pi\rho^*_{He}$ and $\rho_{Hi} = 2\pi\rho^*_{Hi}$, are more convenient and are used everywhere below.)

A case where $\rho_b \gg D_e$ is to be considered a *large body*. The equations that describe the state of a plasma in the vicinity of a large body are extremely complicated, since the boundary conditions at the body's surface (its surface properties) must be taken into account (see Chapter 10). The phenomena around a large body differ in a number of respects from the effects in the vicinity of a *small body*, or *point body*, for which $\rho_b \ll D_e$. In the latter case all that has to be done is to examine the motion of a point charge. In the above zone I artificial bodies are for the most part large, while in zone II at first $\rho_b \simeq D_e$. However, as the distance from the Earth increases, ρ_b progressively becomes less than D_e, until in zone III the bodies are small: $\rho_b \ll D_e$.

The ratios ρ_b/ρ_{Hi} and ρ_b/ρ_{He} (particularly the first), which largely determine the nature of the theoretical problems to be solved (their complexity), as well as the phenomena observed in the body's vicinity naturally also vary considerably with the distance from the Earth. In the majority of cases $\rho_b \ll \rho_{Hi}$ in all the above-indicated zones. However, $\rho_b \gg \rho_{He}, \rho_b \lesssim \rho_{He}$ and $\rho_b \ll \rho_{He}$, in zones I, II, and III, respectively. Hence, the boundary-value problems that have to be solved when studying the phenomena occurring in different plasma regions are diverse in the extreme. Most problematic are cases where the characteristic parameters are commensurable; i.e., $V_0 \simeq v_i, \rho_b \simeq D_e$, and $\rho_b \simeq \rho_{Hi}$.

When we come to the wave processes and resonances taking place in the near-Earth and interplanetary plasma, we find that the division into zones requires a different approach. This applies both to the consideration of the type of possible phenomena and to the frequency ranges of the expected oscillations.

The conditions of wave excitation in a plasma, the nature of plasma instability, and the plasma-oscillation spectra vary considerably, depending on whether the plasma is strongly or weakly magnetized, i.e., depending on the ratio of the energy density $\mu_0 H_0^2/2$ of the external magnetic field $\mathbf{H}_0$ to the density of the gas kinetic energy of the charged particles $Nk(T_e + T_i)$, where k is Boltzmann's constant. These conditions reduce to

$$(V_A/2v_s)^2 \gg, \simeq, \ll 1, \tag{1.6}$$

while for the lengths of the various waves generated in a plasma we have

$$\Lambda^2 = (v_\phi/f)^2 \gg, \simeq, \ll \rho_{He}^2, \rho_{Hi}^2. \tag{1.7}$$

In formulas (1.6) and (1.7), which assume that $T_e \simeq T_i$,

$$V_A = c(\Omega_H/\Omega_0) = c/n_A \tag{1.8}$$

is the Alfvén velocity, where c is the velocity of light, and n_A is the Alfvén refractive index,

$$v_s \simeq (kT_e/M)^{1/2} \tag{1.9}$$

is the nonisothermal velocity of sound, v_ϕ and Λ are the phase velocity and wavelength of the observed plasma oscillations, and

$$\Omega_0 = 2\pi F_0 = (Ne^2/\varepsilon_0 M)^{1/2}, \tag{1.10}$$

where F_0 is the ion Langmuir frequency, otherwise known as the ion plasma frequency; N is the electron or ion number density, these two quantities being equal if, as we assume, the ions are singly charged.

It is readily seen from the tables presented in Chapter 2 that, in the regions with which we are concerned, the plasma is everywhere strongly magnetized: $V_A \gg v_s, \Lambda \gg \rho_{He}, \rho_{Hi}$. Therefore, a number of wave phenomena in the near-Earth plasma are of universal nature. They differ only in terms of their frequency, which often varies by several orders of magnitude. This is due not to any difference in the nature of the physical conditions but rather to a variation in the values of the plasma parameters. For example, in different experiments ion-cyclotron waves were seen to be excited at low altitudes, at $z \simeq 300$ to 400 km, where $F_H \simeq 500$ to 600 Hz, and also at distances of 25 000 to 30 000 km from the Earth, where $F_H \simeq 1$ Hz. Or, for instance, the excitation of Langmuir electron oscillations was recorded with a frequency

$$f_0 = \omega_0/2\pi = (1/2\pi)(Ne^2/\varepsilon_0 m)^{1/2} \simeq 2 \times 10^6 \text{ to } 3 \times 10^6 \text{ Hz}$$
$$\text{or } \simeq 10^4 \text{ to } 2 \times 10^4 \text{ Hz,} \qquad (1.11)$$

respectively, at an altitude of $z \simeq 10^3$ km and at a distance of 10^6 km from Earth in the solar wind. The frequency of the lower-hybrid resonance varies over a still wider range, of the order of 10^4, throughout the near-Earth plasma. Thus, when examining wave processes as a function of frequency, it is correct to classify them according to the type of physical phenomena by which they are elicited, i.e., to use the characteristic frequencies of the different processes as the corresponding criteria. In the following sections we will present such a classification.

2

General properties of the near-Earth and interplanetary plasma

The parameters of the plasma regions under discussion typically vary considerably depending on the time of day, the geographic coordinates, the magnetic and solar activity, and the distance from the Earth's surface. Exceptions are the magnetic field of the earth $\mathbf{H}_0$, the relative variations of which are small right out to distances of several tens of thousands of kilometers, and all the quantities associated with $\mathbf{H}_0$. For example, the electron density N at an altitude $z \simeq 300$ to 400 km may vary tenfold or more from day to night, or with a change in latitude or longitude. The electron temperature T_e varies by a factor of about 5 to 6. At these altitudes the variation of $\Delta H_0/H_0$ is only of the order of 10^{-3} to 10^{-4}. In the interplanetary medium the relative variations of the magnetic field are of course much greater, whereas N is apparently more stable here. Larger relative variations of N and $\mathbf{H}_0$ are observed in the solar wind. The value of N is also extremely unstable in the transition regions of the near-Earth plasma, the plasmapause and magnetopause, where from case to case it may vary by a factor of 10, 100, or even 1000 (see below).

A sufficiently accurate analysis of the various phenomena discussed below, for the purpose of plasma diagnostics or for establishing a closer correspondence between experiment and theory, is naturally possible only if a fairly wide assortment of measurements, including determination of the main plasma parameters, is carried out simultaneously in one and the same experiment. Appropriate research programmes are by now frequently being implemented in modern experiments on artificial Earth satellites and space probes. However, for an analysis of the results of the various theoretical computations, and in order to verify their specific applicability for explaining the processes taking place in a plasma, it is necessary (and often sufficient) to confine oneself to using only approximate values of the fundamental plasma parameters. This having been said, it is important to know the range of their variation under different conditions and also to have an idea of some of the qualitative characteristics of their variation. The pertinent data are given in Fig. 2.1 in very concise form. They

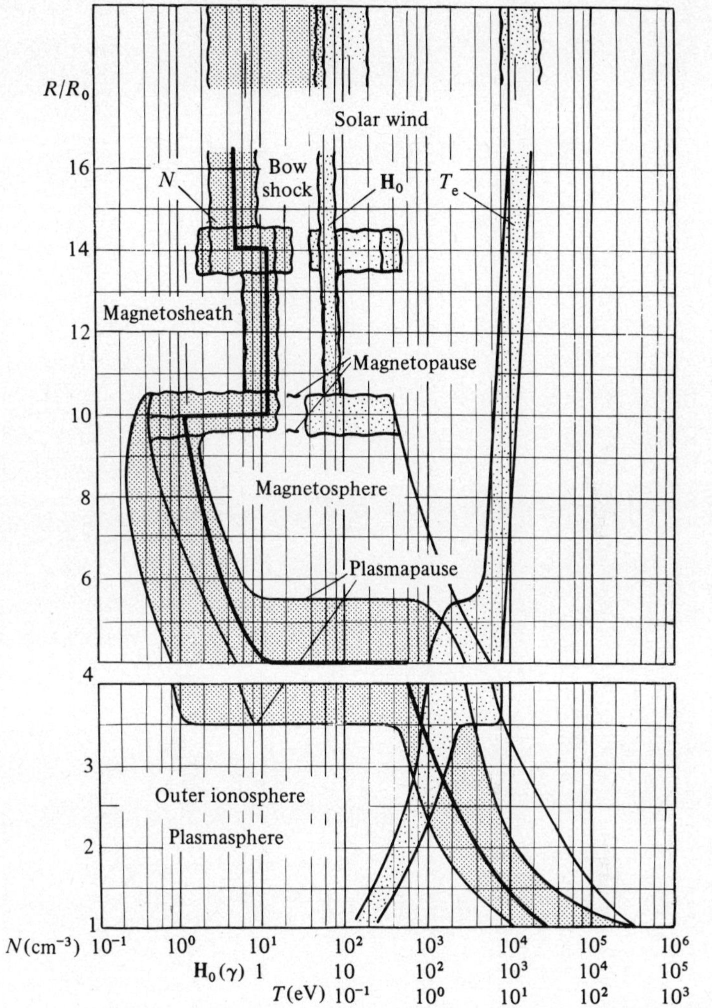

Fig. 2.1. Model of the near-Earth plasma in equatorial regions and at middle latitudes; R/R_0 is the geocentric distance in Earth radii.

summarize a multitude of results of measurements and, naturally, cover only some of the known properties of the near-Earth and interplanetary plasma.

2.1. Structure and properties of the plasma at different distances from the Earth

A model of the near-Earth plasma, describing its properties close to the Earth mainly at middle latitudes, is presented in Fig. 2.1 and Tables 2.1

Table 2.1. *Fundamental parameters of the near-Earth and interplanetary plasma*

Zone	z(km, R_0)	N(cm^{-3})	T_e(K, eV)	H_0(Oe, γ)	v_{ei}(s^{-1})	v_e(cm s^{-1})	v_i(cm s^{-1})	D_e(cm)	NkT(erg cm^{-3})	$\frac{1}{2}\mu_0 H_0^2$(erg cm^{-3})
I	Outer ionosphere (Plasmasphere)									
$V_0 \gg v_i$	300–400	3×10^5–3×10^6	1000–3000	0.45 Oe	4×10^2–2×10^3	1.5×10^7–3×10^7	10^5–2×10^5	0.2–0.7	4×10^{-8}–10^{-6}	8×10^{-3}
$\frac{1}{2}\mu_0 H_0^2 \gg NkT$	500–600	10^5–10^6		0.36	10^2–3×10^2	$\simeq 2.5 \times 10^7$	$\simeq 3 \times 10^5$	$\simeq 0.7$	10^{-8}–3×10^{-7}	5×10^{-3}
$\rho_b \gg \rho_{He}, D_e$	1000	2×10^3–8×10^4	2×10^3–3×10^3	0.29	2–45	2.5×10^7–3×10^7	6×10^5–7×10^5	1–7	5×10^{-10}–3×10^{-8}	3×10^{-3}
$\ll \rho_{Hi}$	2000	10^3–4×10^4	$\simeq 3 \times 10^3$	0.22	0.4–13	$\simeq 3 \times 10^7$	7×10^5	2–10	2×10^{-9}–2×10^{-8}	2×10^{-3}
	5×10^3–6×10^3	5×10^2–5×10^3	4×10^3–5×10^3	0.15	0.1–0.8	3.5×10^7–4×10^7	8×10^5–9×10^5	7–20	3×10^{-10}–3×10^{-9}	9×10^{-4}
II $V_0 \simeq v_i$ $\frac{1}{2}\mu_0 H_0^2 \gg NkT$ $\rho_b \lesssim \rho_{He}, D_e$ $\ll \rho_{Hi}$	Region of the plasmapause $(2.5–6)R_0$:	5×10^{-1}–5×10^2	1–8 eV	300–1000	$\lesssim 10^{-3}$	6×10^7–1.7×10^8	1.5×10^6–4×10^6	10^2–10^3	10^{-12}–6×10^{-9}	4×10^{-7}–4×10^{-6}
III $V_0 \ll v_i$ $\frac{1}{2}\mu_0 H_0^2 \gtrsim NkT$	Magnetosphere $(4–10)R_0$	10^{-1}–10	3–10	40–800	$\lesssim 10^{-5}$	10^8–2×10^8	2.5×10^6–4.5×10^6	10^3–4×10^3	1.5×10^{-12}–1.5×10^{-10}	6×10^{-9}–2×10^{-6}
$\rho_b \gtrsim D_e$	Region of the magnetopause $(9–11)R_0$	2×10^{-1}–10	6–12	4–50	$\lesssim 10^{-5}$	1.5×10^8–2.1×10^8	3.5×10^6–5×10^6	10^3–4×10^3	2×10^{-12}–2×10^{-10}	6×10^{-11}–10^{-8}
$\rho_b \ll \rho_{He}$ $\rho_b \ll \rho_{Hi}$	Magnetosheath	8–15		6–10					2×10^{-10}	
	$(10–15)R_0$		8–15		$\lesssim 10^{-5}$	1.7×10^8–2.3×10^8	4×10^6–5×10^6	$\simeq 10^3$		10^{-10}–4×10^{-10}
		2–20		4–30					5×10^{-11}–5×10^{-10}	
	Interplanetary medium $>(14–16)R_0$	2–10	10–20	5–8	$\lesssim 10^{-5}$	1.9×10^8–2.7×10^8	4.4×10^6–6×10^6	$\simeq 10^3$	3×10^{-11}–3×10^{-10}	$\simeq 10^{-10}$
	Solar wind $\gtrsim 10^6$ km	2–60	$T_e/T_i \simeq$ 8–25(1) 2–6(2)	5–20	$\lesssim 10^{-5}$	1.7×10^8–3×10^8	4×10^6–7×10^6	$\lesssim 10^3$	2×10^{-11}–2×10^{-9}	10^{-10}–10^{-9}

(1) Active solar conditions.
(2) Quiet solar conditions.

Table 2.2. *Fundamental parameters of the near-Earth and interplanetary plasma*

Zone	z(km, R_0)	f_0(Hz)	F_0(Hz)	f_H(Hz)	F_H(Hz)	ρ_{He}(cm)	ρ_{Hi}(cm)	f_L(Hz)	n_A	V_A(cms^{-1})
I	Outer ionosphere (plasmasphere)									
$V_0 \gg v_i$	300–400	5×10^6–1.5×10^7	3×10^4–9×10^4	1.3×10^6	$\simeq 40$	$\simeq 10$–20	8×10^2–1.3×10^3	7×10^3	7.5×10^2–2×10^3	1.5×10^7–4.10^7
$\frac{1}{2}\mu_0 H_0^2 \gg NkT$	500–600	3×10^6–9.10^6	2×10^4–7×10^4	$\simeq 10^6$	$\simeq 70$	$\simeq 25$	4×10^3	8×10^3	3×10^2–10.10^2	3×10^7–10^8
$\rho_b \gg \rho_{He}, D_e$	1000	4×10^5–2×10^6	10^4–5×10^4	7×10^5	400	$\simeq 40$	1.5×10^3	2×10^4	25–125	2×10^8–10^9
$\ll \rho_{Hi}$	2000	3×10^5–1.5×10^6	7×10^3–3×10^4	6×10^5	340	$\simeq 50$	2×10^3	1.5×10^4	20–90	3×10^8–10^9
	5×10^3–6×10^3	2×10^5–6×10^5	5×10^3–1.5×10^4	4×10^5	200	$\simeq 100$	4×10^3	9×10^3	25–75	4×10^8–10^9
II $V_0 \simeq v_i$ $\frac{1}{2}\mu_0 H_0^2 \gg NkT$ $\rho_b \lesssim \rho_{He}, D_e$ $\rho_b \ll \rho_{Hi}$	Region of the plasmapause (2.5–6)R_0	6×10^3–9×10^4	1.5×10^2–2×10^3	8×10^3–3×10^3	5–15	5×10^3–8×10^3	$\simeq 3 \times 10^5$	2×10^2–6×10^2	30–130	2×10^8–10^9
III	Magnetosphere (4–10)R_0	5×10^3–3×10^4	10^2–7×10^2	10^3–2×10^4	0.6–12	10^4–10^5	4×10^5–4×10^6	25–500	60–170	2×10^8–5×10^8
	Region of the magnetopause (9–11)R_0	4×10^3–4×10^4	90–9×10^2	1.2×10^2–10^3	7×10^{-2}–7×10^{-1}	2×10^5–2×10^6	$\simeq 10^7$	3–30	$\simeq 1300$	2×10^7
$V_0 \ll v_i$ $\frac{1}{2}\mu_0 H_0^2 \gtrsim NkT$ $\rho_b \gtrsim D_e$	Magnetosheath (10–15)R_0	10^4–4×10^4	2×10^2–9.10^2	10^2–8×10^2	5×10^{-2}–5×10^{-1}	2×10^5–2×10^6	10^7–10^8	2–20	2×10^3–4×10^3	10^7
$\rho_b \ll \rho_{He}$	Interplanetary medium $>$(14–16)R_0	10^4–3×10^4	2×10^2–7×10^2	1.4×10^2–2×10^2	8×10^{-2}–10^{-1}	10^6–1.5×10^6	$\simeq 10^7$	3–5	9×10^2–2×10^3	2×10^7
$\rho_b \ll \rho_{Hi}$	Solar wind $\gtrsim 10^6$ km	10^4–7×10^4	2×10^2–2×10^3	1.4×10^2–6×10^2	8×10^{-2}–3×10^{-1}	5×10^5–10^6	2×10^7–6×10^7	3–13	2×10^3–6×10^3	0.5×10^7–1.5×10^7

Most of the quantities in this and in the preceding table are given in CGS units. The SI unit for the magnetic energy density is the joule m^{-3} = 10 erg cm^{-3}.

and 2.2. The tables and figure show the range of variation here, as well as in interplanetary space and in the solar wind, of the electron density N in cm^{-3}, the electron temperature T_e in eV (1 eV = 1.16×10^4 K), the intensity of the Earth's magnetic field $\mathbf{H}_0$ in gammas ($\gamma = 10^{-5}$ Oersted = $10^2/4\pi$ A m^{-1}), and other typical plasma parameters.

In the two tables, the first column lists certain approximations and inequalities, the second column lists certain ranges of altitude, while the subsequent columns list certain ranges of plasma parameter values that are covered in the given altitude ranges, and within which the given approximations or inequalities are satisfied. When only one set of figures is given, which is usually the case, it represents the full range of variation of the parameter values for the altitude range concerned.

More detailed dependences of N and $\mathbf{H}_0$ on the distance from Earth are given later in this chapter, in Figs. 2.9, 2.10, and 2.11. In Chapter 7, in connection with the phenomena examined there, we also present some data characterizing the outer ionosphere in the altitude range $z \simeq 200$ to 1000 km (see Table 7.1).

From Fig. 2.1 and Tables 2.1 and 2.2 we see that the ranges of variation of the electron density N, the magnetic field $\mathbf{H}_0$, and the temperature in the solar wind exceed the corresponding variations of these quantities in the interplanetary medium. The relevant data are to be found, in particular, in the following works: Scarf *et al.*, 1970; Scarf *et al.*, 1971; Siscoe *et al.*, 1971; Fredricks *et al.*, 1970; Scarf, 1970; Mariani & Ness, 1969.

It should be noted that the fewest data of all are available for the temperature T_e of the thermal electrons. In the region of the plasmapause the temperature apparently undergoes marked changes and rapidly rises, as does the plasma density (Serbu & Maier, 1970). It is not known whether analogous phenomena are observed in the magnetopause region and in the bow-shock region.

A number of processes in the near-Earth plasma depend on its ion composition. The role played by a small admixture of ions can sometimes be quite considerable. For instance, in Chapter 8 we will see that the energy flux in a multicomponent magnetoplasma has in this case some extremely important characteristics. Data on the relative proportions of the different sorts of ions in the outer ionosphere are given in Table 2.3. Also presented there are values of the effective ion mass

$$1/M_{\text{eff}} = \sum_{s} (\alpha_s/M_s),$$

where $\alpha_s = N_s/N$ is the relative value of the ion density and $\sum \alpha_s = 1$, $\sum N_s = N$.

Table 2.3

z(km)	$N(O_1^+)/N$	$N(H_1^+)/N$	$N(H_e^+)/N$	μ_{eff}^+
300–400	0.95–1.00	2×10^{-2}–10^{-1}	10^{-3}–10^{-1}	10–16
500–600	0.3–1.00	$\simeq (0.1–0.6)$	10^{-3}–10^{-1}	2–4
1000	$\simeq (10^{-2}–10^{-1})$	$\simeq (0.5–1)$	10^{-1}–3×10^{-1}	1–2
2000	$< 10^{-1}$	$\simeq 1$	$\simeq 10^{-2}$	$\simeq 1$
5×10^3–6×10^3	$\ll 10^{-2}$	$\simeq 1$	$\simeq 10^{-2}$	$\simeq 1$

The ion composition in the near-Earth plasma is highly variable, and the tabulated data are extremely approximate, especially at altitudes $z \lesssim 1000$ to 2000 km. There have been observations, for example, of values of $M_{eff} \sim 4$ to 14 for $z \simeq 2500$ to 3000 km (see Ondoh *et al.*, 1974), while at altitudes of several thousand kilometers values of $N(H_e^+) \simeq 10^{-1} N(H_1^+)$ have been recorded (Taylor *et al.*, 1965). However, at still greater distances from the Earth, in the plasmasphere and magnetosphere, most often $N(H_e^+) \simeq 10^{-2} N(H^+)$.

2.2. Distribution of charged-particle density inside and outside the plasmasphere

A characteristic feature of the altitude dependence of the electron density N and the ion density N_i is a rapid, almost step-like, decrease at the boundary of the outer ionosphere (Figs. 2.2 and 2.3; see Angerami & Carpenter, 1966; Chappell *et al.*, 1970a, b, 1971a, b; Shaw & Gurnett, 1972; Taylor *et al.*, 1970). The profiles of $N(H^+)$ and $N(H_e^+)$ are usually similar, whereas the helium-ion density $N(H_e^+)$ is about 10^{-2} times the proton density $N(H^+)$.

The outer part of the ionosphere is generally called the plasmasphere, and its upper boundary is known as the plasmapause. The plasmapause is usually situated nearer the Earth during the night hours (local time), particularly when it is affected by magnetic activity (Fig. 2.4). If the magnetic activity is high, the outer ionosphere becomes, as it were, compressed. The plasmapause is farthest from the Earth at dusk. The diurnal course of the averaged position of the plasmapause has a 'hump' in it (dusk bulge). At dusk the position of the plasmapause reaches values of $L \simeq 7$ (Fig. 2.4 (a)), where L is the McIlwain parameter, which is constant along a line of magnetic force (McIlwain, 1961); roughly speaking, it is the distance from the centre of the Earth, expressed in

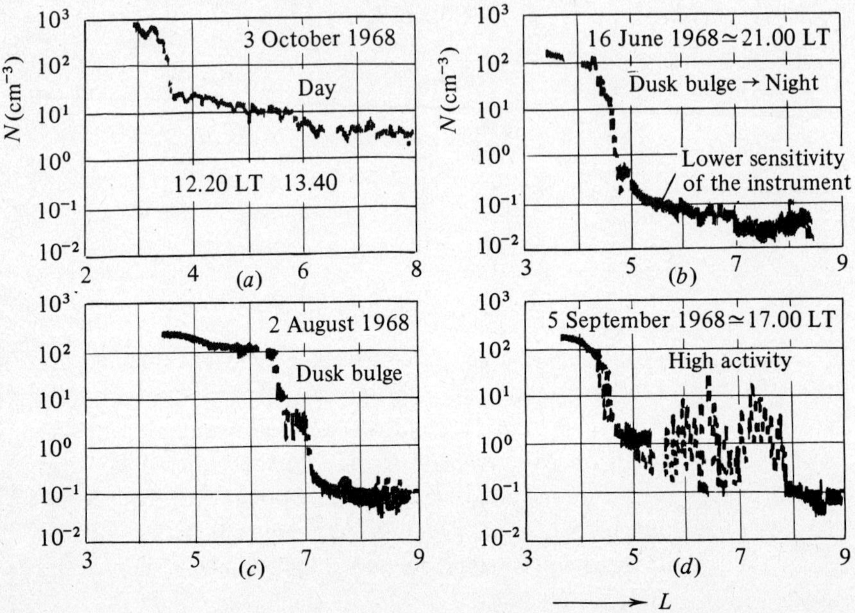

Fig. 2.2. Proton density $N(H^+)$ as a function of L. Data obtained from the results of measurements on OGO 5 with mass spectrometers at geomagnetic latitudes $\phi \lesssim 45$. The value of L is equal to the dimensionless geocentric distance R/R_0 at the geomagnetic equator $\phi_0 = 0$. Times of day are given in local solar time. (Chappell *et al*., 1970a, b, 1971a, b.)

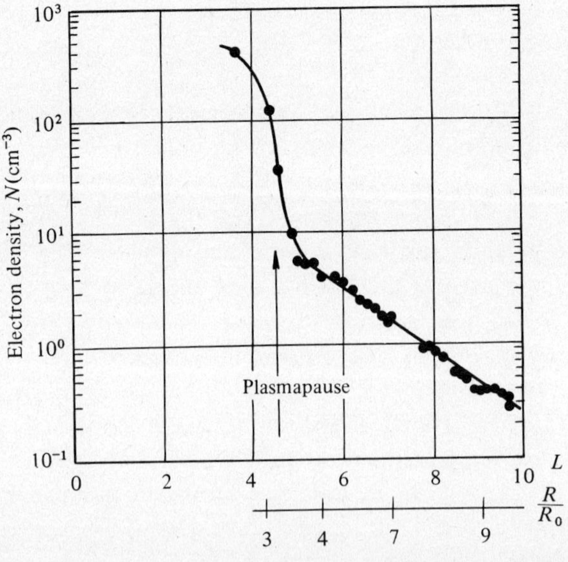

Fig. 2.3. Electron density N as a function of L. The values of R/R_0 refer to the geomagnetic latitudes ϕ at which the measurements were performed. Results of measurements of plasma ω_0 and upper-hybrid ω_U resonance frequencies on IMP 6 (Shaw & Gurnett, 1972).

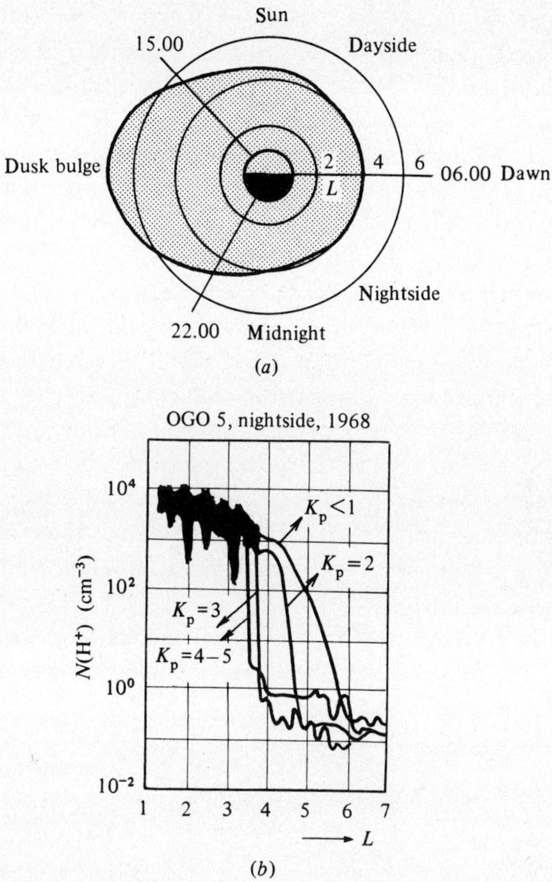

Fig. 2.4. (*a*) Average dependence of the plasmapause position on local time LT. Results of analyzing 150 profiles of proton density $N(H^+)$ obtained from mass-spectrometer measurements on OGO 5; (*b*) dependences of $N(H^+)$ on L during different periods of magnetic activity (K_p index) (Chappell *et al.*, 1971b; Chappell, 1972).

Earth radii, at which the line crosses the magnetic equator. During the daytime hours the plasmapause boundary has a less steep, sometimes quite gently sloping, profile. The region between the plasmapause and the magnetopause (see Fig. 2.1), which in the literature is called the trough, during the daytime (even in periods of moderate activity) in a number of cases becomes filled with isolated formations, the electron density of which attains values of $N \simeq 10$ to $100\,cm^{-3}$; they exceed more than a hundred-fold the values $N \simeq 1\,cm^{-3}$ observed normally (Chappell, 1972, 1974). These formations have a complicated spatial structure with linear dimensions that vary from about 50 to several thousand kilometers (see Chappell *et al.*, 1971b). They apparently arise owing to an influx of plasma from the plasmasphere under the influence of convective electric fields, and in a number of cases they have a wave-like structure. These inhomogeneous cold-plasma formations in the trough markedly affect the stability of the plasma, promoting the excitation of various types of oscillations in it. It is well-known (see Volume 2) that the trough region is indeed 'rich' in various wave processes. In this connection it is important to note that the ratio between the densities of the cold (thermal, $T_e \simeq 1$ to 10 eV) and hot (high energy) particles plays a decisive role in these processes. The effective (Maxwellian) temperature of the latter varies here from several hundred to thousands of electron volts: $T_{eff} \simeq \ldots 10^2$ to $\ldots 10^4$eV. Relatively recent IMP 6 measurements of the densities of thermal and suprathermal charged particles in the plasmatrough between the plasmapause and magnetopause (Fig. 2.5; Gurnett & Frank, 1974a) showed that thermal particles constitute something like 50% of the plasma density around the lower boundary of the trough and approximately 5% at greater distances from the Earth ($\simeq 10$ to 11 R_0).

The data contained in Tables 2.1 and 2.2 and in Figs. 2.1 to 2.4 characterize the properties of the thermal low-energy plasma particles.

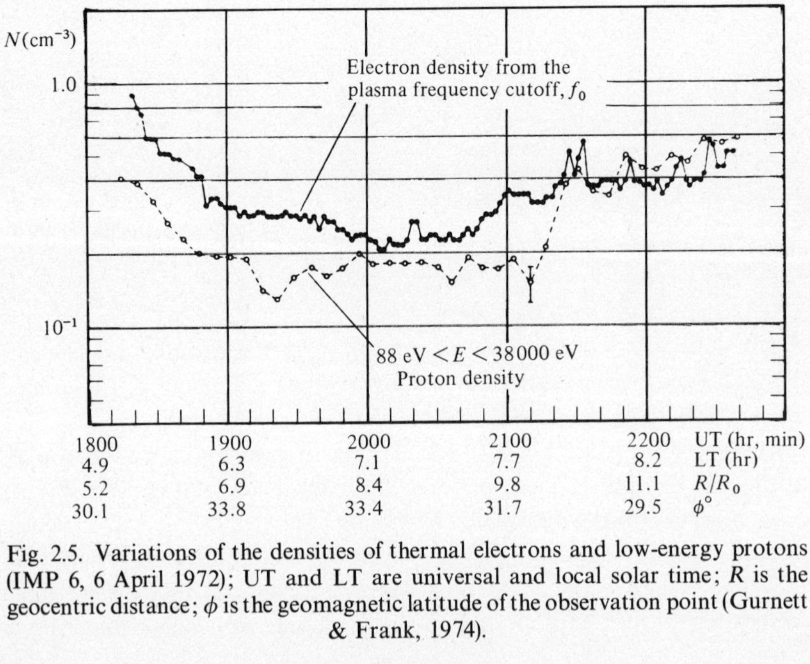

Fig. 2.5. Variations of the densities of thermal electrons and low-energy protons (IMP 6, 6 April 1972); UT and LT are universal and local solar time; R is the geocentric distance; ϕ is the geomagnetic latitude of the observation point (Gurnett & Frank, 1974).

However, in the near-Earth space as a whole, an extremely important part is played by high-energy electrons and protons, and in particular by those trapped by the magnetic field in the so-called radiation belts, at $L \simeq 4$–6. The energies of electrons E_e and protons E_p here reach hundreds of keV and hundreds of MeV, respectively. These particles populate the outer ionosphere, the region of the plasmapause, and the magnetosphere, where they play an important role. The study of the diverse properties of high-energy particles is a specialized, important, and interesting area of the present-day physics of the near-Earth plasma. We refer the reader to books on these aspects (see, for instance, Hess, 1968; Roederer, 1970; Shabanskii, 1972; a collection of papers published in Issledovaniya Kosmicheskogo Prostranstra, 1974), as well as to other papers, some of which are cited below, in order to become acquainted with the properties of these particles. As an illustration we present here some data on radiation-belt-trapped protons in the energy interval $E_p \simeq 20$ to 240 keV, which participate in various wave processes (see Volume 2).

Table 2.4. *Average values of the proton density*, $\bar{N}_p(\mathrm{cm}^{-3})$, in the radiation belt

State of plasma	E_p(keV)	$L=$ 3–3.2	$L=$ 3.5–3.7	$L=$ 4.0–4.2	$L=$ 4.5–5.0	Reference
OGO 3, 10 *June*–23 *July* 1966						
Quiet time	40	0.1	0.1	0.1	0.1	
Prestorm	40	0.3	0.3	0.3	0.3	
Quiet time	20.5	0.2	0.2	0.2	0.2	Frank & Owens, 1970
Prestorm	20.5	0.7	0.7	0.7	0.7	
OGO 3, 7 *July*–8 *Sept*. 1966						
Storm	40	1.1	3.2	2.8	2.6	
Main phase	40	0.1	0.9	1.9	1.5	Frank, 1967, 1970
Explorer 45, *Dec*. 1971						
Prestorm	26.4	0.8	1.8	2.4	1.8	
	42.5	4×10^{-3}	3×10^{-2}	0.5	2.1	Williams *et al.*; 1973
	74.5	10^{-2}	2×10^{-2}	5×10^{-2}	1.1	
	108.5	2×10^{-2}	5×10^{-2}	7×10^{-2}	0.5	Williams & Lyons, 1974
	242.0	0.1	0.2	0.2	0.1	
Explorer 45, 12 *Dec*., 17 *Dec*. 1971						
Storm	35	—	—	—	2.0	
	50	—	—	—	0.5	Williams *et al.*, 1974
	42.5	1	1	3	0.1	
Explorer 45, 24 *Feb*. 1972						
Storm	43.0	—	—	0.1	1.5	Smith & Hoffman, 1974

Table 2.4 gives the average values $\bar{N}_p$ of the density N_p of energetic protons, calculated from the proton fluxes $J_p\,cm^{-2}\,s^{-1}$ measured by OGO 3 and Explorer 45 (Frank, 1970; Frank & Owens, 1970; Williams *et al.*, 1973, 1974; Williams and Lyons, 1974; Smith & Hoffman, 1974; Hultqvist, 1975). The results on which this table is based are typical of the majority of the data found in the literature. In a number of cases the N_p values are considerably smaller, but often they are also commensurable with the N values for the thermal particles (see Figs. 2.1 and 2.5). During periods of intense magnetic activity the particle flux J_p, and accordingly the proton density N_p, increase. Such a case is depicted in Fig. 2.6. The dependence of particle flux J_p on the position of an artificial Earth satellite with respect to the lines of force of the Earth's magnetic field (i.e., on the value of L) shows how sharply J_p increases in the vicinity of $L \simeq 4$ during the main phase of a magnetic storm (Fig. 2.6(a)). This centre of particle-density 'compression' is clearly seen in Fig. 2.6(b), which plots the averaged lines of equal values of the differential fluxes $(\partial J/\partial E)_p$, according to the results of long-term measurements on OGO 3.

Particle fluxes usually have a complex spatial distribution. The largest $\partial J/\partial L$ gradients are observed during strong magnetic disturbances, which contribute to the increased plasma instability during such periods. The spectrogram in Fig. 2.7 is a good illustration of the spatial inhomogeneity of the energetic particles injected into the magnetosphere. It was made

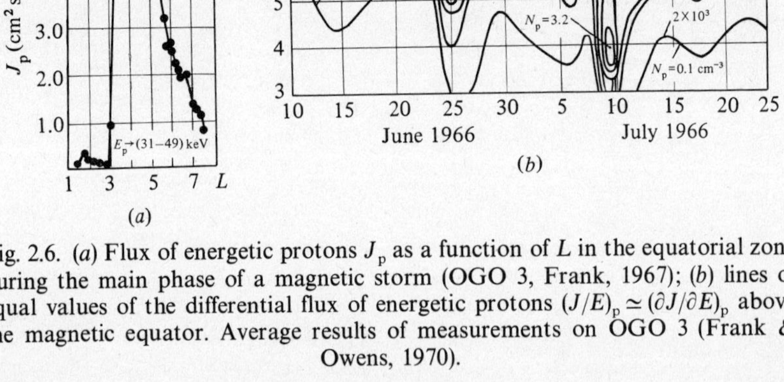

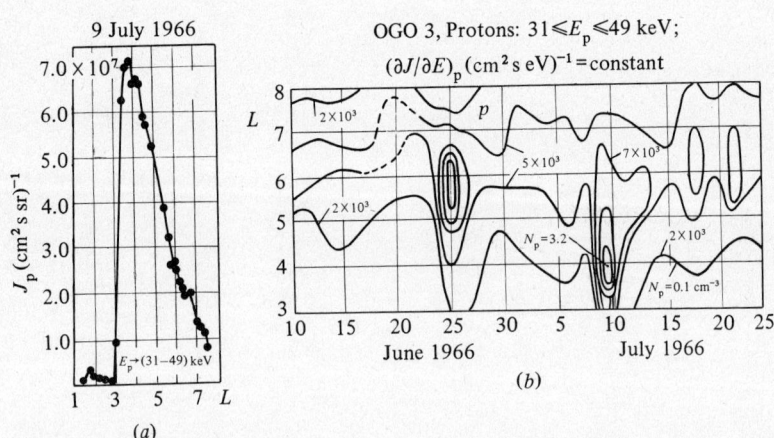

Fig. 2.6. (*a*) Flux of energetic protons J_p as a function of L in the equatorial zone during the main phase of a magnetic storm (OGO 3, Frank, 1967); (*b*) lines of equal values of the differential flux of energetic protons $(J/E)_p \simeq (\partial J/\partial E)_p$ above the magnetic equator. Average results of measurements on OGO 3 (Frank & Owens, 1970).

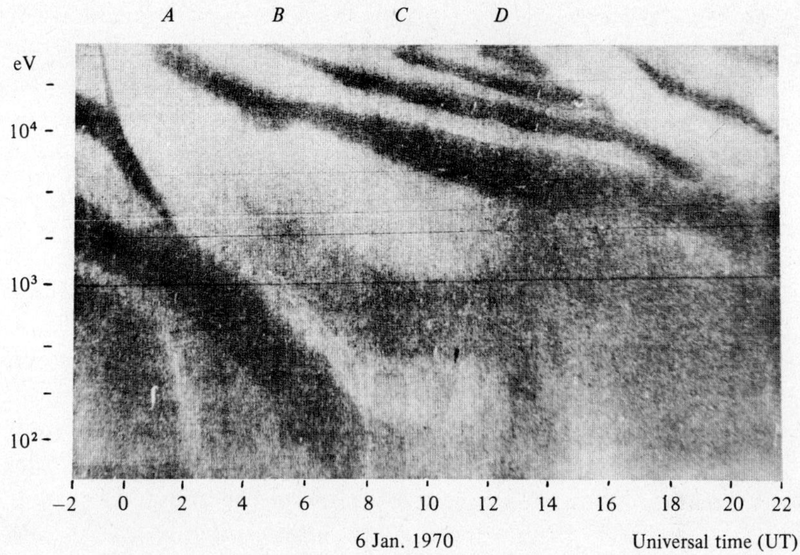

Fig. 2.7. Spectrogram of proton fluxes injected into the magnetosphere on force lines originating in the auroral region, obtained using a spectographic recording technique, making it possible to portray the particle flux as a function of two variables. The dark areas, corresponding to weak particle intensities, and the light, high-intensity areas, give a clear picture of the ionized clouds in the magnetosphere. These spectrograms were recorded on 6 Jan. 1970 aboard ATS 5 in the equatorial zone at $L = 6.8$ (DeForest & McIlwain, 1971).

from data recorded on board the ATS 5 geostationary satellite (DeForest & McIlwain, 1971). The dark bands in the picture correspond to the minimum values of the hot-particle energy E_p. The spatial distribution of the particles is seen to exhibit a cloud-like structure. The structure of the cloud spectrum is stable for a long time, with just the values of the E_p maxima, designated by the letters $A, B, C, \ldots$ in Fig. 2.7, slowly and progressively diminishing. This is apparently the result of energy release by the cloud during its interaction with the surrounding plasma, due to particle diffusion and to a number of other processes, (see DeForest & McIlwain, 1971).

At this point the following should be noted. Hot particles that are trapped in various regions of the near-Earth plasma outside the plasmasphere or that originate in these regions often have a decisive influence on the behaviour of the plasma, on the processes of its instability, and on the nature and type of oscillations and waves excited in it. However, one should not underestimate the importance of thermal particles, the majority of which are included in the outer ionosphere, in bringing about the processes taking place in the plasma regions almost contiguous with the plasmapause. The plasmasphere is a reservoir feeding plasma to the magnetosphere. The overall dynamics of the magnetosphere, the convection of particles and electric fields, and the processes of intermixing (turbulence) are to a significant extent determined by the particles flowing into the magnetosphere from lower-lying regions of the near-Earth plasma. As evidence of this we can cite the good correlation between a number of phenomena observed in the magnetosphere and, for example, the behavior of the F-2 region of the ionosphere, the properties of whistling atmospherics, hydromagnetic whistlers, and so on.

Since the polar zone in general belongs to a special branch of the physics of the near-Earth plasma, here we will not give data describing the latitudinal structure of the plasma, specifically the behaviour of the density of its thermal particles at the transition from the middle latitudes to the polar zones, in the region of the so-called polar cap and polar cusp. In Volume 2 we will present some data characterizing wave processes observed in the magnetosphere whose sources lie in the polar zone. For purposes of illustration, let us just mention that at the higher latitudes the plasmapause is situated considerably lower down than in the equatorial and mid-latitude regions. For instance, according to measurements performed on ISIS 1, already near the magnetoforce line $L \simeq 4$ ($\phi_0 = 60°$ on the Earth's surface) at the transition to the polar cap the proton density at an altitude $z \simeq 3000$ km is reduced about a hundredfold. Rapid temperature changes take place in this same region (Brace & Theis, 1974). OGO 4 measurements also showed that the position of the plasmapause undergoes marked changes at midnight and has a complex structure at altitudes as low as $z \simeq 10^3$ km (Taylor *et al.*, 1971).

From the preceding survey of the spatial changes in the density N of the near-Earth plasma, we see that it is hardly possible to give an analytical description of its dependences on the various parameters. In calculations, therefore, it is often necessary to use empirical or numerical models of the distribution of N for the given conditions. However, in the altitude interval from $z_0 \simeq 10^3$ km to the boundary of the outer ionosphere (the plasmapause) the following formula for the altitude dependence of the electron density

$$N(z) = N_0 \frac{z_0}{z} \exp\left\{ \frac{g_0 M_0 z_0}{2kT_0} \left[\frac{1}{1+(R_0/z)} - \frac{1}{1+(R_0/z_0)} + \ln \frac{1+(R_0/z)}{1+(R_0/z_0)} \right] \right\} \tag{2.1}$$

agrees quite well with results of measurements at middle latitudes (Al'pert,

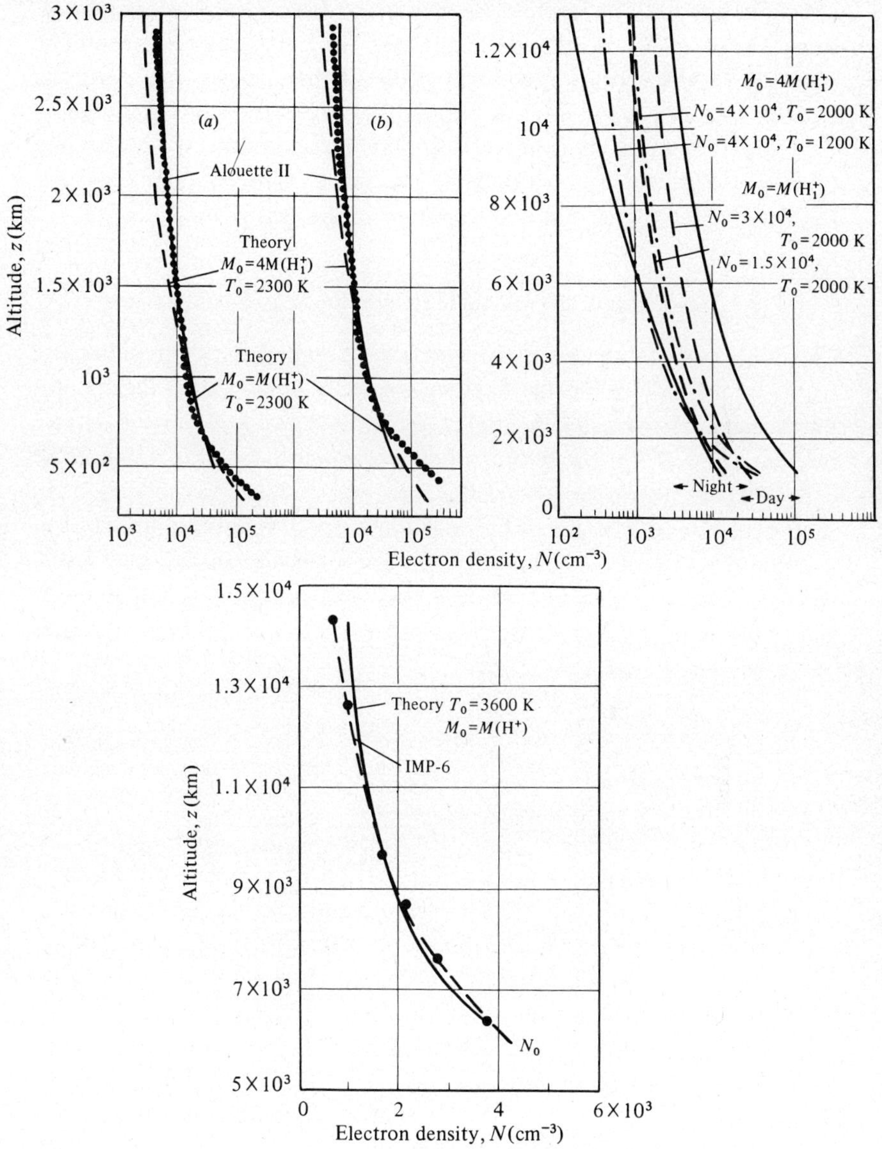

Fig. 2.8. Theoretical and experimental altitude dependences $N(z)$ of the electron density. Next to the theoretical curves are indicated the values of the parameters T_0, M_0, and N_0 at the altitude $z \simeq 1100\,\mathrm{km}$ (see formula (2.1) and Brown, 1973).

1975; see Fig. 2.8). This formula, which was obtained for a hydrostatic model of the outer ionosphere, is used to analyse the properties of whistling atmospherics and hydromagnetic whistlers (see Volume 2; Al'pert & Fligel', 1974, 1977). In formula (2.1), R_0 is the Earth's mean radius. The subscript 0 in the other quantities refers to their values at an altitude $z_0 \simeq 10^3$ km; M_0 is the ion mass, g_0 is the acceleration of gravity, T_0 is the temperature in Kelvin, and k is Boltzmann's constant. When deriving this formula it was assumed that the electron and ion temperatures were the same ($T_e \simeq T_i$) and that only one type of ion was present.

2.3. Structure of the magnetic field in the near-Earth plasma

The distribution of the Earth's magnetic field right up to the region close to the magnetopause on the dayside is usually described by means of a dipole field. The relevant formulas (see below) can be used for diverse calculations and for an analysis of experimental data up to L values of roughly 8 to 10. At $L \simeq 6$ to 7 the force lines of the magnetic field on the Earth's dayside begin to become compressed, this being an effect of the ram pressure of the solar wind. On the nightside, on the other hand, the force lines are stretched out and pass over into the geomagnetic tail (Fig. 2.9), the internal structure of which possesses a number of features

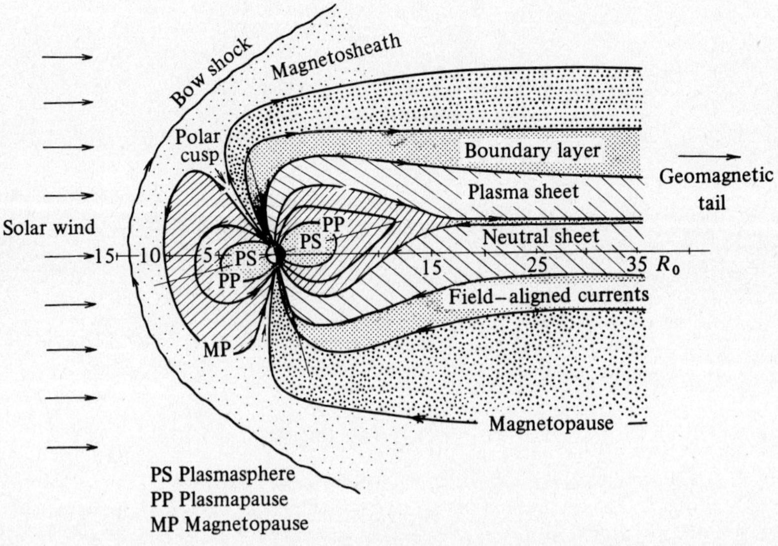

Fig. 2.9. Model of the distribution of magnetic field and plasma in the vicinity of the Earth.

that so far have not received a satisfactory theoretical explanation. According to certain data, in the far region of the tail $\mu_0 H_0 \simeq 10$ to 12γ, while the mean radius of the tail is of the order of 10 to 20 R_0 at distances $R \simeq 30R_0$. The geomagnetic tail is still clearly perceptible for $R \simeq 80R_0$. The thickness of the neutral sheet situated inside the geomagnetic tail is 0.1 to $1R_0$. The neutral sheet separates regions in which the magnetic field points in different directions.

In the outer ionosphere, including part of the region beyond the limits of the plasmapause, a dipole field of the Earth can be used on both sides. It is described by means of the following formulas. Depending on the geocentric distance R/R_0 and the magnetic latitude ϕ, along a force line the magnetic field $H(R)$ is

$$H = H_0(R_0/R)^3(1 + 3\sin^2\phi)^{1/2} \tag{2.2}$$

where

$$R/R_0 = \cos^2\phi/\cos^2\phi_0, \quad H_0 = H(R_0, \phi_0 = 0) \tag{2.3}$$

and ϕ_0 is the geomagnetic latitude of the end of the force line at the Earth's surface ($R = R_0$). The force lines of the dipole magnetic field up to altitudes $z = R_0$ and $z = 9R_0$ are depicted in Figs. 2.10 and 2.11 (Mlodnosky & Helliwell, 1962). Their lengths are indicated in thousands

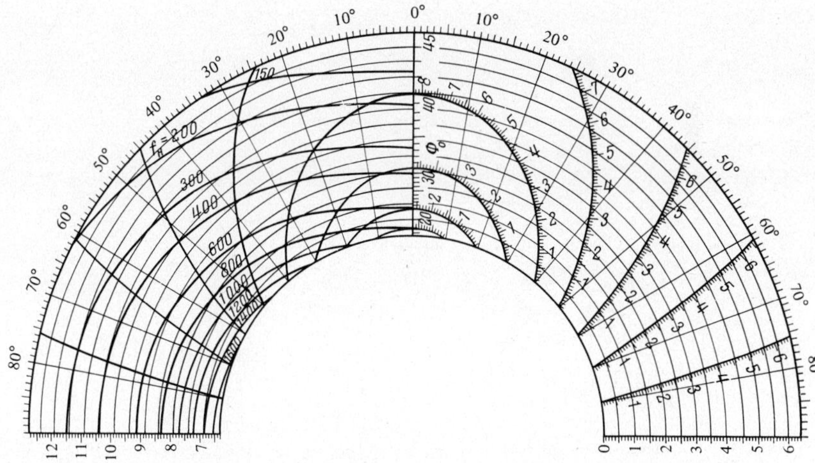

Fig. 2.10. Force lines $\mathbf{H}_0(R_1, \phi)$ of the Earth's dipole magnetic field and lines of equal values of the electron gyrofrequency $f_H(R, \phi)$ in kHz at geocentric distances $R = 1$ to 2 R_0 (where R_0 is the Earth's mean radius). The length $s(R, \phi)$ of each force line, from ground level up to the point (R, ϕ), is given in thousands of kilometers (Mlodnosky & Helliwell, 1962).

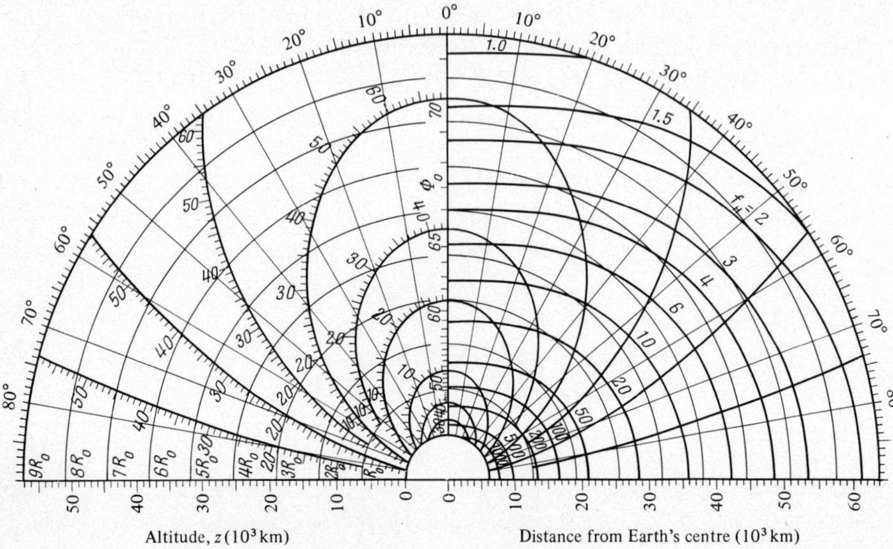

Fig. 2.11. As for Fig. 2.10, but for geocentric distances $R = 1$ to $10\ R_0$ (Mlodnosky & Helliwell, 1962).

of kilometers from the Earth's surface. These lengths were determined by means of formulas obtained from (2.2) and (2.3), namely

$$\mathrm{d}s = \frac{R_0 \cos\phi}{\cos^2\phi_0}(1 + 3\sin^2\phi)^{1/2}\mathrm{d}\phi = -\frac{(1 + 3\sin^2\phi)^{1/2}}{2\sin\phi}\mathrm{d}R, \tag{2.4}$$

$$\begin{aligned} s = {} & \frac{R_0}{2\cos^2\phi_0}\left\{\sin\phi_0(1 + 3\sin^2\phi_0)^{1/2}\right. \\ & \left. + (\sqrt{3})^{-1}\ln\left[\sin\phi_0 + \left(\frac{1 + 3\sin^2\phi_0}{3}\right)^{1/2}\right]\right\} \\ & - \frac{R_0}{2\cos^2\phi_0}\left\{\sin\phi(1 + 3\sin^2\phi)^{1/2}\right. \\ & \left. + (\sqrt{3})^{-1}\ln\left[\sin\phi + \left(\frac{1 + 3\sin^2\phi}{3}\right)^{1/2}\right]\right\}. \end{aligned} \tag{2.5}$$

Above the equator, i.e., at a point $R(L, \phi = 0)$, half of the length of a force line is

$$\begin{aligned} s_0 = {} & \frac{R_0}{2\cos^2\phi_0}\left\{\sin\phi_0(1 + 3\sin^2\phi_0)^{1/2}\right. \\ & \left. + (\sqrt{3})^{-1}\ln\left[(\sqrt{3})\sin\phi_0 + (1 + 3\sin^2\phi_0)^{1/2}\right]\right\} \end{aligned} \tag{2.6}$$

In calculations of the conditions for wave guidance along a magnetic field an important quantity is the force-line curvature $k_H = 1/\rho_H$ (this is the reciprocal of the radius of curvature). For a dipole magnetic field it is readily found that

$$\frac{1}{\rho_H} = \frac{3\cos^2\phi_0}{R_0\cos\phi} \frac{1+\sin^2\phi}{(1+3\sin^2\phi)^{1/2}}. \tag{2.7}$$

The force-line curvature of a field **H** that cannot be described as a dipole field is, especially if the graph $H(R, \phi)$ is specified, easy to find by means of the formula

$$k_H = \frac{1}{\rho_H} = \frac{1}{H}\frac{dH}{dr} = -\frac{1}{H}\mathrm{grad}_\perp H, \tag{2.8}$$

where dr is an element of distance normal to the corresponding force line (Copal Rao & Booker, 1963).

One of the halves of Figs. 2.10 and 2.11 gives curves of equal values of the electron gyrofrequency f_H, assuming that at the equator $\mu_0 H_0 = 0.316$ Gauss, while at the pole $\mu_0 H_0 = 0.628$ Gauss (1 Gauss $= 10^{-4}$ Weber m^{-2}). The dependences of f_H and of the inclination I of the Earth's magnetic field on the latitude ϕ_0 of the foot of the field line at the Earth's surface are presented in Fig. 2.12.

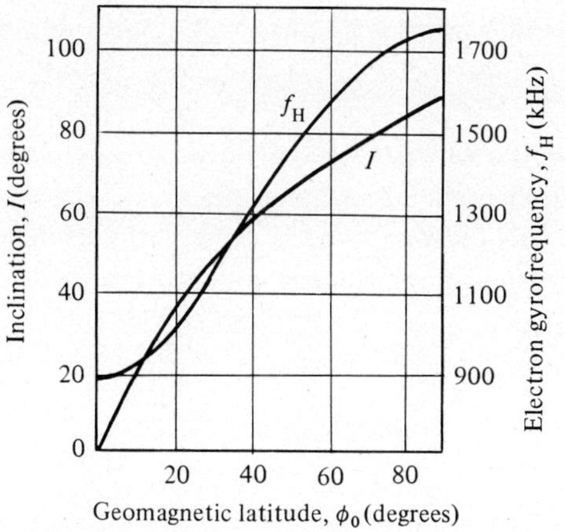

Fig. 2.12. Inclination I of the Earth's magnetic field vector $\mathbf{H}_0$ and the electron gyrofrequency f_H as functions of the geomagnetic latitude ϕ_0 at the Earth's surface.

3

Fundamental equations

The processes taking place in a magnetized plasma are described in the linear approximation by the dispersion equation

$$D(\omega, \mathbf{K}) = A\tilde{n}^4 + B\tilde{n}^2 + C = 0 \tag{3.1}$$

where

$$\tilde{n} = n - \mathrm{i}\kappa = (c/v_\phi) - \mathrm{i}\kappa \tag{3.2}$$

is the complex refractive index of the plasma, n is its real part, corresponding to the phase velocity v_ϕ of the waves, and κ is their spatial attenuation factor (Silin & Rukhadze, 1961; Stix, 1962; Akhiezer *et al.*, 1964, 1975; Ginzburg, 1967; Ginzburg & Rukhadze, 1970).

It is also generally convenient to use a complex frequency

$$\tilde{\omega} = \omega + \mathrm{i}\gamma. \tag{3.3}$$

Here, depending on whether γ is positive or negative, the oscillation amplitude is attenuated or augmented in the course of time. Our description of harmonic waves in the form $\exp(\mathrm{i}\tilde{\omega}t)$ defines the quantity γ as the time attenuation coefficient or (*decay rate*) of the oscillations if $\gamma > 0$ and as the *growth rate* if $\gamma < 0$. When the attenuation is small the spatial attenuation factor of the wave κ is related to γ by the equation

$$\gamma \simeq (\omega/c)(\partial\omega/\partial K)\kappa, \tag{3.4}$$

where $K = \omega n/c$ is the real part of the wave number.

In the general case, the coefficients A, B, and C in the dispersion equation (3.1) depend on the components of the permittivity tensor of the plasma: namely

$$\{\varepsilon_{sp}\} = \begin{Bmatrix} \varepsilon_{11} & \varepsilon_{12} & \varepsilon_{13} \\ \varepsilon_{21} & \varepsilon_{22} & \varepsilon_{23} \\ \varepsilon_{31} & \varepsilon_{32} & \varepsilon_{33} \end{Bmatrix} \tag{3.5}$$

as follows:

$$\left.\begin{aligned} A &= \varepsilon_{11}\sin^2\theta + 2\varepsilon_{13}\sin\theta\cos\theta + \varepsilon_{33}\cos^2\theta, \\ B &= -[\varepsilon_{11}\varepsilon_{33} + (\varepsilon_{22}\varepsilon_{33} + \varepsilon_{33}^2)\cos^2\theta - \varepsilon_{13}^2 \\ &\quad + (\varepsilon_{11}\varepsilon_{22} + \varepsilon_{12}^2)\sin^2\theta - 2(\varepsilon_{12}\varepsilon_{23} - \varepsilon_{13}\varepsilon_{22})\sin\theta\cos\theta], \\ C &= \varepsilon_{33}(\varepsilon_{11}\varepsilon_{22} + \varepsilon_{12}^2) + \varepsilon_{11}\varepsilon_{23}^2 + 2\varepsilon_{12}\varepsilon_{13}\varepsilon_{23} - \varepsilon_{22}\varepsilon_{13}^2, \end{aligned}\right\} \quad (3.6)$$

where θ is the angle between the wave vector $\mathbf{K}$ and the vector of the external magnetic field $\mathbf{H}_0$. The values of ε_{cq} are determined from a self-consistent solution of the linearized kinetic equations together with the Maxwell equations for the conditions posed by the particular problem; the subscripts s and p refer to the axes of a right-handed rectangular coordinate system, which we have labelled 1, 2, and 3, with the 3-axis parallel to the magnetic field $\mathbf{H}_0$.

For a plasma consisting of two kinds of particles (electrons e and one kind of ion i), the system of kinetic equations in the unsteady-state case has the form

$$\left.\begin{aligned} &\frac{\partial f_{\mathrm{i}}}{\partial t} + \mathbf{v}\frac{\partial f_{\mathrm{i}}}{\partial \mathbf{r}} + \frac{e\mathbf{E}}{M}\frac{\partial f_{\mathrm{i}}}{\partial \mathbf{v}} + \frac{\mu_0 e}{M}(\mathbf{V}_{\mathrm{L}} \times \mathbf{H}_0)\frac{\partial F_{\mathrm{i}}}{\partial \mathbf{v}} \\ &\frac{\partial f_{\mathrm{e}}}{\partial t} + \mathbf{v}\frac{\partial f_{\mathrm{e}}}{\partial \mathbf{r}} - \frac{e\mathbf{E}}{m}\frac{\partial f_{\mathrm{e}}}{\partial \mathbf{v}} - \frac{\mu_0 e}{m}(\mathbf{V}_{\mathrm{L}} \times \mathbf{H}_0)\frac{\partial F_{\mathrm{e}}}{\partial \mathbf{v}} \end{aligned}\right\} \quad (3.7)$$

and, for example, when examining the flow of a plasma around a body moving in it, the system (3.7) has to be solved simultaneously with the Poisson equation

$$\Delta\phi = -\frac{e}{\varepsilon_0}\left(\int f_{\mathrm{i}} d^3 v - \int f_{\mathrm{e}} d^3 v\right), \quad \mathbf{E} = -\operatorname{grad}\phi. \quad (3.8)$$

Here and in the expression for the Lorentz force in (3.7), $\mathbf{V}_{\mathrm{L}} = \mathbf{v} + \mathbf{V}_0$, where $\mathbf{V}_0$ is the velocity of the body.

In (3.7) and (3.8) t is time, $\mathbf{r}$ is a vector defining the position of a particle, $\mathbf{v}$ is its velocity, ϕ and $\mathbf{E}$ are the electric potential and the electric field respectively, while $f_{\mathrm{e}}(\mathbf{r}, \mathbf{v}, t)$ and $f_{\mathrm{i}}(\mathbf{r}, \mathbf{v}, t)$ are the distribution functions of the electrons and ions, which in the general case depend on the spatial coordinates, the velocity, and time. A plasma may have an ordered velocity V_{p} relative to the observation point. In a problem concerning the motion of a body in a quiescent plasma $V_{\mathrm{p}} = -V_0$. If the ordered velocity V_{p} characterizes a particle flux traversing the plasma, then the corresponding equations must be supplemented with components describing the distribution functions of the particle fluxes, which are in fact the external sources acting upon the plasma.

Depending on the various conditions of the problem, the elements of tensor ε_{sp} will have one form or another, which determines the nature of the phenomena observed in the plasma, and in particular its oscillation spectra. Naturally, concrete problems require formulations of specific conditions, for instance: the temperature properties of the particles, the boundary conditions at the surface of a moving body, the nature of the sources (external electric fields, incident waves, oncoming fluxes), etc. In addition, in problems relating, for example, to plasma flow around a body (Al'pert *et al.*, 1963, 1965), the collision integrals y_e and y_i, which take into account the effect of particle collisions on the distribution function, have to be added to the right-hand side of (3.7). In a number of cases the effect of the collision integral can be neglected. However, when examining the scattering of radio waves in a body's wake or wave spectra, for instance, it is of fundamental importance to take into account collisions, since otherwise the formulas obtained may diverge (see Al'pert *et al.*, 1963). Specific cases that have been studied experimentally or theoretically will be presented in the following sections. Here let us just mention some of the important general properties of a plasma and of the quantities characterizing it.

In undisturbed plasma regions the distribution functions f_e and f_i are Maxwellian and depend only on the particle velocities $\mathbf{v}$. For example, at fairly large distances from a moving body, where the plasma is only slightly disturbed,

$$\left.\begin{aligned} f_{i0} &= N_{i0}\left(\frac{M}{2\pi kT}\right)^{3/2}\exp\left[-\frac{M(\mathbf{v}+\mathbf{V}_0)^2}{2kT}\right],\\ f_{e0} &= N_{e0}\left(\frac{m}{2\pi kT}\right)^{3/2}\exp\left[-\frac{m(\mathbf{v}+V_0)^2}{2kT}\right],\end{aligned}\right\} \tag{3.9}$$

N_{e0} and N_{i0} being the undisturbed electron and ion densities. In disturbed regions the corresponding particle densities are defined by integrals:

$$\left.\begin{aligned} N_i(\mathbf{r},t) &= \int f_i(\mathbf{r},\mathbf{v},t)\mathrm{d}^3v,\\ N_e(\mathbf{r},t) &= \int f_e(\mathbf{r},\mathbf{v},t)\mathrm{d}^3v.\end{aligned}\right\} \tag{3.10}$$

In studies of plasma flow around a body, if the problems are solved in a coordinate system moving with the body, then the dependences of all the quantities on time $\partial f/\partial t = 0$, and the problem becomes a steady-state one. Naturally, in the absence of ordered plasma motion (see (3.9)), we have

$\mathbf{V}_p = 0$. If particle fluxes (beams of electrons or ions) impinging on the plasma are present, for instance with a Maxwellian distribution, then N_{i0} or N_{e0} in (3.9) for a beam must be replaced by the particle densities of the beams, and $\mathbf{V}_0$ by the beam velocity $\mathbf{V}_b$. Note that the Maxwellian velocity-distribution function is distinguished by the fact that, when $\mathbf{V}_0 = 0$, we have $\partial f_0/\partial v = 0$ if $\mathbf{v} = 0$ and $\partial f_0/\partial v < 0$ if $\mathbf{v} \neq 0$, throughout the entire velocity region $|\mathbf{v}| > 0$. This means that in an *equilibrium plasma* γ is always greater than 0 (see (3.3)); thus there can be no process of oscillation growth (wave excitation) in it. It has been shown that, in general, for an arbitrary form of the distribution function:

$$\gamma \simeq -\left(\frac{\partial f}{\partial v}\right) \tag{3.11}$$

(Landau, 1946). Therefore, if $\partial f/\partial v < 0$, then $\gamma > 0$. However, there may be distribution functions that, within a certain interval of variation of the particle velocities, have values of $\partial f/\partial v > 0$. For example, in the case of a beam of particles incident upon a plasma, their ordered velocity may in a certain velocity interval exceed the equilibrium thermal velocities of the plasma. In this velocity region the general distribution function will have an ascending branch with a derivative $\partial f/\partial v > 0$, where $\gamma < 0$. This leads to the possibility of an increase in the plasma oscillations and to the excitation of waves in the plasma. It is this, properly speaking, that constitutes *beam instability* of a plasma, leading to an accumulation of resonance oscillations in it (see Chapter 6).

There is one circumstance that greatly simplifies the solution of a certain type of problem in the near-Earth and interplanetary plasma, for example, problems related to plasma flow around a moving artificial body whose velocity $V_0 \ll v_e$. In this case, in the region where we may disregard the influence of absorption and emission of particles by the body's surface (see Chapter 10), the electrons have a Maxwell–Boltzmann distribution, i.e.,

$$f_e(\mathbf{v}) = N_{e0}(m/2\pi kT)^{3/2} \exp\left[e\varphi/kT - m(\mathbf{v} + \mathbf{V}_0)^2/2kT\right]. \tag{3.12}$$

Therefore, the second equation in system (3.7) disappears, and the Poisson equation (3.8) becomes simplified and can be rewritten as

$$\Delta\phi = -(e/\varepsilon_0)\left(\int f_i d^3 v - N_{e0} \exp(e\phi/kT)\right). \tag{3.13}$$

Naturally, just as in the system of equations (3.7) and (3.8), in (3.13) the potential $\phi = \phi(\mathbf{r}, t)$, i.e., it depends on $\mathbf{r}$ and t if an unsteady-state problem is being considered, whereas $\phi = \phi(\mathbf{r})$ if the problem is steady-state, that is, if it is assumed in (3.7) that $\partial f/\partial t = 0$.

A plasma that can be described on the basis of kinetic theory, this determined by whether the physical phenomena to be considered depend on the velocities of particles of various kinds, is known to possess frequency and spatial dispersion. *Frequency dispersion* is manifested in the fact that different physical parameters of the described phenomena are functions of the frequency f (the angular frequency, or pulsatance, is $\omega = 2\pi f$), i.e., the state of the plasma at a given moment depends on how the processes took place previously, and *temporal inertia* of the plasma exists. *Spatial dispersion* of a plasma, on the other hand, means that various processes in it are nonlocal in terms of space, the state of the plasma at a given point depending essentially on the influence exerted by the entire region surrounding it. As a result, the various quantities describing the phenomena occurring in the plasma depend on the wave number $K = 2\pi/\lambda$. There then exists *spatial inertia* of the plasma, related to a transfer of the 'effect' from one point to another. Specifically, the elements of the tensor ε_{sp} of the plasma permittivity are therefore functions both of ω and of $\mathbf{K}$.

In solving various problems it is often necessary to use the complex conductivity tensor $\{\sigma_{sp}\}$, which is related as follows to the elements of the tensor ε_{sp}:

$$\varepsilon_{sp}(\omega, K) = \delta_{sp} - (\mathrm{i}/\varepsilon_0\omega_0)\sigma_{sp}(\omega, K). \tag{3.14}$$

In (3.14) the elements of σ_{sp} depend on the polarizability of the various kinds of particles and are found from the solution of the system of equations (3.7) and (3.8). The numbers δ_{sp}, known as Kronecker constants, are equal to unity if $s = p$ and to zero if $s \neq \mathrm{p}$.

The complex refractive index (3.2) may be written as

$$\tilde{n}(\omega, \mathbf{K}) = n(\omega, \mathbf{K}) - \mathrm{i}\kappa(\omega, \mathbf{K}). \tag{3.15}$$

The frequencies determined from (3.1) are also functions of $\mathbf{K}$, i.e.,

$$\tilde{\omega}(\mathbf{K}) = \omega(\mathbf{K}) + \mathrm{i}\gamma(\mathbf{K}). \tag{3.16}$$

Therefore, the dispersion equations are often reduced to the form

$$F(\omega, \mathbf{K}) = 0 \text{ or } \omega = \omega(\mathbf{K}). \tag{3.17}$$

This not only is a convenient form in which to write them, but sometimes also makes it easier to grasp the inherent significance of the phenomena in question.

Resonance properties of a plasma manifest themselves, theoretically speaking, as follows. An analysis of the integrals defining the tensor of the permittivity of the plasma shows that they always have singular points, which determine the conditions that produce the strongest interaction between the plasma particles and the wave field, and there may be a *resonance accumulation* of oscillations in the plasma, when during the wave-particle interaction the wave frequency coincides with the *natural frequencies* of the oscillators (i.e. the resonating plasma particles). In such a case, naturally, the resonance frequencies of the oscillators are shifted Dopplerwise with respect to the frequencies they possess in their own stationary coordinate system, these being, for example, the frequencies $s\omega_H$, where $s = \pm 1, 2, 3\ldots$ Corresponding resonance conditions arise when the wave field interacts with both the electrons and the ions of the plasma, and they have the following form:

$$\omega = \mathbf{K}\cdot\mathbf{v}_{\parallel}, \quad \omega = \mathbf{K}\cdot\mathbf{v}_{\parallel} + s\omega_H, \quad \omega = \mathbf{K}\cdot\mathbf{v}_{\parallel} - s\Omega_H, \tag{3.18}$$

where $\mathbf{K}\cdot\mathbf{v}_{\parallel} = Kv_{\parallel}\cos\theta, \theta$ being the angle between $\mathbf{K}$ and $\mathbf{H}_0$, and $\mathbf{v}_{\parallel}$ is the longitudinal (along the vector $\mathbf{H}_0$) component of the mean velocity of the particles. Note that for particle velocities close to c relativistic effects have to be taken into account, so in (3.18) the gyrofrequencies ω_H and Ω_H must be replaced by $\omega_H(1 - v^2/c^2)^{1/2}$ and $\Omega_H(1 - v^2/c^2)^{1/2}$, where v is the total particle velocity.

The first of conditions (3.18) describes the *Čerenkov–Vavilov effect* and defines the conditions for the so-called Čerenkov attenuation or, conversely, the excitation of plasma oscillations. If the wave phase velocity ω/K is greater than the longitudinal component of the particle velocity $v_{\parallel}$, then the oscillations become damped, since the particles receive more energy from the field than they release. The reverse phenomenon, Čerenkov excitation, takes place for $\omega/K < v_{\parallel}$, when in the course of particle–wave interaction the particles derive less energy from the waves than they give off. Here it is obvious that the condition $\omega = Kv_{\parallel}\cos\theta$ can be fulfilled only if $\cos\theta > 0$. This means that the Čerenkov radiation is in the direction in which the particle is moving.

The other two conditions of (3.18) describe the *magnetobraking* (cyclotron) excitation or attenuation of waves, also depending on whether the phase velocity of the wave is greater or smaller than the particle velocity. The physical significance of the terms $\mathbf{K}\cdot\mathbf{v}_{\parallel}$ in (3.18) is that they define the Doppler shifts of the frequencies of the excited oscillations. When $s > 0$ the Doppler effect is normal: the wave phase velocity $v_{\phi} > v_{\parallel}$. When $s < 0$ the Doppler effect is anomalous: the phase velocity $v_{\phi} < v_{\parallel}$. It is readily seen that, depending on the sign of s, the condition of gyroresonance is satisfied for either $\cos\theta > 0$ or $\cos\theta < 0$. In the first case $\theta < \pi/2$, i.e., the wave emission is in the direction of particle motion, this corresponding to the anomalous Doppler effect, as in the case of Čerenkov radiation. In the second case $\theta > \pi/2$, i.e., the direction of emission is opposite to the direction of particle motion, this corresponding to the normal Doppler effect.

Another fundamental property of a *kinetic equilibrium* plasma, i.e., a plasma in which the effect of the thermal motion of the particles is taken into account, is as follows. When collisions between particles are disregarded, i.e., in a collisionless plasma, the *oscillatory processes become attenuated* due to their interaction with the particles ($\gamma < 0$). The absence of this attenuation, which is known as *Landau attenuation* or *Landau damping* (Landau, 1946), and also as Čerenkov or gyroresonance attenuation, is in fact one of the main qualitative distinctions of a *cold* plasma. Here it is important to note that resonance phenomena are also observed in a cold plasma, i.e., when the theory is constructed on the basis of hydrodynamic (or quasihydrodynamic) equations or on the basis of macroscopic equations.

4

Refractive indexes for a cold magnetoplasma

4.1. Permittivity-tensor elements for a collisional plasma

In the general case, the dispersion equation (3.1) defines the two complex refractive indexes

$$\tilde{n}_{\pm}^2 = \tilde{n}_{12}^2 = (n_{12} - \mathrm{i}\kappa_{12})^2 = \frac{-B \pm \sqrt{(B^2 - 4AC)}}{2A}. \tag{4.1}$$

When the influence of the thermal motion of particles is disregarded, i.e., when it is assumed that the plasma is cold, the temperatures of its particles being equal to zero and fluxes being absent from it, the quantities A, B, and C in (4.1) are described by the formulas

$$\left.\begin{aligned} A &= \varepsilon_1 \sin^2\theta + \varepsilon_3 \cos^2\theta, \\ B &= -\varepsilon_1\varepsilon_3(1 + \cos^2\theta) - (\varepsilon_1^2 - \varepsilon_2^2)\sin^2\theta, \\ C &= \varepsilon_3(\varepsilon_1^2 - \varepsilon_2^2). \end{aligned}\right\} \tag{4.2}$$

In a collisionless plasma the tensor elements have the form

$$\left.\begin{aligned} &\varepsilon_1 = \varepsilon_{11} = \varepsilon_{22} = 1 - \frac{\omega_0^2}{\omega^2 - \omega_{\mathrm{H}}^2} - \frac{\Omega_{01}^2}{\omega^2 - \Omega_{\mathrm{H}1}^2} - \frac{\Omega_{02}^2}{\omega^2 - \Omega_{\mathrm{H}2}^2} - \ldots, \\ &\mathrm{i}\varepsilon_2 = \varepsilon_{12} = \varepsilon_{21} = \mathrm{i}\frac{\omega_0^2\omega_{\mathrm{H}}}{\omega(\omega^2 - \omega_{\mathrm{H}}^2)} - \mathrm{i}\frac{\Omega_{01}^2\Omega_{\mathrm{H}1}}{\omega(\omega^2 - \Omega_{\mathrm{H}1}^2)} \\ &\qquad - \mathrm{i}\frac{\Omega_{02}^2\Omega_{\mathrm{H}2}}{\omega(\omega^2 - \Omega_{\mathrm{H}2}^2)} - \ldots, \\ &\varepsilon_3 = \varepsilon_{33} = 1 - \frac{\omega_0^2}{\omega^2} - \frac{\Omega_{01}^2}{\omega^2} - \frac{\Omega_{02}^2}{\omega^2} - \ldots, \\ &\varepsilon_{13} = \varepsilon_{31} = \varepsilon_{23} = \varepsilon_{32} = 0. \end{aligned}\right\} \tag{4.3}$$

On the right-hand side of (4.3), the subscripts 1, 2... refer to the different kinds of ion. The refractive indexes are naturally real quantities.

It is important to stress that the formulas (4.1) and (4.2) have a wider field of applicability than just a cold plasma. They can be used if the conditions

$$\left.\begin{array}{l}\left.\begin{array}{l}\left(\dfrac{K_\perp v_e}{\omega_H}\right)^2=\left(\dfrac{\rho_{He}}{\Lambda_{\perp e}}\right)^2\ll 1,\\[2ex] \left(\dfrac{K_\perp v_i}{\Omega_H}\right)^2=\left(\dfrac{\rho_{Hi}}{\Lambda_{\perp i}}\right)^2\ll 1,\end{array}\right\}\\[4ex] \left.\begin{array}{l}\left(\dfrac{\omega-s\omega_H}{K_{\|}v_e}\right)=\left(1-\dfrac{s\omega_H}{\omega}\right)\dfrac{v_{\phi\|e}}{v_e}\gg 1,\\[2ex] \qquad\qquad\left(1-\dfrac{s\Omega_H}{\omega}\right)\dfrac{v_{\phi\|i}}{v_i}\gg 1,\end{array}\right\}\end{array}\right\} \tag{4.3a}$$

are satisfied where $s=0,\pm 1,\pm 2,\ldots$, and the subscripts $\perp$ and $\|$ signify the transverse and longitudinal (along $\mathbf{H}_0$) components of the wave vector $\mathbf{K}$ and of the wavelengths Λ. These conditions mean that the *approximation of a cold plasma* is valid when the wavelengths of low-frequency (LF) and high-frequency (HF) oscillations are greater than the Larmor radii of the ions and electrons, respectively, and the phase velocities of the waves are large in comparison with the thermal velocities of these particles. This approximation does not apply, however, under conditions quite close to the gyroresonances of the ions and electrons.

The complex refractive index is naturally determined by tensor elements including the effects of collisions between electrons and ions $\nu_{e,is}=\nu_{es}$, between ions and electrons $\nu_{is,e}=\nu_{se}$, and between different kinds of ions $\nu_{is,p}=\nu_{sp}$ and $\nu_{ip,s}=\nu_{ps}$. Collisions ν_{ee} and ν_{ii} between electrons and between ions of one kind also play an important (and under certain conditions a predominant) role (see below). However, it goes without saying that a consistent computation of their effect is possible only using kinetic theory.

The general formula that is commonly used for the frequency of collisions between charged particles of different kinds has the form

$$\nu_{sp}=\frac{4\sqrt{(2\pi)}e_s^2e_p^2N_p}{3(M_sk^3T^3)^{1/2}}\left(\frac{M_s}{M_s+M_p}\right)^{1/2}L(\ldots), \tag{4.4}$$

where e_s, e_p and M_s, M_p are, respectively, the charges and masses of the corresponding kinds of particles and $L(\ldots)$ is the so-called Coulomb logarithm (see Eq. (4.5) below). Here for electrons $(m,-e)$ we use the subscript e everywhere (for example, v_e), and for ions (M,e) the subscripts s and p (for example v_s, v_p).

Apparently, beginning with the paper by Smith & Brice, 1964, incorrect formulas for the tensor elements in a collisional plasma were used in the literature. In Smith & Brice, 1964, as in subsequent studies (see, for instance, Gurnett *et al.*, 1965; Jones, 1969, 1970), the masses m and M_s in formulas (4.3), in the expressions for the Langmuir (plasma) frequencies ω_0, Ω_{0s} and for the gyrofrequencies ω_H, Ω_{Hs} are replaced by the 'effective' masses $m_{eff} = m(1 - i\nu_e/\omega)$ and $M_{s\,eff} = M_s(1 - i\nu_{is}/\omega)$. Here the values of ν_e and ν_{is} are determined as the 'effective numbers of collisions' of particles of type s. However, formulas with effective masses are erroneous, and their use does not merely result in quantitative errors. In Section 4.4 we will present the results of numerical calculations illustrating this point. It is readily shown that only at frequencies $\omega \gg \omega_L \equiv \omega_H \cos\theta$, where the influence of ion motion is negligible, will a term of type m_{eff}, i.e., the effective mass of the electrons, appear in the formulas for $(n - i\kappa)$.

Formulas with effective masses are obtained by solving equations of motion in which the relative velocities of different kinds of charged particles are not taken into account. Here the inherent meaning of the collisions characterized by ν_e and ν_{is} is not revealed. Effective masses have been used to calculate the complex refractive indexes (see Jones, 1969 under conditions close to the ion gyrofrequencies with $\nu_e = \nu_{ee} + \sum \nu_{es}$ and $\nu_{is} = \sum (\nu_{iis} + \nu_{sp})$. However, there is no basis for such a definition of ν_e and ν_{is}.

4.1.1. Tensor elements taking into account collisions between different kinds of charged particles

If a collisional plasma is multicomponent, i.e., if it consists of several kinds of ions, then the formulas for the tensor elements are extremely cumbersome, their structure lacking symmetry and becoming more and more complicated as the number of ions increases. New formulas therefore have to be derived for each number of ions. This in general makes it very problematic to perform kinetic calculations, and these are carried out only for certain specific cases.

Below we present formulas for $\varepsilon_1, \varepsilon_2$ and ε_3 for a collisional cold plasma containing only one or two kinds of ions. They were obtained in the hydrodynamic approximation, assuming also that the plasma does not contain negative ions (Al'pert, 1980a). Along with these we give formulas for three kinds of attenuation factors κ_c, κ_{ii} and κ_{ee} and the corresponding expressions for the tensor elements in the frequency range $0 < \omega \sim \Omega_{Hs}$ under conditions where $n \gg \kappa$. These were obtained from kinetic calculations with the aid of the collision integral (Akhiezer *et al.*, 1976). Under these conditions the refractive indexes n_{12} of a cold plasma will have a greater degree of accuracy. A combination of results of calculations in the hydrodynamic and kinetic approximations thus makes it possible to determine the total attenuation of ELF waves in a number of cases. The attenuation factors κ_{ii} and κ_{ee} are determined by collisions ν_{ii} and ν_{ee} between charged particles of the same kind i.e., they are purely kinetic in nature. On the other hand, the attenuation factor κ_c depends on the collision frequencies $\nu_{es}, \nu_{se}, \nu_{sp}$ and ν_{ps} and is determined solely by collisions between different kinds of charged particles. The formulas for κ_c and for κ_{12} are very similar.

The fact that a plasma is multicomponent is of great significance and leads, as will be seen, to new effects. In order to understand these and to analyze the results of a number of experiments, we can often simply confine ourselves, in the plasma regions of interest to us, to formulas which take into account the presence of only one or two kinds of ions ($s = 1, 2$). When deriving the tensor elements, and everywhere below, we will use the frequencies of the collisions of electrons with ions and between different kinds of ions, these being determined with the aid of the formulas

$$\left.\begin{aligned}\nu_{e,i1} &= \nu_{e1} = \pi\frac{e^4}{(kT)^2}v_e N_1 \ln\left(0.37\frac{kT}{e^2N^{1/3}}\right) = \frac{5.5N_1}{T^{3/2}}\ln\left(220\frac{T}{N^{1/3}}\right),\\ \nu_{e,i2} &= \nu_{e2} = \nu_{e1}N_2/N_1, \quad \nu_{1e} = \nu_{2e} = \nu_{e1}N/N_1,\end{aligned}\right\} \tag{4.5a}$$

$$\left.\begin{aligned}\nu_{i1,i2} = \nu_{12} &= 2\sqrt{(2\pi)}\frac{e_1^2e_2^2}{(kT)^{3/2}}N_2\left[\frac{M_2}{M_1(M_1+M_2)}\right]^{1/2}\\ &\quad\times\ln\left(0.37\frac{kT}{e^2N^{1/3}}\right)\\ &= \frac{5.5N_2}{T^{3/2}}\left[\frac{mM_2}{M_1(M_1+M_2)}\right]^{1/2}\ln\left(220\frac{T}{N^{1/3}}\right),\\ \nu_{21} &= \nu_{12}\frac{N_1M_1}{N_2M_2}, \quad (e_1 = e_2 = e, \quad N_1 + N_2 = N).\end{aligned}\right\} \tag{4.5b}$$

Here the frequencies of collisions between particles of the same kind are

$$\nu_{ee} = \frac{N}{(\sqrt{2})N_s}\nu_{e,is} \text{ and } \nu_{ii,s} = \frac{1}{\sqrt{2}}\nu_{e,is}\left(\frac{m}{M_s}\right)^{1/2}. \tag{4.5c}$$

Let us note here that in (4.5) the frequencies of collisions are determined somewhat differently than in (4.4), this being related to the form in which the equations of motion (4.7) have been written, so as to ensure that they obey the law of conservation of momentum (see below).

In the formulas (4.5) N is the electron density, N_1, N_2 and M_1, M_2 are, respectively, the densities and masses of the different kinds of ions ($s=1, p=2$); $v_e=(8kT/\pi m)^{1/2}$ is the mean thermal velocity of the electrons (Chapman & Cowling, 1939; Cowling, 1945; Nicolet, 1953; Ginzburg, 1967). It should be borne in mind that the quantities entering into the Coulomb logarithm, as well as the numerical coefficients of formulas (4.5), are accurate only to a multiplier $\simeq 1$ and that they correspond to the case where the ions are singly ionized, i.e., $e_1=e_2=e$. It is readily perceived from (4.5) that the ratio $\nu_{12}/\nu_{ei} \sim (m/M)^{1/2}$; this fact may be useful when deriving approximate formulas for $\tilde{n}^2$ in a multicomponent plasma. However, in view of the awkwardness of the general formulas, the corresponding expressions require accurate manipulation of small terms, of the same order of smallness, for example, as terms of the form $(\Omega_{Hs}/\omega_H)^{1/2}$ or Ω_{0s}/ω_0.

Some of the formulas for $\tilde{n}_{12}^2$ given below can also be obtained taking into account collisions between electrons and neutral particles, characterized by

$$\nu_{en} = \tfrac{4}{3}\sigma_{en} N_n v_e, \tag{4.6}$$

since in the lower regions of the ionosphere they can still play an appreciable role. In formula (4.6), σ_{en} is the effective cross-section and N_n is the number density of neutral particles. Curves for σ_{en} as a function of temperature, for the particles playing a role in these regions of the ionosphere, are given in Fig. 4.1. The quantity ν_{en} will be used, in particular, in Chapter 7, when describing nonlinear effects in a plasma.

To determine the permittivity tensor for a cold plasma consisting of electrons and two kinds of ions, we use the equations of motion of the

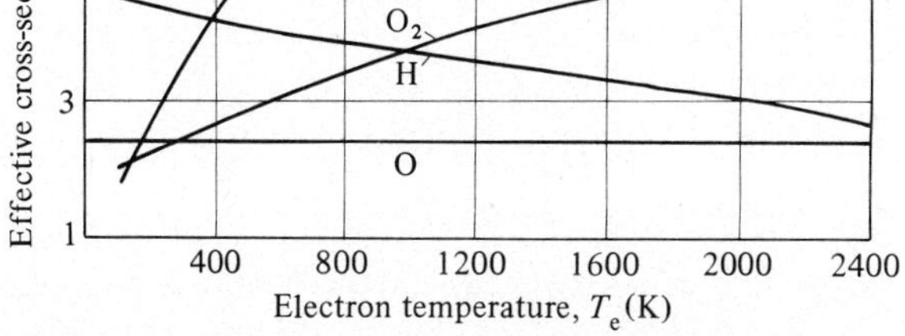

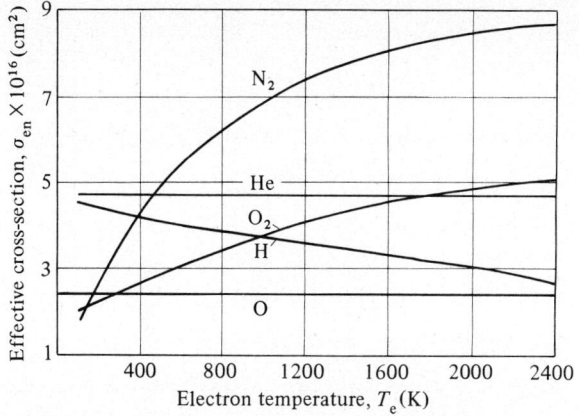

Fig. 4.1. Effective cross-sections σ_{en}, for collisions of electrons with different kinds of neutral particles, as functions of the temperature T_e.

charged particles, written as

$$\left.\begin{aligned} m\ddot{\mathbf{r}}_e &= -e\mathbf{E} - \mu_0 e\dot{\mathbf{r}}_e \times \mathbf{H}_0 - m\nu_{e1}(\dot{\mathbf{r}}_e - \mathbf{r}_{i1}) - m\nu_{e2}(\dot{\mathbf{r}}_e - \dot{\mathbf{r}}_{i2}) \\ M_1\ddot{\mathbf{r}}_{i1} &= e\mathbf{E} + \mu_0 e\dot{\mathbf{r}}_{i1} \times \mathbf{H}_0 + m\nu_{1e}(\dot{\mathbf{r}}_e - \dot{\mathbf{r}}_{i1}) - M_1\nu_{12}(\dot{\mathbf{r}}_{i1} - \mathbf{r}_{i2}), \\ M_2\ddot{\mathbf{r}}_{i2} &= e\mathbf{E} + \mu_0 e\dot{\mathbf{r}}_{i2} \times \mathbf{H}_0 + m\nu_{2e}(\dot{\mathbf{r}}_e - \dot{\mathbf{r}}_{i2}) + M_2\nu_{21}(\dot{\mathbf{r}}_{i1} + \dot{\mathbf{r}}_{i2}), \end{aligned}\right\} \quad (4.7a)$$

and the equation for the total current

$$\mathbf{j} = -e(N\dot{\mathbf{r}}_e - N_1\dot{\mathbf{r}}_1 - N_2\dot{\mathbf{r}}_2) = \{\sigma_{sp}\}\mathbf{E}. \quad (4.7b)$$

The collision frequencies appearing in (4.7*a*) are defined in (4.5); $\dot{\mathbf{r}}_e$, $\dot{\mathbf{r}}_{11}$ and $\dot{\mathbf{r}}_{12}$ are the ordered particle velocities determined by the action of the electric field $\mathbf{E}$ of the electromagnetic wave $\mathbf{E} = \mathbf{E}_0 \exp[i(\omega t - \mathbf{K}\cdot\mathbf{r})]$ and the external constant magnetic field $\mathbf{H}_0$. In (4.7*b*), $\{\sigma_{sp}\}$ is the tensor of the complex conductivity. It is related to the permittivity tensor as follows (see Eq. (3.14)):

$$\{\varepsilon_{sp}\} = \begin{Bmatrix} 1 & 0 & 0 \\ 0 & 1 & 0 \\ 0 & 0 & 1 \end{Bmatrix} - \frac{i}{\omega\varepsilon_0}\{\sigma_{sp}\} \quad (4.7c)$$

Thus, in order to determine the tensor elements σ_{sp} and ε_{sp}, we need to solve Eqs. (4.7*a*) for the velocities $\dot{\mathbf{r}}_{e1,2}$, i.e., to establish their connection with E_x, E_y, and E_z. To do this, we can also use the following relation which ensues directly from (4.7*a*) and (4.7*b*):

$$\mu_0\mathbf{j} \times \mathbf{H}_0 = i\omega e(Nm\dot{\mathbf{r}}_e + N_1M_1\dot{\mathbf{r}}_1 + N_2M_2\dot{\mathbf{r}}_2) \quad (4.7d)$$

If only one kind of ion is present, so that $M_2 = 0$ and $N_2 = 0$, then assuming that $M_1 = M$ and $N_1 = N$ and bearing in mind that $\nu_{e1} = \nu_{1e} = \nu_{ei}$, the corresponding calculations lead us to the following formulas for the permittivity-tensor elements:

$$\left.\begin{aligned} \varepsilon_1 &= 1 - \frac{\omega_0^2 + \Omega_0^2}{P(\omega)}(\omega^2 - \Omega_H\omega_H - i\nu_{ei}\omega), \\ \varepsilon_2 &= \omega\frac{\omega_0^2\omega_H - \Omega_0^2\Omega_H}{P(\omega)}, \\ \varepsilon_3 &= 1 - \frac{\omega_0 + \Omega_0^2}{\omega^2(1 - i\nu_{ei}/\omega)}, \end{aligned}\right\} \quad (4.8)$$

where

$$P(\omega) = [(\omega + \omega_H)(\omega - \Omega_H) - i\nu_{ei}\omega][(\omega - \omega_H)(\omega + \Omega_H) - i\nu_{ei}\omega]. \quad (4.9)$$

In (4.8) and (4.9) the multipliers $(1 + m/M)$, which in the given case are not significant, are omitted from the collision frequencies ν_{ei} (Al'pert, 1980).

If two kinds of ions are present in the plasma, taking into account that

$(N_1+N_2)=N$, then the following formulas are obtained for the tensor elements ε_1, ε_2 and ε_3:

$$\varepsilon_1=\varepsilon_{11}=\varepsilon_{22}=1-\frac{\omega_0^2}{\omega^2}\frac{\{A\cdot D\}+\dfrac{N_2\omega_{\rm H}\Omega_{\rm H1}}{N\omega^2}\{B\cdot F\}}{\{A^2\}-\dfrac{\omega_{\rm H}^2}{\omega^2}\{B^2\}},$$

$$i\varepsilon_2=\varepsilon_{12}=-\varepsilon_{21}=i\frac{\omega_0^2}{\omega^2}\frac{\dfrac{N_2}{N}\dfrac{\Omega_{\rm H1}}{\omega}\{A\cdot F\}+\dfrac{\omega_{\rm H}}{\omega}\{B\cdot D\}}{\{A^2\}-\dfrac{\omega_{\rm H}^2}{\omega^2}\{B^2\}}, \qquad (4.10)$$

$$\varepsilon_3=\varepsilon_{33}=1-\frac{\omega_0^2}{\omega^2}\frac{D}{\{A\}+\dfrac{\omega_{\rm H}}{\omega}\dfrac{\Omega_{\rm H2}}{\omega}\{C\}},$$

$$\varepsilon_{13}=\varepsilon_{31}=\varepsilon_{32}=\varepsilon_{23}=0,$$

where

$$\left.\begin{aligned}
\{A\}&=\left\{\left(1-\frac{\Omega_{\rm H1}+\Omega_{\rm H2}}{\omega}\frac{\omega_{\rm H}}{\omega}+\frac{\Omega_{\rm H1}\Omega_{\rm H2}}{\omega^2}\right)-\frac{\nu_{\rm e1}}{\omega}\frac{N}{N_1}\left(\frac{\nu_{12}}{\omega}+\frac{\nu_{\rm e1}}{\omega}\frac{m}{M_1}\frac{N_2}{N_1}\right)\right.\\
&\quad\times\left(1+\frac{M_1N_1}{M_2N_2}+\frac{mN}{M_2N_2}\right)-i\left\{\frac{\nu_{12}}{\omega}\left(1+\frac{M_1N_1}{M_2N_2}-\frac{N}{N_2}\frac{\Omega_{\rm H2}}{\omega}\frac{\omega_{\rm H}}{\omega}\right)\right.\\
&\quad\left.\left.+\frac{\nu_{\rm e1}}{\omega}\frac{N}{N_1}\left(1-\frac{\Omega_{\rm H1}\Omega_{\rm H2}}{\omega^2}+\frac{m}{M_1}+\frac{m}{M_2}\right)\right\}\right\},\\
\{B\}&=\left\{\left(1+\frac{\Omega_{\rm H1}\Omega_{\rm H2}}{\omega^2}-\frac{m}{M_1}-\frac{m}{M_2}\right)\right\}-i\left\{\frac{\nu_{12}}{\omega}\left(1+\frac{M_1N_1}{M_2N_2}+\frac{mN}{M_2N_2}\right)\right.\\
&\quad\left.-\frac{\nu_{\rm e1}}{\omega}\frac{2N}{N_1}\frac{m^2}{M_1M_2}\right\},\\
\{C\}&=\left\{\left(1+\frac{M_2}{M_1}-\frac{m}{M_1}\right)-i\left(\frac{\nu_{12}}{\omega}\frac{N}{N_2}+\frac{\nu_{\rm e1}}{\omega}\frac{m}{M_1}\frac{N}{N_1}\right)\right\},\\
\{D\}&=\left\{\left(1+\frac{m}{M_1}\frac{N_1}{N}+\frac{m}{M_2}\frac{N_2}{N}\right)-i\left(\frac{\nu_{12}}{\omega}+\frac{\nu_{\rm e1}}{\omega}\frac{m}{M_1}\frac{N_2}{N_1}\right)\right.\\
&\quad\left.\times\left(1+\frac{M_1N_1}{M_2N_2}+\frac{mN}{M_2N_2}\right)\right\},\\
\{F\}&=\left\{1+\frac{M_1N_1}{M_2N_2}+\frac{mN}{M_2N_2}\right\}.
\end{aligned}\right\} \qquad (4.10a)$$

Neglecting terms in (4.10) and (4.10*a*) of the order of m/M and $(m/M)^{1/2}$, we get:

$$\left.\begin{aligned}
\{A\} &= \left\{\left(1 - \frac{\Omega_{H1} + \Omega_{H2}}{\omega}\frac{\omega_H}{\omega}\right) - \frac{\nu_{12}}{\omega}\frac{\nu_{e1}}{\omega}\frac{N}{N_1}\left(1 + \frac{M_1 N_1}{M_2 N_2}\right)\right\} \\
&\quad - \mathrm{i}\left\{\frac{\nu_{12}}{\omega}\left(1 + \frac{M_1 N_1}{M_2 N_2} - \frac{N}{N_2}\frac{\Omega_{H2}}{\omega}\frac{\omega_H}{\omega}\right)\right. \\
&\quad \left. + \frac{\nu_{e1}}{\omega}\frac{N}{N_1}\left(1 - \frac{\Omega_{H1}\Omega_{H2}}{\omega^2}\right)\right\}, \\
\{B\} &= \left\{\left(1 + \frac{\Omega_{H1}\Omega_{H2}}{\omega^2}\right) - \mathrm{i}\frac{\nu_{12}}{\omega}\left(1 + \frac{M_1 N_1}{M_2 N_2}\right)\right\}, \\
\{C\} &= \left\{\left(1 + \frac{M_2}{M_1}\right) - \mathrm{i}\left(\frac{\nu_{12}}{\omega}\frac{N}{N_2}\right)\right\}, \\
\{D\} &= \left\{1 - \mathrm{i}\frac{\nu_{12}}{\omega}\left(1 + \frac{M_1 N_1}{M_2 N_2}\right)\right\}, \\
\{F\} &= \left\{1 + \frac{M_1 N_1}{M_2 N_2}\right\},
\end{aligned}\right\} \qquad (4.10b)$$

together with a simple expression for the tensor element ε_3, namely

$$\varepsilon_3 = 1 - (\omega_0^2/\omega^2)[1 - \mathrm{i}(\nu_{ei}/\omega)(N/N_1)]^{-1}. \qquad (4.10c)$$

At this point it is important to note the following. It emerges from an analysis of the formulas (4.10) that in the frequency range $0 < \omega \ll (\Omega_H \omega_H)^{1/2}$ the role of collisions between different kinds of ions is greater than that of collisions between electrons and ions.

4.1.2. Effect of collisions between like particles (electrons, or ions of one kind) in the ELF range $0 \leqslant \omega \simeq \Omega_{HS}$

A calculation of the permittivity tensor taking into account the effect of electron–electron (ν_{ee}) and ion–ion (ν_{ii}) collisions requires a sufficiently accurate and complete solution of the corresponding problem based on an analysis of the kinetic equation, including the collision integral. In problems of this type in a number of cases subtle effects are revealed which often cannot be predicted by qualitative *a priori* considerations.

The effect of mutual collisions between electrons in the high-frequency range $\omega \gg \omega_L \simeq (\omega_H \Omega_H)^{1/2}$ is insignificant and can virtually be neglected at frequencies sufficiently remote from the electron gyrofrequency ω_H. The corresponding formulas for the complex tensor elements, obtained on the basis of kinetic theory (Gurevich, 1958a, 1959), are given below (see (7.19)–(7.22) and Fig. 7.9). As we know, however, the effects of ν_{ee} and ν_{ii} on the attenuation of ELF waves $0 \leqslant \omega \lesssim \Omega_{Hs}$, where the principal role is played by the motion of ions, has been studied only very recently (Akhiezer *et al.*, 1976). On the basis of qualitative considerations we could assume that ion–ion collisions must play a role in a plasma consisting of electrons and one kind of ion. The following qualitative considerations lead to just such a conclusion. An analysis of the quantity κ_{12} in a cold plasma having several kinds of ions showed that collisions between the different kinds of ions, specified by ν_{sp}, may prove to be more significant than collisions between electrons and ions, specified by ν_{ei}. If is therefore natural to assume that, on going to the limit, where only one kind of ion remains in the plasma, the effect of ν_{ii} must manifest itself.

The attenuation factor of ELF waves $0 \leqslant \omega \sim \Omega_{Hs}$, including the beginning of the range of VLF waves, where $\Omega_{Hs} \lesssim 0 \ll \omega_L$, can, assuming the conditions

$$\left.\begin{aligned} &n \gg \kappa, \quad \frac{\nu_{es}}{\Omega_H} \ll \frac{v_e}{c} n\cos\theta, \quad \frac{\nu_{sp}}{\Omega_H} \ll \frac{v_{is}}{c} n\cos\theta, \quad \frac{v_{is}}{c} n\sin\theta \ll \frac{\Omega_{Hs}}{\omega}, \\ &\frac{v_e}{c} n\sin\theta \ll \frac{\omega_H}{\omega}, \quad \frac{v_{is}}{c} n\cos\theta \ll 1, \quad \left(\frac{v_{is}}{V_A}\right)^2 \simeq \frac{4\pi N_s M_s}{H_0^2}\frac{2kT}{M_s} \ll 1, \end{aligned}\right\} \tag{4.11}$$

are satisfied, be expressed as the sum of three components:

$$(\kappa)_{12} = (\kappa_c)_{12} + (\kappa_{ii})_{12} + (\kappa_{ee})_{12}, \tag{4.11a}$$

where V_A is the Alfvén velocity, and all the other quantities in (4.11) have been defined above. For the definitions of the terms VLF and ELF, see the classification of frequency ranges given below in Section 4.3.2.

The quantity (κ_c) in (4.11*a*) is determined by the frequencies ν_{es} and ν_{sp} of the collisions between different kinds of charged particles. Thus, this type of attenuation arises due to the *frictional forces* between them, so (κ_c) must coincide with the values of κ_{12} obtained using the formulas worked out for a cold plasma. The quantity (κ_{ii}) defines the attenuation due to ion–ion collisions and it appears as a result of *ion viscosity*. As shown by an analysis of the formulas given below, it plays a role in a plasma containing only one kind of ion. The attenuation factor $(\kappa_{ee}) \sim \nu_{ee}$, on the other hand, arises due to *electron viscosity* and the motion of the electrons relative to the ions along the magnetic field $\mathbf{H}_0$.

The values of all these attenuation factors are given by the following formulas:

$$(\kappa_e)_{12}=\frac{\varepsilon_{es1}[(1+\cos^2\theta)n_{12}^2-2\varepsilon_1]-2\varepsilon_2\varepsilon_{es2}}{2n_{12}[2n_{12}^2\cos^2\theta-\varepsilon_1(1+\cos^2\theta)]},\tag{4.11b}$$

$$(\kappa_{ii})_{12}=\frac{\varepsilon_{ii1}(n_{12}^2-\varepsilon_1)+\varepsilon_{ii2}(n_{12}^2\cos^2\theta-\varepsilon_1)-2\varepsilon_2\varepsilon_{ii12}}{2n_{12}[2n_{12}^2\cos^2\theta-\varepsilon_1(1+\cos^2\theta)]},\tag{4.11c}$$

$$\left.\begin{aligned}(\kappa_{ee})_{12}=&\frac{\nu_{ee}}{\omega}\sin^2\theta\left\{\frac{1}{\sqrt{2}}\frac{\omega^2}{\omega_0^2}(\varepsilon_1 n_{12}^2-\varepsilon_1^2+\varepsilon_2^2)n_{12}\right.\\&+\frac{\omega}{\omega_H}\frac{v_e^2}{c^2}n_{12}\left[\frac{1+\sqrt{2}}{5}\frac{\omega_0^2}{\omega\omega_H}(n_{12}^2\cos^2\theta-\varepsilon_1)\right.\\&\left.\left.-\frac{4+7\sqrt{2}}{5}\varepsilon_2 n_{12}^2\cos^2\theta\right]\right\}\\&\times[2n_{12}^2\cos^2\theta-\varepsilon_1(1+\cos^2\theta)]^{-1}.\end{aligned}\right\}\tag{4.11d}$$

Here the expression for κ_{ee} is written assuming fulfillment of the condition

$$v_e/c \ll 1/n_{12}\cos\theta.\tag{4.11e}$$

In formulas (4.11*b*–*d*) the quantities n_{12} and ε_1, ε_2 are the real values of the refractive indexes and of the dielectric tensor elements for a cold plasma. The other tensor elements in (4.11*b*) and (4.11*c*) may be determined for a multicomponent plasma ($s=1,2,\dots$) with the aid of the following formulas:

$$\left.\begin{aligned}\varepsilon_{es1}=&\sum_{(s)}\frac{m}{M_s}\frac{N}{N_s}\frac{\nu_{es}}{\omega}\frac{\Omega_{0s}^2}{\Omega_{Hs}^2}\frac{\omega^2(\omega^2+\Omega_{Hs}^2)}{(\omega^2-\Omega_{Hs}^2)^2}+\sum_{(s)}\left\{\frac{\nu_{s(s+1)}}{\omega}\frac{M_s}{M_{(s+1)}}\Omega_{0s}^2\right.\\&\times\left[\frac{\omega^2+\Omega_{Hs}^2}{(\omega^2-\Omega_{Hs}^2)^2}\frac{M_{(s+1)}}{M_s}-\frac{2(\omega^2+\Omega_{Hs}\Omega_{H(s+1)})}{(\omega^2-\Omega_{Hs}^2)(\omega^2-\Omega_{H(s+1)}^2)}\right.\\&\left.\left.+\frac{M_s}{M_{(s+1)}}\frac{\omega^2+\Omega_{H(s+1)}^2}{(\omega^2-\Omega_{H(s+1)}^2)^2}\right]\right\},\end{aligned}\right\}\tag{4.11e}$$

$$\left.\begin{aligned}\varepsilon_{es2}=&-2\sum_{(s)}\frac{m}{M_s}\frac{N}{N_s}\frac{\nu_{es}}{\Omega_{Hs}}\frac{\omega^2\Omega_{0s}^2}{(\omega^2-\Omega_{Hs}^2)^2}-2\sum_{(s)}\left\{\frac{\nu_{s(s+1)}}{\Omega_{Hs}}\right.\\&\times\Omega_{0s}^2\left[\frac{\Omega_{Hs}^2}{(\omega^2-\Omega_{Hs}^2)^2}-\frac{\Omega_{H(s+1)}^2+\Omega_{Hs}\Omega_{H(s+1)}}{(\omega^2-\Omega_{Hs}^2)(\omega^2-\Omega_{H(s+1)}^2)}\right.\\&\left.\left.+\frac{M_s}{M_{(s+1)}}\frac{\Omega_{H(s+1)}^2}{(\omega^2-\Omega_{H(s+1)}^2)^2}\right]\right\},\end{aligned}\right\}\tag{4.11f}$$

$$\left.\begin{aligned}\varepsilon_{\mathrm{ii}1} = \sum_{(\mathrm{s})}\Bigg\{&\left[\frac{2}{5}\frac{\nu_{\mathrm{iis}}}{\omega}\frac{\omega^2\Omega_{0\mathrm{s}}^2}{(\omega^2-\Omega_{\mathrm{Hs}}^2)^2}\left(\frac{v_{\mathrm{is}}}{c}\right)^2 n_{12}^2\right]\\ &\times\left[3\cos^2\theta\frac{(\omega^2+\Omega_{\mathrm{Hs}}^2)^2+4\omega^2\Omega_{\mathrm{Hs}}^2}{(\omega^2-\Omega_{\mathrm{Hs}}^2)^2}\right.\\ &\left.+\sin^2\theta\frac{4\omega^4+31\omega^2\Omega_{\mathrm{Hs}}^2+28\Omega_{\mathrm{Hs}}^4}{(\omega^2-4\Omega_{\mathrm{Hs}}^2)^2}\right]\Bigg\},\end{aligned}\right\} \tag{4.11g}$$

$$\left.\begin{aligned}\varepsilon_{\mathrm{ii}2} = \sum_{(\mathrm{s})}\Bigg\{&\left[\frac{2}{5}\frac{\nu_{\mathrm{iis}}}{\omega}\frac{\omega^2\Omega_{0\mathrm{s}}^2}{(\omega^2-\Omega_{\mathrm{Hs}}^2)^2}\left(\frac{v_{\mathrm{is}}}{c}\right)^2 n_{12}^2\right]\\ &\times\left[3\cos^2\theta\frac{(\omega^2+\Omega_{\mathrm{Hs}}^2)^2+4\omega^2\Omega_{\mathrm{Hs}}^2}{(\omega^2-\Omega_{\mathrm{Hs}}^2)^2}\right.\\ &\left.+\sin^2\theta\frac{3\omega^6+40\omega^4\Omega_{\mathrm{Hs}}^2+4\omega^2\Omega_{\mathrm{Hs}}^4+16\Omega_{\mathrm{Hs}}^6}{\omega^2(\omega^2-4\Omega_{\mathrm{Hs}}^2)^2}\right]\Bigg\},\end{aligned}\right\} \tag{4.11h}$$

$$\left.\begin{aligned}\varepsilon_{\mathrm{ii}12} = -\sum_{(\mathrm{s})}\Bigg\{&\left[\frac{2}{5}\frac{\nu_{\mathrm{iis}}\Omega_{\mathrm{Hs}}\omega^2\Omega_{0\mathrm{s}}^2}{(\omega^2-\Omega_{\mathrm{Hs}}^2)^2}\left(\frac{v_{\mathrm{is}}}{c}\right)^2 n_{12}^2\right]\\ &\times\left[12\cos^2\theta\frac{\omega^2+\Omega_{\mathrm{Hs}}^2}{(\omega^2-\Omega_{\mathrm{Hs}}^2)^2}\right.\\ &\left.+\sin^2\theta\frac{19\omega^4+28\omega^2\Omega_{\mathrm{Hs}}^2+16\Omega_{\mathrm{Hs}}^4}{\omega^2(\omega^2-\Omega_{\mathrm{Hs}}^2)^2}\right]\Bigg\}.\end{aligned}\right\} \tag{4.11i}$$

An analysis of formulas (4.11*b*–*d*) leads us to a number of important conclusions:

(1) Numerical calculations show that the attenuation factor $(\kappa_{\mathrm{c}})_{12}$ (see (4.11*a*)) is very close to the corresponding values of κ_{12} for a cold plasma. When $\theta = 0$ and the plasma consists of one kind of ion, the corresponding formulas are identical. Thus in this case it follows from (4.1.3) that

$$(\kappa_{\mathrm{c}})_{12} = \left(\frac{m}{M}\frac{\nu_{\mathrm{ei}}}{\Omega_{\mathrm{H}}}\frac{\Omega_0}{\Omega_{\mathrm{H}}}\right)\left(\frac{\omega}{\Omega_{\mathrm{H}}}\right)(1 \mp \omega/\Omega_{\mathrm{H}})^{-3/2}. \tag{4.11j}$$

An analogous formula for κ_{12} is obtained for a cold plasma when $\omega^2 \ll \omega_{\mathrm{H}}^2$ (see (4.19)).

Another property of κ_{c} consists in the following. If we transform κ_{c} into the sum of two terms:

$$(\kappa_{\mathrm{c}}) = \kappa_{\mathrm{c}}(\nu_{\mathrm{sp}}) + \kappa_{\mathrm{c}}(\nu_{\mathrm{es}}), \tag{4.11k}$$

then it is readily shown that

$$\kappa_c(\nu_{es})/\kappa_c(\nu_{sp}) \simeq (m/M_s)^{1/2}, \qquad (4.11l)$$

i.e., the effect of collisions between the different kinds of ions markedly predominates in a multicomponent plasma. It has already been mentioned that this property of the attenuation factor also comes to the fore in an analysis of the attenuation factors κ_{12} of a multicomponent cold plasma.

(2) In a plasma consisting of one kind of ion an important role may be played by ion–ion collisions. In a multicomponent plasma, however, the quantity κ_{ii} is small in comparison with the other components of the attenuation factor. The results of numerical calculations of the ratio $(\kappa_{ii})_1/(\kappa_c)_1$, presented in Fig. 4.2 for the case of one kind of ion, show that near the frequency Ω_H, where, however, the conditions $n_1^2 \gg \kappa_1^2$ are still satisfied quite well, the value of κ_{ii} becomes considerably greater than the attenuation factor for a cold plasma. The ratio $(\kappa_{ii})_1/(\kappa_c)_1$ is more or less proportional to Ω_0^2/Ω_H^2. This means that as the density of the plasma increases the role of (κ_{ii}) increases rapidly.

(3) In the case of quasi-transverse wave propagation, where the angle

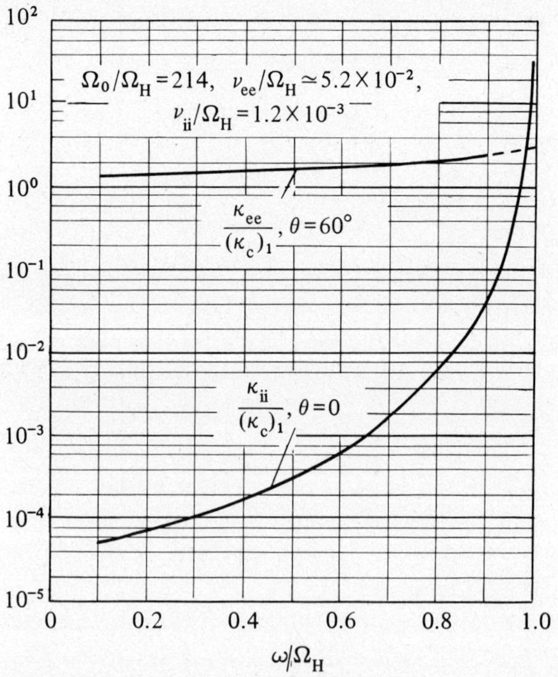

Fig. 4.2. Frequency-dependences of the attenuation-factor ratios: $\kappa_{ee}(\nu_{ee})$ and $\kappa_{ii}(\nu_{ii})$ for the ion branch of the waves n_1, divided by $\kappa_{ei}(\nu_{ei}, \nu_{sp})$ (Al'pert, 1980a).

θ between the wave vector **K** and the magnetic field $\mathbf{H}_0$ is appreciable, the attenuation factor $(\kappa_{ee})_1$, caused by the collisions ν_{ee} between electrons, can exceed both $(\kappa_c)_1$ and $(\kappa_{ii})_1$. This is also seen from the computational results presented in Fig. 4.2.

In conclusion, let us note that if condition (4.11*e*) is satisfied then the Čerenkov attenuation κ_{ee} of the ELF waves can be disregarded, being exponentially small. In the general case

$$\left.\begin{aligned}\kappa_e = &\frac{\pi^{1/2}\sin^2\theta}{4\ \ Z_e n_{12}}\exp(-Z_e^2)[\omega_0^2(n_{12}^2\cos^2\theta - \varepsilon_1)(\omega_H^2\cos^2\theta)^{-1}\\ &+\omega^2 n_{12}^2(\varepsilon_1 n_{12}^2 - \varepsilon_1^2 + \varepsilon_2^2)(\omega_0^2|1 + i\pi^{1/2}Z_e W(Z_e)|)^{-1}]\\ &\times[2n_{12}^2\cos^2\theta - \varepsilon_1(1+\cos^2\theta)]^{-1},\end{aligned}\right\}\quad(4.11m)$$

where $W(Z_e)$ is Kramp's function (plasma dispersion function, see (5.26) below), $Z_e = \omega/Kv_e\cos\theta$.

The data examined in this section thus lead to the important general conclusion that in a number of cases electron–electron and ion–ion collisions play a decisive role in the attenuation of ELF waves. Yet this fact is not usually taken into account when analysing the results of the various experiments; hence, incorrect conclusions, both quantitative and qualitative, may well be drawn.

4.2. Formulas for $\tilde{n}_{12}^2$ in the presence of ions of one kind over the entire frequency range $0 \leqslant \omega \to \infty$

Detailed studies of the formulas for $\tilde{n}_{12}^2$ have been made by Ratcliffe, 1959, by Budden, 1961, by Booker & Dyce, 1965, by Rawer & Suchy, 1967 and by Booker, 1975. Generally speaking, however, the formula for the complex refractive index is inconvenient for numerical calculations, even for a plasma with only one kind of ion. Two simpler, overlapping quantitative formulas, making it possible to study the main characteristics of $\tilde{n}_{12}$, are obtained if the entire frequency range is broken up into two parts:

$$0 \leqslant \omega^2 \ll \omega_H^2 \quad \text{and} \quad \Omega_H\omega_H \ll \omega^2 \to \infty. \qquad (4.12)$$

In the lower frequency range $0 \leqslant \omega^2 \ll \omega_H^2$, granted fulfilment of the inequalities

$$\Omega_H^2 \ll \Omega_0^2, \quad \nu\omega \ll \omega_0^2, \qquad (4.12a)$$

which are satisfactorily complied with in the plasma regions of interest to us, the square of the refractive index is quite accurately described by the formula

$$\tilde{n}_{12}^2 = 1 + \frac{2\Omega_0^2/\Omega_H^2}{2\left(1 - \frac{\omega^2}{\Omega_H \omega_H}\right) - \left(1 + \frac{\omega^2}{\Omega_0^2}\right)\sin^2\theta \pm \left[\left(1 - \frac{\omega^2}{\Omega_0^2}\right)^2 \sin^4\theta + 4\frac{\omega^2}{\Omega_H^2}\cos^2\theta\right]^{1/2} + 2i\frac{m}{M}\frac{\nu\omega}{\Omega_H^2}} \tag{4.13}$$

Here, near the ion gyroresonance, approximately for $\omega/\Omega_H \gtrsim 0.995$, when $\theta \neq 0$ and $\theta = \pi/2$, formula (4.13) for n_1 is no longer valid.

From (4.13) it follows that, when $\nu = 0$ and $\theta = 0$,

$$n_{12}^2 = n_\pm^2 = 1 + \frac{\Omega_0^2}{\Omega_H^2}\left[1 \mp \frac{\omega}{\Omega_H}\left(1 \pm \frac{\omega}{\omega_H}\right)\right]^{-1} = 1 + \frac{\Omega_0^2}{\Omega_H^2(1 \mp \omega/\Omega_H)}, \tag{4.14}$$

while for $\theta = \pi/2$

$$n_1^2 = n_+^2 = 1 - \frac{\omega_0^2}{\omega^2}, \quad n_2^2 = n_-^2 = 1 + \frac{\Omega_0^2}{\Omega_H^2\left[1 - \omega^2\left(\frac{1}{\Omega_0^2} + \frac{1}{\Omega_H \omega_H}\right)\right]}. \tag{4.15}$$

Thus it is seen that in the frequency range $\omega^2 \ll \omega_H^2$ formula (4.13) has two infinities. One of these corresponds to the ion gyroresonance $\omega = \Omega_H$ for $\theta = 0$ of the wave branch n_1, which is called the ordinary wave. The second infinity corresponds to the lower-hybrid resonance $\omega_L = (\Omega_H \omega_H)^{1/2}(1 + \omega_H^2/\omega_0^2)^{-1/2}$ of the branch n_2, which is called the extraordinary wave. When $\nu \neq 0$, the refractive indexes attain maxima at, or very close to, these resonance frequencies; their values are

$$\begin{aligned} n_{+\infty}(\theta = 0) &= n_{1\infty} \simeq \kappa_{1\infty} \simeq \frac{\omega_0}{(2\nu\Omega_H)^{1/2}}, \\ n_{-\infty}(\theta = \pi/2) &= n_{2\infty} \simeq \kappa_{2\infty} \simeq \frac{\omega_0}{(2\nu\omega_L)^{1/2}}. \end{aligned} \tag{4.16}$$

Formula (4.13) describes, according to the classification to be presented below, the properties of very-low-frequency (VLF) and extremely-low-frequency (ELF) waves, as well as the beginning of the range of low-frequency (LF) waves.

The upper of the two broad frequency ranges defined in (4.12) must now be considered. When $\omega^2 \gg \Omega_H \omega_H$, the ranges of LF and high-frequency (HF) waves are described well by the familiar Appleton–Hartree formula

(Appleton, 1928; Hartree, 1931). This formula could, by the way, also be called the Appleton–Lassen formula (Lassen, 1927; see Rawer & Suchy, 1967). It has the form

$$\tilde{n}_{12}^2 = 1 - \frac{2\dfrac{\omega_0^2}{\omega^2}\left(1 - \dfrac{\omega_0^2}{\omega^2} - \mathrm{i}\dfrac{\nu}{\omega}\right)}{2\left(1 - \dfrac{\omega_0^2}{\omega^2} - \mathrm{i}\dfrac{\nu}{\omega}\right)\left(1 - \mathrm{i}\dfrac{\nu}{\omega}\right) - \dfrac{\omega_H^2}{\omega^2}\sin^2\theta \pm \left[\dfrac{\omega_H^4}{\omega^4}\sin^4\theta + 4\dfrac{\omega_H^2}{\omega^2}\cos^2\theta\left(1 - \dfrac{\omega_0^2}{\omega^2} - \mathrm{i}\dfrac{\nu}{\omega}\right)^2\right]^{1/2}}. \tag{4.17}$$

When the collision frequency $\nu = 0$, the curve for the refractive index n_2^2 (vs. ω) has two infinities. If $\theta = 0$, one of them corresponds to the electron gyroresonance $\omega = \omega_H$. When $\theta = \pi/2$, the second infinity at the frequency $\omega_U = (\omega_H^2 + \omega_0^2)^{1/2}$ corresponds to the upper-hybrid resonance of an extraordinary slow wave (see below). At the resonance frequencies, when $\nu \neq 0$, on condition that $\nu^2 \ll \omega_H^2$ the refractive index and the attenuation factor are, respectively,

$$\left.\begin{aligned} n_{-\infty}(\theta = 0) = n_{2\infty} \simeq \kappa_{2\infty} \simeq \frac{\omega_0}{(2\nu\omega_H)^{1/2}}, \\ n_{-\infty}(\theta = \pi/2) = n_{2\infty} \simeq \kappa_{2\infty} \simeq \frac{\omega_0}{(2\nu\omega_U)^{1/2}}. \end{aligned}\right\} \tag{4.18}$$

The resonance frequencies in the entire frequency range for an arbitrary angle $\theta \neq 0$ will be examined below.

Compact formulas for $\tilde{n}_{12}^2$ can be obtained for the entire frequency range in the two limiting cases, namely $\theta = 0$ and $\theta = \pi/2$. They make it easy to reveal, apart from the resonances, certain additional properties of the refractive indexes.

In the frequency range $0 \leqslant \omega \to \infty$ for $\theta = 0$:

$$\begin{aligned} \tilde{n}_{12}^2 = (\varepsilon_1 \pm \varepsilon_2) &= 1 - \frac{\omega_0^2 + \Omega_0^2}{(\omega \pm \omega_H)(\omega \mp \Omega_H) - \mathrm{i}\nu\omega} \\ &= \frac{(\omega - \omega_+)(\omega - \omega_-) - \mathrm{i}\nu\omega}{(\omega \pm \omega_H)(\omega \mp \Omega_H) - \mathrm{i}\nu\omega}, \end{aligned} \tag{4.19}$$

whereas for $\theta = \pi/2$

$$n_1^2 = \varepsilon_3 = 1 - \frac{\omega_0^2 + \Omega_0^2}{\omega^2 - \mathrm{i}\nu\omega} = \frac{(\omega^2 - \omega_0^2) - \mathrm{i}\nu\omega}{\omega^2 - \mathrm{i}\nu\omega}, \tag{4.20}$$

$$\tilde{n}_2^2 = \frac{(\varepsilon_1^2 - \varepsilon_2^2)}{\varepsilon_1}$$
$$= \frac{(\omega - \omega_+)^2(\omega^2 - \omega_-^2) + 2i\nu\omega\omega_0^2}{(\omega^2 - \omega_L^2)(\omega^2 - \omega_U^2) - \nu^2\omega^2 + i\nu\omega[(\omega_0^2 + \Omega_0^2) - 2(\omega^2 - \Omega_H\omega_H)]} \tag{4.21}$$

In (4.21) ω_L and ω_U are, respectively, the lower-hybrid and upper-hybrid frequencies already obtained from (4.13) and (4.17):

$$\omega_U \simeq (\omega_H^2 + \omega_0^2)^{1/2}, \quad \omega_L \simeq (\Omega_H\omega_H)^{1/2}\left(\frac{\omega_H^2}{\omega_0^2} + 1\right)^{-1/2} \tag{4.22}$$

They satisfy the equation

$$\omega^4 - \omega^2(\omega_0^2 + 2\Omega_H\omega_H + \omega_H^2) + \Omega_H\omega_H(\Omega_H\omega_H + \omega_0^2) = 0, \tag{4.23}$$

which, like (4.19), (4.20), and (4.21), is accurate to terms of the order of m/M if the condition $\omega_H^2 \gg \nu^2$ is satisfied.

The frequencies ω_- and ω_+ in (4.21) determine the zero points $n_{12}^2 = 0$ for $\nu = 0$. They satisfy the equation

$$\omega_\mp^2 \pm \omega_\mp\omega_H - \omega_0^2 - \Omega_H\omega_H = 0, \tag{4.24}$$

from which it follows, omitting terms of the order of m/M, that

$$\omega_- = -\frac{\omega_H}{2} + \left(\omega_0^2 + \frac{\omega_H^2}{4}\right)^{1/2}, \quad \omega_+ = \frac{\omega_H}{2} + \left(\omega_0^2 + \frac{\omega_H^2}{4}\right)^{1/2}. \tag{4.25}$$

Thus, the refractive index of a plasma containing just one kind of ion goes to zero only in the range of HF waves; the frequency values corresponding to the zero points depend little on the effect of forced ion motion. When $\theta = 0$, the branches of the so-called ordinary and extraordinary fast waves have zero points $n_1^2(\omega_-) = 0$ and $n_2^2(\omega_+) = 0$ (see Figs. 4.4 and 4.5 below). For $\theta = \pi/2$ both extraordinary HF waves have zero points $n_2^2(\omega_-) = 0$ and $n_2^2(\omega_+) = 0$. However, the refractive index n_1^2 of the ordinary fast wave also has a third zero value, $n_1^2(\omega_0) = 0$. It is readily shown that the frequencies $\omega = \omega_-$ and $\omega = \omega_+$ determining the zero values of n_{12}^2 do not depend on the angle θ. For $0 < \theta < \pi/2$ the quantities n_{12}^2 always have three zero points: $n_1^2(\omega_- = n_2^2(\omega_0) = n_1^2(\omega_+) = 0$. The frequencies ω_- and ω_+ are called the *cutoff frequencies*, while the frequency ω_0 is the *Langmuir resonance* (or *plasma*) *frequency* of the electrons (see below).

Naturally, when $\nu^2 \ll \omega_H^2$ the formulas (4.19), (4.20), and (4.21) directly determine the corresponding resonance values $n_{1\infty}$ and $n_{2\infty}$ (see (4.16) and (4.18)). In formulas (4.19)–(4.20), in the terms proportional to ν, the multipliers $(1 + m/M)$ are omitted; here the collision frequency $\nu = \nu_{ei} + \nu_{en}$, i.e., it is obtained taking into account collisions of electrons with neutral particles.

In the HF range, when the collision frequency is taken into account, the quantity n^2 displays an interesting peculiarity. When $\omega^2 = \omega_0^2$ and

$$\nu = \nu_c(\omega_H/2)(\sin^2\theta/\cos\theta), \tag{4.26}$$

the two refractive indexes are equal: $n_1^2 = n_2^2$, i.e., formula (4.2) has a multiple root. This follows directly from (4.17), since for fulfilment of the specified conditions the square-root term in its denominator equals zero. Hence, when $\nu = \nu_c$, which is known as the *critical collision frequency*, the plasma loses its birefringent properties, and there exists in it one wave with one type of polarization (see below). The value ν_c constitutes a specific kind of boundary, across which the properties of waves propagated in the plasma change.

When

$$\nu_c^2 \ll \nu_{ei}^2 + ((\omega^2 - \omega_0^2)/\omega)^2, \tag{4.27}$$

conditions are created in the plasma for the so-called *quasi-longitudinal* or *quasi-parallel* ($\theta = 0$) *propagation* of HF waves (see (4.31)). In this case

$$\tilde{n}_{12}^2 \simeq 1 - \frac{\omega_0^2}{\omega[(\omega - i\nu_{ei}) + \omega_H\cos\theta]}. \tag{4.28}$$

Both waves are circularly polarized, having opposite senses of rotation (see (4.45) and (4.46)).

If the inverse of (4.27) is satisfied, i.e., if

$$\nu_c^2 \gg \nu_{ei}^2 + [(\omega^2 - \omega_0^2)/\omega]^2 \tag{4.29}$$

then conditions are created in the plasma for *quasi-transverse* or *quasi-perpendicular* ($\theta = \pi/2$) *propagation* of HF waves and

$$\left.\begin{aligned} \tilde{n}_1^2 &= 1 - \frac{\omega_0^2}{\omega(\omega - i\nu_{ei}) + (\omega^2 - \omega_0^2 - i\nu_{ei}\omega)\cot^2\theta}, \\ \tilde{n}_2^2 &= 1 - \frac{\omega_0^2(\omega^2 - \omega_0^2 - i\nu_{ei}\omega)}{\omega(\omega - i\nu_{ei})(\omega^2 - \omega_0^2 - i\nu_{ei}\omega) - \omega^2\omega_H^2\sin^2\theta}. \end{aligned}\right\} \tag{4.30}$$

These two waves are linearly polarized in mutually perpendicular directions. A general analysis of the polarization coefficients of the different types of waves will be given below (see (4.40)–(4.46)).

Compact formulas for $\tilde{n}_{12}^2$ are also obtained in the case of quasi-longitudinal propagation of electromagnetic waves, which is of interest, for example, when studying the guidance of various types of waves by the Earth's magnetic field or the trapping of waves in magnetic-field-aligned ducts.

For the LF branch of the waves

$$\omega_L \ll \omega \lesssim \omega_H \tag{4.31}$$

the condition for quasi-longitudinality has the form

$$\left(\frac{\sin^2\theta}{2\cos\theta}\right)^2 \ll \left|\frac{\omega^2 - \omega_0^2 - i\nu\omega}{\omega\omega_H}\right|^2. \tag{4.31a}$$

In this case, in the frequency range considered, only $n_2^2 > 0$, as is directly seen from the formula

$$\tilde{n}_{12}^2 = 1 - \frac{\omega_0^2}{\omega[(\omega - i\nu) \pm \omega_H \cos\theta]}. \tag{4.32}$$

Naturally, formula (4.3.2) is identical to (4.28).

In the ELF and VLF ranges the condition for quasi-longitudinal propagation has the form

$$\left(\frac{\sin^2\theta}{2\cos\theta}\right)^2 \ll \left(\frac{\omega}{\Omega_H}\right)^2 \frac{|\Omega_0^2 + i(m/M)\nu\omega|^2}{|\omega^2 - \Omega_0^2 - \Omega_H^2 + i(m/M)\nu\omega|^2}. \tag{4.33}$$

If the condition (4.33) is satisfied, when $0 \leqslant \omega^2 \ll \Omega_H \omega_H$, the refractive index is described by the formula

$$\tilde{n}_{12}^2 = 1 + \frac{\Omega_0^2}{\Omega_H \cos\theta(\Omega_H \cos\theta \mp \omega) + i(m/M)\nu\omega}. \tag{4.34}$$

For arbitrary values of the angle θ, i.e., when the condition for quasi-longitudinal propagation is not satisfied in this frequency range,

$$\tilde{n}_{12}^2 \simeq \frac{2\Omega_0^2}{\Omega_H^2\left\{(1+\cos^2\theta) - \frac{\omega^2}{\Omega_0^2}\sin^2\theta \mp \left[\left(1 - \frac{\omega^2}{\Omega_0^2}\right)^2 \sin^4\theta + 4\frac{\omega^2}{\Omega_H^2}\cos^2\theta\right]^{1/2} + 2i\frac{m}{M}\nu\omega\right\}}. \tag{4.35}$$

In the vicinity of the ion gyrofrequency, i.e., for $\omega \simeq \Omega_H$, we have

$$\tilde{n}_1^2 \simeq \frac{\Omega_0^2(1 + \cos^2\theta)}{2\Omega_H(\Omega_H - \omega)\cos^2\theta + i(m/M)\nu\omega(1 + \cos^2\theta)}, \tag{4.36}$$

$$\tilde{n}_2^2 \simeq \frac{\Omega_0^2}{\Omega_H^2(1 + \cos^2\theta) + i(m/M)\nu\omega}, \tag{4.37}$$

whereas for $\omega/\Omega_H \ll 1$, when $\sin^4\theta/4\cos^2\theta \gg \omega^2/\Omega_H^2$, we have

$$\tilde{n}_1^2 \simeq \frac{\Omega_0^2}{\Omega_H^2 \cos^2\theta + i(m/M)\nu\omega}, \tag{4.38}$$

$$\tilde{n}_2^2 \simeq \frac{\Omega_0^2}{\Omega_H^2 + i(m/M)\nu\omega}. \tag{4.39}$$

Formulas (4.38) and (4.39) represent the refractive indexes of the so-called Alfvén and modified Alfvén waves.

4.3. Properties of the different branches of waves

In a cold plasma the quantities A, B, and C defining the refractive index, (4.1), do not depend on the wave vector **K**. Therefore the dispersion equation, (3.1), which in the general case, taking into account spatial dispersion, is transcendental, turns out to be an algebraic equation. The latter is a fourth-order equation in $\tilde{n}(\omega, \theta)$. It determines two values of $\tilde{n}_{12}^2(\omega, \theta)$, which describe the properties of two elliptically polarized waves – an ordinary wave and an extraordinary wave – which have different phase velocities and different polarization senses, i.e., different directions of rotation of the electric-field vector **E** of the wave. However, relative to the square of the frequency ω^2, Eq. (3.1) is a fifth-degree equation and determines *five wave branches* in the frequency range $0 \leqslant \omega \to \infty$.

4.3.1. Polarization coefficients

One of the elliptically polarized waves generated in a plasma exhibits left-handed rotation, i.e., the vector **E** rotates counterclockwise. The second wave exhibits right-handed rotation, the vector **E** rotating clockwise. Both waves are transverse, i.e., they are electromagnetic waves. Under certain conditions they become linearly polarized in one direction or in two mutually perpendicular directions. In a right-handed coordinate system $Oxyz$ so chosen that the wave vector **K** is directed along the z-axis and the vector of the constant magnetic field $\mathbf{H}_0$ makes an angle θ with it, the polarization of the two waves in a multicomponent collisional plasma is described by the formula

$$\rho_{12} = \left(\frac{E_y}{E_x}\right)_{12} = i\frac{\varepsilon_2 \cos\theta\, \tilde{n}_{12}^2}{(\varepsilon_1^2 - \varepsilon_2^2) - \varepsilon_1 \tilde{n}_{12}^2}$$

$$= i\frac{\varepsilon_1\varepsilon_3 - n_{12}^2(\varepsilon_1 \sin^2\theta + \varepsilon_3 \cos^2\theta)}{\varepsilon_2\varepsilon_3 \cos\theta}. \tag{4.40}$$

The quantity ρ_{12} is called the *polarization coefficient*. The two formulas (4.40) are equivalent and can be converted into one other. By eliminating $\tilde{n}_{12}^2$ from (4.40) we readily obtain a quadratic equation determining the two values of ρ_{12}, namely

$$\rho^2 + 2\mathrm{i}\rho\left\{\frac{\sin^2\theta}{\cos\theta}\,\frac{\varepsilon_1^2-\varepsilon_2^2-\varepsilon_1\varepsilon_3}{2\varepsilon_2\varepsilon_3}\right\}+1=0 \tag{4.41}$$

Obviously, (4.41) will have identical roots $\rho_1=\rho_2=\pm 1$ if

$$\frac{\varepsilon_1^2-\varepsilon_2^2-\varepsilon_1\varepsilon_3}{\varepsilon_2\varepsilon_3}=\mp\mathrm{i}\frac{2\cos\theta}{\sin^2\theta}. \tag{4.42}$$

Thus, (4.42) specifies the conditions on which both waves are linearly polarized.

In a *collisional plasma* (4.42) is also a condition that the square-root term in equation (4.1) be equal to zero. This means that, when (4.42) is satisfied, the refractive indexes of the two waves are the same.

Two equations

$$\left.\begin{aligned}\mathrm{Im}\left(\frac{\varepsilon_1^2-\varepsilon_2^2-\varepsilon_1\varepsilon_3}{\varepsilon_2\varepsilon_3}\right)&=\pm\frac{2\cos\theta}{\sin^2\theta},\\ \mathrm{Re}\left(\frac{\varepsilon_1^2-\varepsilon_2^2-\varepsilon_1\varepsilon_3}{\varepsilon_2\varepsilon_3}\right)&=0,\end{aligned}\right\} \tag{4.43}$$

derived by taking the imaginary (Im) and real (Re) parts of (4.42) determine, for the given values of the collision frequencies and of the other plasma parameters, the frequency $\omega=\omega_{\mathrm{cr}}$ and the angle $\theta=\theta_{\mathrm{cr}}$ at which the branches $n_1(\omega,\theta)$ and $n_2(\omega,\theta)$ intersect. The frequency ω_{cr} is called the intersection frequency, or crossover frequency and exists only in a multicomponent plasma (see Section 4.4). Both waves are linearly polarized at the points of intersection. Upon passing through these points, the polarization senses of the waves described by the refractive indexes n_1 and n_2 change. The wave exhibiting right-handed rotation becomes left-handed polarized, and vice versa. It can be shown that this polarization reversal occurs in a collisional plasma for angles $\theta\geqslant\theta_{\mathrm{cr}}$ and that at the point of intersection $Re(\varepsilon_2)=0$.

In the vicinity of the frequency ω_{cr} different types of waves may interact strongly. Therefore, when considering the phenomena taking place in this region, a strict study must be made of the system of coupled equations describing the fields of these waves (Försterling, 1942; Budden, 1961). In the vicinity of the crossover frequency the directions of the energy fluxes of both waves show important characteristics (see Chapter 8). Note also that certain phenomena occurring in the frequency range $(\omega_{\mathrm{cr}}\rightarrow\Omega_{\mathrm{H}})$ were first detected by artificial Earth satellites (see Chapter 14, Volume 2).

In a *collisionless plasma*, where $\nu = 0$. the condition (4.42) naturally cannot be satisfied in the form in which it is written, since all the tensor elements are real values. In this case, for an analysis of the conditions at which the branches n_{12}^2 of the different types of waves intersect and their polarization senses change, the equality to zero of the square-root term in (4.1) may be rewritten as

$$(\varepsilon_1^2 - \varepsilon_2^2 - \varepsilon_3\varepsilon_1)^2 \sin^4\theta + 4(\varepsilon_2\varepsilon_3)^2 \cos^2\theta = 0 \qquad (4.42a)$$

It follows from (4.42*a*) that it is possible for simultaneously the refractive indexes to be equal ($n_1 = n_2$) and both waves to be linearly polarized (see the first formula of (4.40)), provided that two conditions are satisfied:

$$\varepsilon_2 = 0, \quad \theta = 0. \qquad (4.44)$$

or

$$\varepsilon_1 - \varepsilon_2 - \varepsilon_3\varepsilon_1 = 0, \quad \theta = \pi/2 \qquad (4.44a)$$

When the first equation of (4.44) holds, both waves are linearly polarized for any value of θ (see (4.40)). Hence, in a collisionless plasma the crossover frequencies ω_{cr} are always determined from the condition $\varepsilon_2 = 0$. However, the branches n_1^2 and n_2^2 intersect only when $\theta = 0$. For arbitrary values of $\theta \neq 0$, if $\nu = 0$, the change in the wave-polarization senses takes place at those frequencies at which the equation $\varepsilon_2 = 0$ holds; however, at these points $n_1(\omega, \theta) \neq n_2(\omega, \theta)$ (see Figs. 4.8, 4.9 below). The first of the equations (4.44*a*) determines the frequencies $\omega = \omega_{\mathrm{T}1}, \omega_{\mathrm{T}2}, \ldots, \omega_{\mathrm{T}}$ where $n_1^2(\omega_{\mathrm{T}1}, \ldots) = n_2^2(\omega_{\mathrm{T}1}, \ldots) < 0$ and both of the waves change their sense of polarization (see Figs. 4.5, 4.6, and 4.8). The polarization of LF and HF waves, i.e., in the frequency range where the influence of ions can be disregarded, is described by the formula

$$\rho_{12} = \mathrm{i}\frac{2(\omega^2 - \omega_0^2 - \mathrm{i}\nu\omega)\cos\theta}{\omega\omega_{\mathrm{H}}\sin\theta \pm [\omega^2\omega_{\mathrm{H}}^2\sin^4\theta + 4(\omega^2 - \omega_0^2 - \mathrm{i}\nu\omega)^2\cos^2\theta]^{1/2}}. \qquad (4.45)$$

From (4.40) and (4.45) it is readily seen that

$$\rho_1\rho_2 = 1, \qquad (4.46)$$

i.e., the rotation senses of the two types of elliptically polarized waves are opposite, and their polarization ellipses are mutually perpendicular. For $\theta = 0$ the values of $\rho_{12} = \pm\mathrm{i}$, i.e., the waves are circularly polarized, whereas for $\theta = \pi/2$ and $\nu = 0$ we have $\rho_{12} = 0, \infty$, i.e., both waves are linearly polarized in mutually perpendicular directions. An interesting case of reversal of the polarization senses of HF waves was described above (see Eqs. (4.26)–(4.30)).

4.3.2. Refractive indexes. Resonances. Classification of the waves

The general properties of the various types of oscillations and waves observed in a magnetoplasma, as well as the properties of their refractive indexes, can be revealed quite fully without taking the collision frequencies into account.

In a two-component plasma, consisting of electrons and one kind of ion, the dispersion equation, (3.1), is of the fifth degree in ω^2; thus it determines *five wave branches*. Three of these are described by the refractive index $n_1^2(\omega,\theta)$ and two by the refractive index $n_2^2(\omega,\theta)$ (see Figs. 4.3–4.6). In the two limiting cases, namely for $\theta=0$ and $\theta=\pi/2$, the n_{12}^2 curves have four branches. If $\theta=0$, one branch n_1^2 degenerates, i.e., the divergence $n_1^2\to\infty$, bordering on the upper-hybrid resonance, disappears. In this case resonances are possible only at the ion and electron gyrofrequencies Ω_H and ω_H. When $\theta=\pi/2$, the n_1^2 branch, corresponding to the gyroresonance $\omega\to\Omega_H$, goes to infinity. In this case n^2 has three n_2^2 branches with two resonance values $n_2^2\to\infty$ at the lower-hybrid and

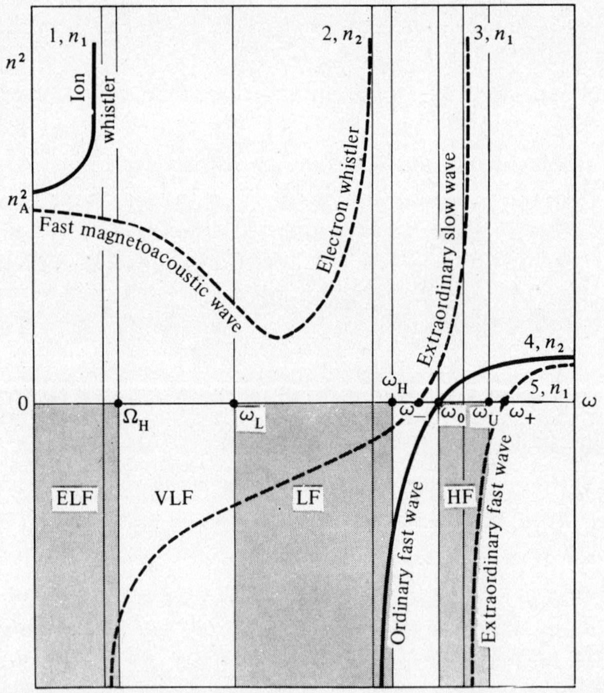

Fig. 4.3. The five branches of the refractive indexes n_{12}^2 of a cold two-component magnetoplasma for $0<\theta<\pi/2$. The resonance regions $\omega_3(\theta), \omega_2(\theta)$, and $\omega_1(\theta)$ in the ELF, LF, and HF ranges are demarcated (see Fig. 4.7).

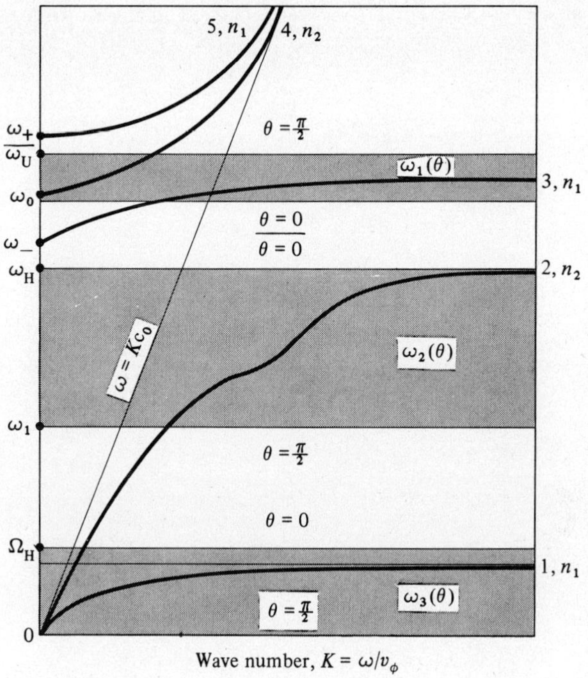

Fig. 4.4. The five branches of the ω(K) relations for a cold two-component magnetoplasma.

upper-hybrid frequencies ω_L and ω_U (see (4.21)) and one n_1^2 branch (see Figs. 4.5 and 4.6). At the Langmuir frequency ω_0 the refractive index $n_1^2(\omega_0)=0$. The Langmuir frequency is also a resonance frequency at which longitudinal high-frequency oscillations can be excited for $\theta=0$, just as at the other resonance frequencies (see below). The dependences $n_+^2=n_1^2$ and $n_-^2=n_2^2$ for $\omega_0<\omega_H$ and $\omega_0>\omega_H$ and for the values $\theta=0$, $0<\theta<\pi/2$ and $\theta=\pi/2$ (Figs. 4.5 and 4.6) make it possible to trace the evolution of the n_{12}^2 curves as the values of the various parameters are changed. When examining Figs. 4.3–4.6, it is to be borne in mind that in them the range of frequency changes by several orders of magnitude (approximately by a factor of $10M/m$), so that, analogously, the values of n_{12}^2 with which we have to deal in the plasma regions of interest to us also change by several orders of magnitude, even taking collisions into account. These figures therefore do not depict the true course of the n_{12}^2 curves. At this point let us note the following properties of $n_\pm^2=n_{12}^2$, in addition to those already described.

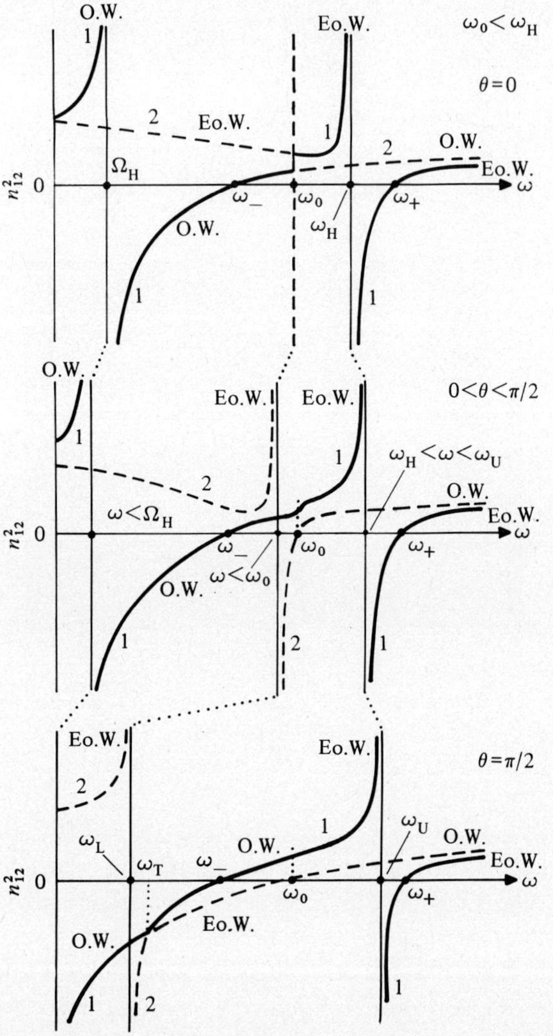

Fig. 4.5. Schematic curves of the refractive indexes n_{12}^2 of the ordinary wave (O.W.) and of the extraordinary wave (Eo.W.) for $\theta = 0, 0 < \theta < \pi/2$ and $\theta = \pi/2$ in the case where the Langmuir frequency ω_0 is lower than the electron gyrofrequency ω_H.

(1) The value of $n_2(\omega = 0)$ does not depend on the angle θ; it equals the refractive index of the so-called modified Alfvén wave (see (4.56) below) and of the Alfvén wave for $\theta = 0$.

(2) In the interval of angles $0 < \theta < \pi/2$ the refractive index n_1^2 $(\omega = \omega_0) = 1$, i.e., it coincides with the values $n_1^2(\omega \to \infty) = 1$ and n_2^2 $(\omega \to \infty) = 1$. If $\theta = \pi/2$, then $n_2^2(\omega = \omega_0) = 0$.

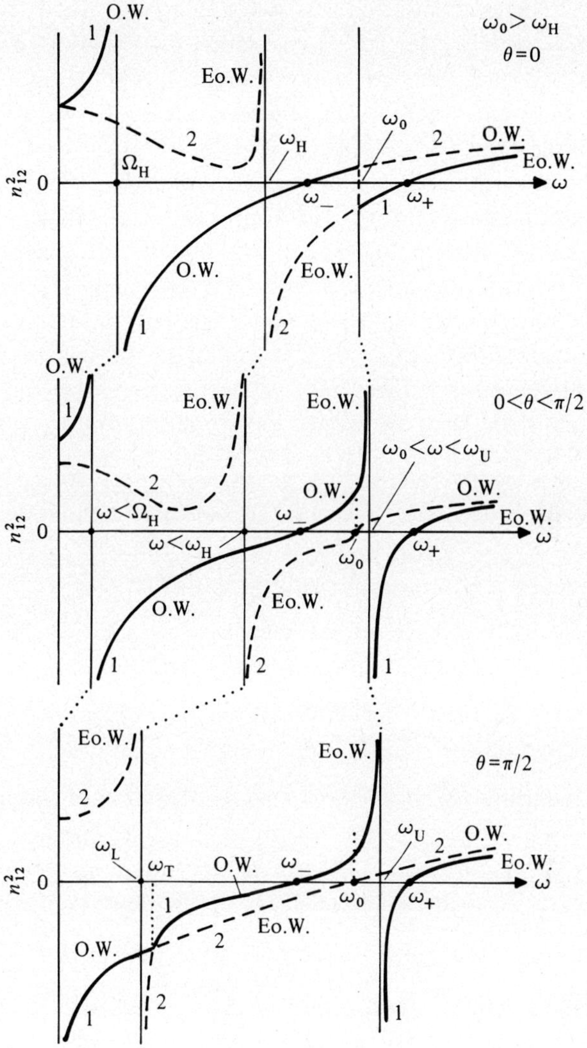

Fig. 4.6. As for Fig. 4.5, but with $\omega_0 > \omega_H$.

(3) For $\omega_0 > \omega_H$ the refractive index $n_2^2 \to \infty$ in the frequency range $\omega_0 \leqslant \omega \leqslant \omega_U$ at a certain value of the angle θ in the interval from 0 to $\pi/2$. If $\omega_0 < \omega_H$, the corresponding frequency range is $\omega_H \leqslant \omega \leqslant \omega_U$.

(4) For $\omega_0 > \omega_H$ or $\omega_0 < \omega_H$ the refractive index $n_2^2 \to \infty$, correspondingly, in the frequency ranges $\omega_L \leqslant \omega \leqslant \omega_H$ or $\omega_L \leqslant \omega \leqslant \omega_0$, in the interval of angles θ from 0 to $\pi/2$. The behavior of the divergences (infinities) of n_2^2 is associated with the resonance properties of the plasma, which will be described shortly, following paragraph (6).

(5) In the interval of angles $0<\theta\leqslant\pi/2$, when ω goes continuously through ω_0 and $\omega_{cr12}, \omega_{cr23}, \ldots$ (see Section 4.4), the polarization senses of both of the waves change: right-handed rotation becomes left-handed and vice versa and when $\omega=\omega_0$ and $\omega=\omega_{cr}, \ldots$ the waves are linearly polarized (see Figs. 4.5, 4.6, 4.8).

(6) When $\theta=\pi/2$ both of the waves are linearly polarized, and when ω goes continuously through $\omega=\omega_{cr12}, \omega_{cr23}, \ldots$ where $n_1=n_2$ and through $\omega=\omega_{T1}, \omega_{T2}, \ldots \omega_T$ where $n_1^2=n_2^2<0$, the senses of polarization of both of the waves change. But for simplicity the names of the extra-low-frequency (ELF) ordinary and extraordinary waves in the ranges $\omega_{T12}\leqslant\omega\leqslant\omega_{cr12}, \omega_{T23}\leqslant\omega\leqslant\omega_{cr23}$ and the names of the high-frequency (HF) waves in the range $\omega_T\leqslant\omega\leqslant\omega_0$ are not changed (see Figs. 4.5, 4.6, and 4.8). They are named as at the frequencies $\omega>\omega_{cr23}$ and $\omega<\omega_0$.

In a cold collisionless plasma the dispersion equation, (3.1), has divergences $n^2\to\infty$ when $A\to 0$ (see (4.1)). In this case

$$n_+^2=n_1^2=-C/B, \quad n_-^2=n_2^2=-B/A \tag{4.47}$$

Since $B\neq 0$, only the refractive index $n_-^2=n_2^2$ satisfies this condition, when $A=0$. This leads to a third-degree equation in ω^2:

$$1-\frac{\omega_0^2\cos^2\theta}{\omega^2}+\frac{\omega_0^2\sin^2\theta}{\omega^2-\omega_H^2}-\frac{\Omega_0^2\cos^2\theta}{\omega^2}-\frac{\Omega_0^2\sin^2\theta}{\omega^2-\Omega_H^2}=0. \tag{4.48}$$

It determines the three *resonance branches* $\omega(\theta)$ for a cold plasma. A kinetic study shows that these branches determine the conditions on which natural longitudinal plasma oscillations can be excited along the wave vector. Since spatial dispersion is absent in the approximation of a cold plasma considered here, the group velocity of these oscillations $\dfrac{\partial\omega}{\partial\mathbf{K}}=0$. Thus, in this approximation we cannot speak of a resonance excitation of longitudinal waves, because the oscillations cannot emerge from the region in which they are excited.

The three resonance branches for a cold plasma $\omega_1(\theta)$, $\omega_2(\theta)$ and $\omega_3(\theta)$ are shown schematically in Fig. 4.7. One of these, namely the *high-frequency* (HF) branch $\omega_1(\theta)$, is only slightly influenced by the ions, because $\omega_0, \omega_H\gg\Omega_0, \Omega_H$. It is described by the formula

$$\omega_1^2(\theta)=\tfrac{1}{2}\{(\omega_0^2+\omega_H^2)+[(\omega_0^2+\omega_H^2)^2-4\omega_0^2\omega_H^2\cos^2\theta]^{1/2}\}. \tag{4.49}$$

If $\omega_0>\omega_H$, then $\omega_1(\theta)$ changes within the following limits:

$$\begin{aligned} &\omega_1(\theta)=\omega_0 \qquad \text{for } \theta=0,\\ &\omega_1(\theta)=\omega_U=(\omega_H^2+\omega_0^2)^{1/2} \quad \text{for } \theta=\pi/2. \end{aligned} \tag{4.50}$$

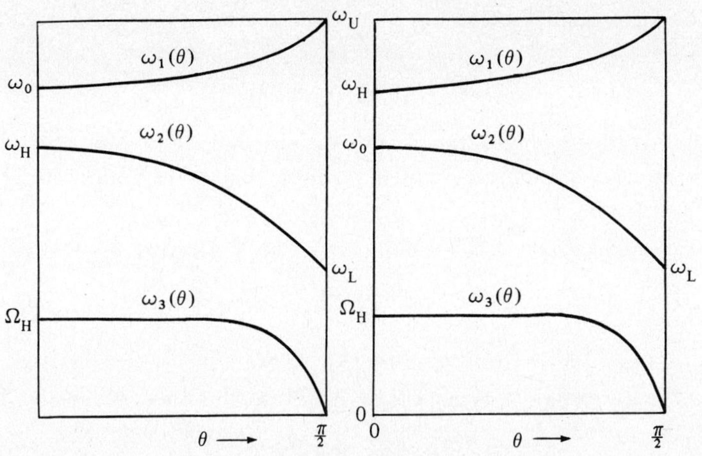

Fig. 4.7. The three resonance branches $\omega_{123}(\theta)$ for a two-component cold plasma.

The frequency $\omega_1(\pi/2) = \omega_U$. This is the *upper-hybrid resonance frequency* obtained above. Note here that taking spatial dispersion into account markedly affects the behavior of n_2^2 in the vicinity of the resonance frequencies $\omega_{123}(\theta)$. However, the values of $\omega(\theta)$ are not changed much (see Akhiezer *et al.*, 1964).

When $\omega_0 < \omega_H$, the branch $\omega_1(\theta)$ changes in the interval of angles $0 \leqslant \theta \leqslant \pi/2$ within the limits $\omega_H \leqslant \omega \leqslant \omega_U$. The values $\omega = \omega_H$ or $\omega = \omega_0$, respectively, when $\omega_0 > \omega_H$ or $\omega_0 < \omega_H$, determine for $\theta = 0$ the beginning of another resonance branch $\omega_2(\theta)$. This branch is described in the frequency range $\omega_H^2, \omega_0^2 \gtrsim \omega^2 \gg \Omega_H \omega_H$ by formula (4.49) with a minus sign before the square root term. Its continuation in the frequency range $\Omega_H \omega_H \ll \omega^2 \ll \omega_H^2$, as well as the third resonance branch $\omega_3(\theta)$ at frequencies such that $0 \leqslant \omega \leqslant \Omega_H$, are found from the equation

$$\omega^4 - \omega^2 \frac{\omega_0^2(\cos^2\theta + m/M)}{\omega_0^2 + \omega_H^2 \sin^2\theta} + \frac{m^2}{M^2} \frac{\omega_0^2 \cos^2\theta}{\omega_0^2 + \omega_H^2 \sin^2\theta} = 0, \qquad (4.51)$$

which is obtained from (4.13) for these frequency ranges.

The largest of the roots of (4.51) describes the angular dependence of the low-frequency part of the second resonance branch, namely

$$\omega_2(\theta) = \omega_H \frac{(\cos^2\theta + m/M)^{1/2}}{[1 + (\omega_H^2/\omega_0^2)\sin^2\theta]^{1/2}}. \qquad (4.52)$$

To the accuracy of terms of the order of m/M, it follows from (4.52) that $\omega^2(\pi/2)$ is the *lower-hybrid resonance frequency* ω_L (see (4.22)). The second root of (4.51) describes the angular dependence of the third resonance branch for a cold plasma:

$$\omega_3(\theta) \simeq \frac{\Omega_H \cos\theta}{(\cos^2\theta + m/M)^{1/2}}. \tag{4.53}$$

Thus, Equations (4.49) and (4.51) overlap each other and describe the branches over the entire frequency range, just as Equations (4.13) and (4.17) do for $n_{12}^2(\omega, \theta)$.

The *low-frequency* (LF) branch $\omega_2\theta$ varies within the following limits:

$$\begin{aligned} \omega_2(\theta) &= \omega_H \text{ or } \omega_0, \quad \text{for } \theta = 0, \\ \omega_2(\theta) &= \omega_L, \quad \text{for } \theta = \pi/2, \end{aligned} \tag{4.54}$$

while the *extremely low-frequency* (ELF) branch varies within the limits

$$\begin{aligned} \omega_3(\theta) &= \Omega_H, \quad \text{for } \theta = 0, \\ \omega_3(\theta) &= \left(\frac{\Omega_H \cos\theta}{\Omega_H^2 + \Omega_0^2}\right)\omega_0 \to 0, \quad \text{for } \theta = \pi/2. \end{aligned} \tag{4.55}$$

The $\omega_1(\theta), \omega_2(\theta)$, and $\omega_3(\theta)$ curves show that, in a cold plasma, resonance oscillations are absent in three frequency ranges: $\Omega_H \leqslant \omega \leqslant \omega_L$, $\omega_H \leqslant \omega \leqslant \omega_0$ and $\omega > \omega_U$.

As will be seen below, between the ion gyrofrequency Ω_H and the lower-hybrid frequency ω_L, oscillations are excited in a nonisothermal plasma, taking into account the thermal motion of particles, when $T_e \gg T_i$. It is appropriate to call the oscillations in this frequency range *very-low-frequency* (VLF) *resonance oscillations*. Note that in the VLF region, i.e., for $\Omega_H \leqslant \omega \leqslant \omega_L$, some very interesting effects arise in the near-Earth plasma due to the influence of the different kinds of ions: waves are trapped and cut off, their trajectories are of a complex type, and so on (see Volume 2). In the frequency range bordering on HF oscillations, i.e., between ω_H and ω_0, there may also be resonances which are also revealed theoretically when taking into account the spatial dispersion of the plasma. At frequencies $\omega > \omega_U$ the high-frequency resonance properties of a plasma also manifest themselves in special cases only in a warm plasma (see Volume 2).

To conclude this section, let us describe in addition some properties of the five branches of the refractive index $n_{12}^2(\omega)$ of a cold plasma.

Branch 1 in Figs. 4.3 and 4.4 is the ordinary ion wave $n_1^2(\omega)$. It ends at the ion gyrofrequency $\omega = \Omega_H$, where its refractive index $n_1^2 \to \infty$. For $\omega > \Omega_H$ the amplitude of the ion wave decays rapidly. This attenuation of the wave is 'kinetic' in nature and is associated with the ion-cyclotron resonance. This branch is sometimes called the *ion-whistler wave* (see Volume 2). For $\omega \to 0$ (i.e., $\omega \ll \Omega_H$), it describes the *Alfvén wave*:

$$\omega_{10} = \frac{K V_A \cos\theta}{[1 + (c/V_A)^2]^{1/2}}, \quad V_A = \frac{c}{n_A}, \quad n_A = \frac{\Omega_0}{\Omega_H},$$

$$n_{10}^2 = \frac{n_A^2}{(1 - \omega/\Omega_H)\cos^2\theta}, \quad \kappa_{10} = \frac{n_{10}}{2} \frac{\nu\omega}{\Omega_H \omega_H \cos^2\theta}, \tag{4.56}$$

where, as assumed above, $\Omega_0^2 \gg \Omega_H^2$. If $\theta = \pi/2$, then

$$n_{10} \simeq \kappa_{10} \simeq \omega_0/(2\nu\omega)^{1/2}, \tag{4.56a}$$

whereas for $\omega \to 0$, the Alfvén wave disappears: $n_1, \kappa_1 \to \infty$ (see Figs. 4.5 and 4.6).

In (4.56) V_A is the Alfvén velocity, n_A is the Alfvén refractive index (see (1.8)), and κ_{10} is the collisional attenuation factor. Here, as everywhere below, the subscript 0 indicates the value of the parameter in a cold plasma. For the cyclotron wave, when $\omega \to \Omega_H$ for $\nu = 0$, we have

$$n_{10}^2 = n_A^2(1 + \cos^2\theta)/(1 - \omega^2/\Omega_H^2)\cos^2\theta. \tag{4.56b}$$

The range of frequencies spanned by branch 1 corresponds to the ELF range of resonances $\omega_3(\theta)$, described by formulas (4.51) and (4.53) (see Fig. 4.7).

Branch 2 in Figs. 4.3 and 4.4 describes the properties of waves in the frequency range $0 < \omega \lesssim \omega_H$. It includes *fast magnetoacoustic waves* and the *whistler mode*, i.e., the extraordinary waves n_{20}. This branch stops at the gyrofrequency $\omega \simeq \omega_H$ due to cyclotron resonance of electrons, and also due to the effect of the spatial dispersion of the plasma. Branch 2 therefore spans three frequency ranges: ELF, VLF, and LF waves. Only one resonance region of a cold plasma corresponds to it, namely the LF branch $\omega_2(\theta)$ (see Fig. 4.7). In the limit, for $\omega \ll \Omega_H$, the index $n_2(\omega, \theta)$ describes the properties of the so-called *modified Alfvén wave*:

$$n_{20} = \frac{n_A}{(1 + \omega/\Omega_H)^{1/2}}, \quad \kappa_{20} = \frac{n_A \nu\omega}{2\Omega_H \omega_H}, \quad \omega_{20} = \frac{K_0 V_A}{[1 + (V_A/c)^2]^{1/2}}. \tag{4.57}$$

For $\omega \simeq \Omega_H$, we have

$$n_{20}^2 = \frac{n_A^2}{(1 + \omega/\Omega_H)} \frac{2}{(1 + \cos^2\theta)} \simeq \frac{n_A^2}{1 + \cos^2\theta}. \tag{4.57a}$$

For $\theta = 0$, it follows that $n_{20} = n_A/\sqrt{2}$. If $\theta = \pi/2$, we have $n_{20} = n_A$, i.e., to the accuracy of terms of the order of ν/ω_H, in the interval $0 \leqslant \theta \leqslant \pi/2$, the index n_{20} changes little. For an increase in frequency, the branch $n_2(\theta)$ breaks off when $\theta = \pi/2$ at the lower-hybrid frequency (see Figs. 4.5 and 4.6).

In the frequency range $0 < \omega \ll \omega_H, \omega_0$, when $\Omega_0^2 \gg \Omega_H^2$, the refractive indexes of the two wave branches, namely the Alfvén branch (for $\omega > \Omega_H$, ion-whistler) and the modified Alfvén branch (for $\omega > \Omega_H$, fast magneto-acoustic) (see Fig. 4.3), can be described by the following dispersion equations:

$$n_{12}^2 = \frac{1}{\cos^2\theta}\{\varepsilon_1(1+\cos^2\theta) \pm [\varepsilon_1^2(1+\cos^2\theta)^2 - 4(\varepsilon_1^2 - \varepsilon_2^2)\cos^2\theta]^{1/2}\} \tag{4.58}$$

$$\omega_{12}^2(K) = \tfrac{1}{2}K^2V_A^2\{1+\cos^2\theta(1+K^2c^2/\Omega_0^2) \mp [[1+\cos^2\theta(1+K^2c^2/\Omega_0^2)]^2 - 4\cos^2\theta]^{1/2}\}, \tag{4.59}$$

where the wave number $K = (\omega_{12}/c)n$. For values of $\omega \ll \Omega_H$ and when $\omega \to \Omega_H$, (4.59) naturally gives us the formulas (4.56) and (4.57).

In the frequency range $\omega \gg \omega_L$, the branch $n_2(\omega, \theta)$ describes the properties of an electron-whistler wave, or electron whistler. The term 'whistler wave' is applied to waves in the LF range $\omega_L \ll \omega \lesssim \omega_H$, for which

$$\left.\begin{aligned} n_{20}^2 &= \frac{\omega_0^2}{(\omega_H\cos\theta - \omega)\omega}, \\ \kappa_{20} &= \frac{n_{20}}{2}\frac{\nu}{(\omega_H\cos\theta - \omega)}, \\ \omega_{20}(\mathbf{K}) &= K^2c^2\frac{\omega_H\cos\theta}{\omega_0^2 + K^2c^2}. \end{aligned}\right\} \tag{4.60}$$

The values of κ are given in Eqs. (4.56)–(4.59) for the region of the near-Earth plasma, where collisional attenuation of the waves plays a role. This corresponds to altitudes $z < 1000$–3000 km. At higher altitudes the main role begins to be played by collisionless attenuation (see the next section).

Branches 3, 4, and 5 of $n_{12}(\omega, \theta)$ (see Figs. 4.3 and 4.4) correspond to two extraordinary HF waves and one ordinary HF wave. They are suitably described by formula (4.17). Their characteristic frequencies have been discussed above. The frequency range in which these waves are able to be propagated in a plasma, i.e., when $n^2 > 0$, varies within the limits $\omega_- \leqslant \omega \to \infty$ and includes the HF resonance region $\omega_0 \leqslant \omega \leqslant \omega_U$ for a cold plasma.

4.4. Refractive indexes for a multicomponent plasma

Let us consider the properties of the refractive index of a plasma containing several kinds of ions. In the near-Earth plasma and in the wave phenomena observable in it which are of interest to us here (see Volume 2), protons (H^+), helium ions (He^+), and atomic oxygen ions (O^+) play a role in the outer ionosphere at altitudes $z \lesssim 300$ to 1500 km. Above this, the O^+ ions rapidly disappear, and protons (H^+) predominate. He^+ ions are present to a lesser extent at great heights. However, they still have an effect on various processes out to remote distances from the Earth's surface.

In the presence of several kinds of ions the relationships involving the refractive indexes $n_{12}(\omega, \theta)$ vary significantly only in the range of frequencies comparable with the ion gyrofrequencies Ω_{Hs}. Thus, when expanding the formula (4.1) in order to study the properties of the different n_{12}^2 branches, we can use the condition $\omega^2 \ll \omega_H^2$, the role of the multicomponent nature of the plasma being important only in the VLF and ELF ranges, including part of the vicinity of the lower-hybrid frequency.

4.4.1. Properties of $\mathbf{n}_{12}^2$ in a collisionless plasma

The formulas for the refractive indexes of a multicomponent plasma for arbitrary values of θ are complicated. When $\theta = 0$ and $n_{12}^2 = (\varepsilon_1 \pm \varepsilon_2)$, these formulas are considerably simpler. With the aid of formula (4.3), it is easy to show that

$$\begin{aligned} n_1^2 &= 1 - \frac{\omega_0^2}{\omega\omega_H} + \frac{\Omega_{01}^2}{\omega(\Omega_{H1} - \omega)} + \frac{\Omega_{02}^2}{\omega(\Omega_{H2} - \omega)} + \dots, \\ n_2^2 &= 1 + \frac{\omega_0^2}{\omega\omega_H} - \frac{\Omega_{01}^2}{\omega(\Omega_{H1} + \omega)} - \frac{\Omega_{02}^2}{\omega(\Omega_{H2} + \omega)} - \dots, \end{aligned} \tag{4.61}$$

To be specific, let us assume here that $\Omega_{H1} < \Omega_{H2} < \Omega_{H3}$ are the gyrofrequencies and that $\Omega_{01}, \Omega_{02}, \Omega_{03}$ are the Langmuir frequencies, respectively, for the ions O^+, He^+, and H^+. In the case of three kinds of ions ($s = 1, 2, 3$), the refractive index $n(\omega)$ naturally has an infinity at each of the three ion gyrofrequencies, and, correspondingly, three branches. Note, too, that when $\omega \to 0$ the formulas (4.61) have an apparent infinity. We can, however, readily show that, for instance, for $\omega = 0$ and $\theta = 0$

$$n_1^2 = n_2^2 = 1 + \omega_0^2/\omega_H^2 + \sum_s \Omega_{0s}^2/\Omega_{Hs}^2. \tag{4.61a}$$

For arbitrary non-zero values of the angle θ, naturally the values of n_{12}^2 will also be finite for $\omega = 0$.

As pointed out above (see Section 4.3), an analysis of the general formula for the refractive index shows that in a collisionless plasma intersections between the n_{12}^2 branches are possible only for $\theta=0$. The values of ω corresponding to these intersections, known as the *crossover frequencies* $\omega_{cr12}=\omega_{12}$, $\omega_{cr23}=\omega_{23}\ldots$ (see formulas (4.62) and (4.63), are defined by the condition $\varepsilon_2=0$. Of course, in the vicinity of the crossover frequencies, the ordinary and extraordinary waves may, as noted previously, interact strongly, and a transformation from one type of wave to the other may even occur. An investigation of this process requires a careful analysis of the corresponding system of coupled differential equations (see Budden, 1961 and Försterling, 1942).

In a collisionless plasma for $\theta\neq 0$ at the crossover frequencies ω_{12}, $\omega_{23},\ldots$, we have $n_1^2\neq n_2^2$, However, the two waves are linearly polarized for $\omega=\omega_{cr}$, just as in the case of $\theta=0$. Upon transition through these points, the senses of polarization of the corresponding waves change (the direction of rotation of their electric-field vector **E**). From the condition $\varepsilon_2=0$ for a quasi-neutral plasma containing two kinds of ions, we obtain for the crossover frequency

$$\omega_{12}=(\alpha_2\Omega_{H1}^2+\alpha_1\Omega_{H2}^2)^{1/2}, \tag{4.62}$$

where $\alpha_1=N_1/N$ and $\alpha_2=N_2/N$ are the densities of the different kinds of ions and N is the electron density ($\alpha_1+\alpha_2=1$ and $N_1+N_2=N$). In a plasma containing three kinds of ions ($s=1,2,3$), the formulas for the crossover frequencies ω_{12} and ω_{23} are determined with the aid of an equation that is quadratic in ω^2:

$$(\omega^2-\omega_{12}^2)(\omega^2-\omega_{23}^2)=\omega^4-\omega^2(\omega_{12}^2+\omega_{23}^2)+\omega_{12}^2\omega_{23}^2=0, \tag{4.63}$$

where

$$\begin{aligned}(\omega_{12}^2+\omega_{23}^2)&=\Omega_{H1}^2(1-\alpha_1)+\Omega_{H2}^2(1-\alpha_2)+\Omega_{H3}^2(1-\alpha_3),\\(\omega_{12}^2\omega_{23}^2)&=\alpha_1(\Omega_{H2}^2\Omega_{H3}^2)+\alpha_2(\Omega_{H3}^2\Omega_{H1}^2)+\alpha_3(\Omega_{H1}^2\Omega_{H2}^2).\end{aligned} \tag{4.63a}$$

Figure 4.8 shows schematic curves of the refractive indexes $n_1^2(\omega,\theta)$ and $n_2^2(\omega,\theta)$ for a collisionless plasma. The curves pertain to the values $\theta=0, 0<\theta<\pi/2$ and $\theta=\pi/2$ and to the frequency range $0\leq\omega\sim\Omega_{H3}$. With an increase in frequency to $\omega>$ or $\gg\Omega_{H3}$, the branch n_2^2 (on the right-hand side of the figure) links up with the n_2^2 branch describing the properties of a fast magnetoacoustic wave (whistler mode), which has an infinity for $\omega\to\omega_H$. When $\theta=\pi/2$, the corresponding branch also becomes an n_2^2 branch exhibiting an infinity at the lower-hybrid frequency.

Figure 4.9 shows the results of numerical calculations of the refractive indexes n_{12}^2 of a three-component plasma, for $v=0$ and various values

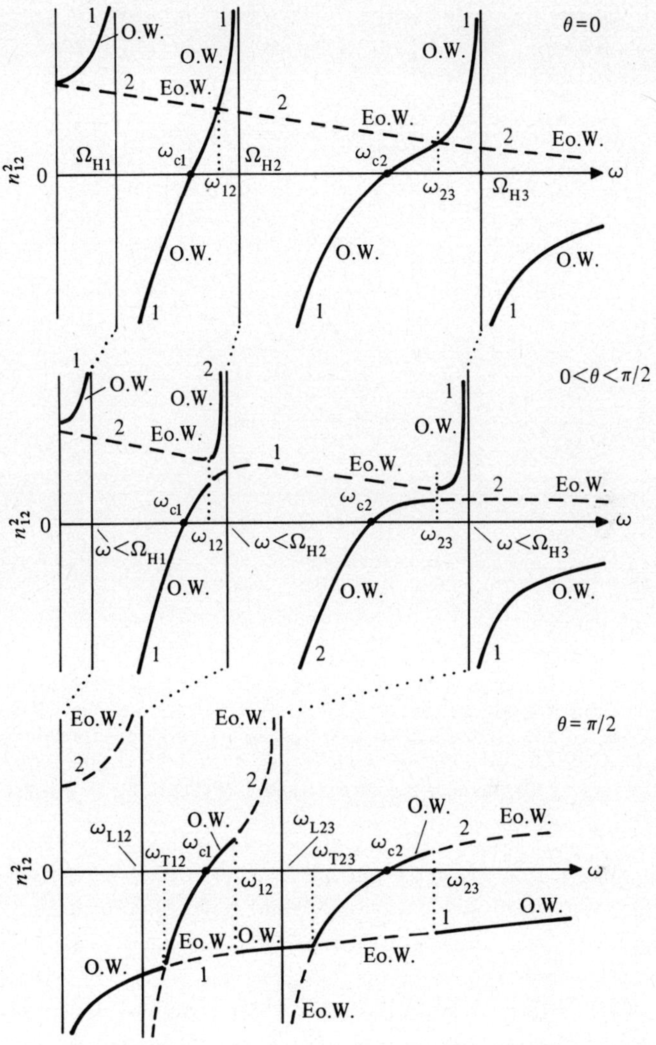

Fig. 4.8. Schematic curves of the refractive indexes n_{12}^2 of a plasma consisting of three kinds of ions for $\theta = 0, 0 < \theta < \pi/2$, and $\theta = \pi/2$.

of θ (from 0 to 85°). The $n_{12}^2(\omega, \theta = \text{constant})$ curves in this figure illustrate the quantitative extent and the nature of the variations in the refractive indexes with a change in θ at heights z from 500 to 600 km in the outer ionosphere.

For $\theta = \pi/2$ in a multicomponent plasma, resonances appear at the hybrid frequencies. One of these corresponds to the lower-hybrid frequency $\omega_L = (\Omega_H \omega_H)^{1/2}$ examined above. This frequency depends considerably on

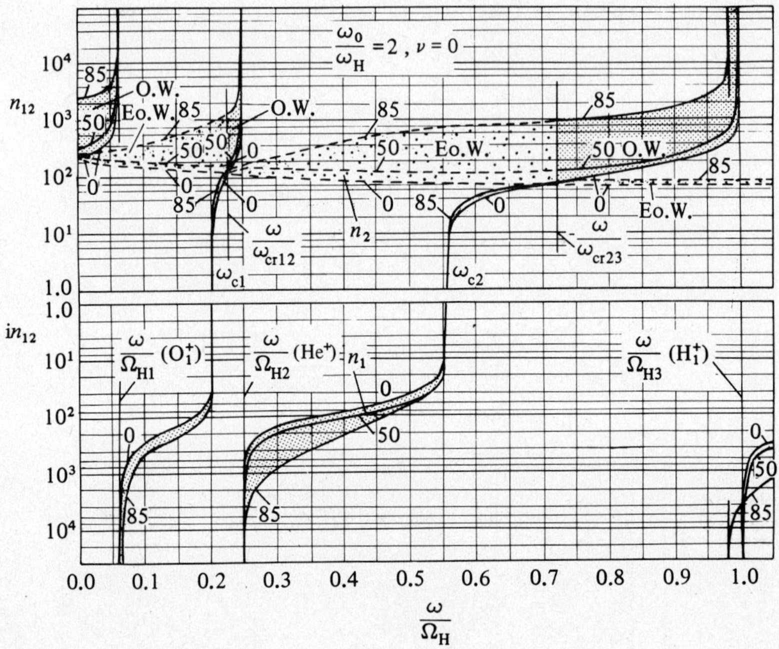

Fig. 4.9. Results of numerical calculations of n_{12}^2 for a collisionless plasma ($\nu = 0$), consisting of three kinds of ions (O^+, He^+, and H^+), for various values of the angle θ (given by the numbers next to the curves). Upon transition through the crossover frequencies ω_{12} and ω_{23}, the types of waves (ordinary and extraordinary) change (Al'pert, 1980a).

the electron gyrofrequency ω_H, and also, in this frequency range, on the total effect that the ions of various kinds have on the motion of the electrons. The other hybrid resonances are connected mainly with the motion of the ions and lie in the range of the ion gyrofrequencies. As we saw back in Section 4.1, the condition for resonance when $\theta = \pi/2$ is $\varepsilon_1 = 0$. For a plasma with two kinds of ions, provided the condition $\Omega_0^2 \gg \Omega_H^2$ is satisfied (see above), it follows from formula (4.1) that the resonance frequencies are found by solving the equation

$$\omega^4 - \omega^2 \frac{\omega_H^2(\Omega_{01}^2 + \Omega_{02}^2)}{(\omega_0^2 + \omega_H^2)} + \frac{\omega_H^2(\Omega_{01}^2\Omega_{H2}^2 + \Omega_{02}^2\Omega_{H1}^2)}{(\omega_0^2 + \omega_H^2)} = 0. \quad (4.64)$$

One of the roots of (4.64), as is easy to show, is considerably larger than the other. Actually, to the accuracy of terms of the order of m/M, we obtain from (4.64) an expression for one of the resonance frequencies, namely the *lower-hybrid frequency*

$$\omega_L = \omega_H \left(\frac{\Omega_{01}^2 + \Omega_{02}^2}{\omega_H^2 + \omega_0^2} \right)^{1/2}. \tag{4.65}$$

The second root of (4.64) defines the so-called *ion–ion hybrid frequency*

$$\omega_{L12} = \left(\frac{\Omega_{01}^2 \Omega_{H2}^2 + \Omega_{02}^2 \Omega_{H1}^2}{(\Omega_{01}^2 + \Omega_{02}^2)} \right)^{1/2} = \Omega_{H2} \left(\frac{1 + (\alpha_2/\alpha_1)(M_2/M_1)}{1 + (\alpha_2/\alpha_1)(M_1/M_2)} \right)^{1/2}, \tag{4.66}$$

where, as previously, $\alpha_1 + \alpha_2 = 1$. If three kinds of ions are present ($s = 1, 2, 3$), it follows from (4.1) that the lower-hybrid frequency

$$\omega_L = \omega_H \left(\frac{\Omega_{02}^2 + \Omega_{02}^2 + \Omega_{03}^2}{\omega_H^2 + \omega_0^2} \right)^{1/2}. \tag{4.65a}$$

The two ion–ion hybrid frequencies are determined in this case from an equation of the following form, quadratic in ω^2:

$$(\omega^2 - \omega_{L12}^2)(\omega^2 - \omega_{L23}^2) = \omega^4 - \omega^2(\omega_{L12}^2 + \omega_{L23}^2) + \omega_{L12}^2 \omega_{L23}^2 = 0 \tag{4.67}$$

where

$$(\omega_{L12}^2 + \omega_{L23}^2) = \frac{\Omega_{01}^2(\Omega_{H2}^2 + \Omega_{H3}^2) + \Omega_{02}^2(\Omega_{H3}^2 + \Omega_{H1}^2) + \Omega_{03}^2(\Omega_{H1}^2 + \Omega_{H2}^2)}{(\Omega_{01}^2 + \Omega_{02}^2 + \Omega_{03}^2)}, \tag{4.68}$$

$$(\omega_{L12}^2 \omega_{L23}^2) = \frac{\Omega_{01}^2 \Omega_{H2}^3 \Omega_{H3}^2 + \Omega_{02}^2 \Omega_{H3}^2 \Omega_{H1}^2 + \Omega_{03}^2 \Omega_{H1}^2 \Omega_{H2}^2}{(\Omega_{01}^2 + \Omega_{02}^2 + \Omega_{03}^2)}. \tag{4.68a}$$

It can be shown that in a plasma consisting of an arbitrary number of ions the lower hybrid frequency ω_L is given by the formula

$$\omega_L^2 = \frac{m}{M_{eff}} \frac{\omega_H^2 \cdot \omega_0^2}{(\omega_H^2 + \omega_0^2)}, \tag{4.69}$$

where the quantity M_{eff}, defined by

$$1/M_{eff} = \sum_s (\alpha_s/M_s), \tag{4.70}$$

could be called the effective mass of the mixture of the s kinds of ion, while the coefficients $\alpha_s = N_s/N$ are the relative ion densities with $\sum \alpha_s = 1$. Formulas (4.65) and (4.65 a) follow directly from (4.70).

If the following condition is satisfied:

$$\frac{4\Omega_{01}^2 \Omega_{03}^2 \Omega_{H2}^2 (\Omega_{H2}^2 - \Omega_{H3}^2)}{\Omega_{01}^2(\Omega_{H2}^2 - \Omega_{H3}^2) + \Omega_{02}^2(\Omega_{H1}^2 - \Omega_{H3}^2) + \Omega_{03}^2(\Omega_{H1}^2 - \Omega_{H2}^2)} \ll 1, \tag{4.71}$$

which may well be the case in the outer ionosphere, then the ion–ion hybrid frequencies (4.67) are given by the symmetrical formulas

$$\left.\begin{aligned}\omega_{L12}^2 &= \frac{\Omega_{01}^2\Omega_{H2}^2+\Omega_{02}^2\Omega_{H1}^2+\Omega_{03}^2\Omega_{H1}^2}{(\Omega_{01}^2+\Omega_{02}^2+\Omega_{03}^2)},\\ \omega_{L23}^2 &= \frac{\Omega_{02}^2\Omega_{H3}^2+\Omega_{03}^2\Omega_{H2}^2+\Omega_{01}^2\Omega_{H3}^2}{(\Omega_{01}^2+\Omega_{02}^2+\Omega_{03}^2)}.\end{aligned}\right\} \tag{4.72}$$

The zero points $n_1^2(\omega)=0$ in the ELF range, at $\theta=0$ and $\theta=\pi/2$, are defined by one and the same condition: $(\varepsilon_1+\varepsilon_2)=0$ (see (4.19) and (4.20)). The frequencies satisfying this condition are known as the *cutoff frequencies.* In a plasma consisting of two kinds of ions, there is only one cutoff frequency in the ELF range:

$$\omega_{c1}=\alpha_1\Omega_{H2}+\alpha_2\Omega_{H1}. \tag{4.73}$$

As previously, in (4.73), we have $\alpha_1+\alpha_2=1$, and the condition $\omega_{c1}\Omega_H \ll \Omega_0^2$ is satisfied, which is equivalent to the condition $\Omega_H^2 \ll \Omega_0^2$ used above. As pointed out earlier, this condition is satisfied in the plasma regions being considered here. It should be noted that in the ELF range also the cutoff frequencies do not depend on the angle θ, nor do they in the HF range $\Omega_H\omega_H \ll \omega \to \infty$. If three kinds of ions are present, the two values of the cutoff frequency in the ELF range are found from the quadratic equation

$$(\omega-\omega_{c1})(\omega-\omega_{c2})=\omega^2-(\omega_{c1}+\omega_{c2})+\omega_{c1}\omega_{c2}=0, \tag{4.74}$$

where

$$\left.\begin{aligned}(\omega_{c1}+\omega_{c2}) &= \Omega_{H1}(1-\alpha_1)+\Omega_{H2}(1-\alpha_2)+\Omega_{H3}(1-\alpha_3),\\ (\omega_{c1}\omega_{c2}) &= \alpha_1\Omega_{H2}\Omega_{H3}+\alpha_2\Omega_{H3}\Omega_{H1}+\alpha_3\Omega_{H1}\Omega_{H2}.\end{aligned}\right\} \tag{4.75}$$

In Section 4.2 (see formula (4.25)) we saw that in the HF range $\omega \gg (\Omega_H\omega_H)^{1/2}$ the refractive indexes have three more zero points, two of which correspond to the cutoff frequencies ω_- and ω_+.

4.4.2. Relationships for n_{12} and κ_{12} in a collisional plasma

The effects of the collisions naturally are these: the infinities of n_{12}^2 disappear, the refractive indexes n_1^2, n_2^2 are not cut off, and the attenuation factors κ_1 and κ_2 are non-zero over the entire frequency range.

At the resonance frequencies the quantities n_1 and κ_1 attain maxima. In the frequency range where for $\nu=0$ it is the case that $n_1^2<0$, i.e. in an opaque region for the wave, we have $\kappa_1 \gg n_1$. With an increase in ω, upon transition through the cutoff region $\omega \simeq \omega_c$ (see (4.73) and (4.74)), the attenuation factor κ_1 descreases rapidly, while n_1, on the other hand, increases rapidly. Such a pattern of the frequency dependences of n and κ is observed clearly over all the frequency ranges in the vicinity of the resonance frequencies, and for the predominantly ELF waves that are of interest to us in this subsection ($0 \leqslant \omega \sim \Omega_{Hs}$), upon transition through all the ion gyrofrequencies and ion–ion hybrid frequencies Ω_{Hs}, ω_{L12}, and ω_{L23}.

In a narrow frequency region around the gyroresonances, the coefficients

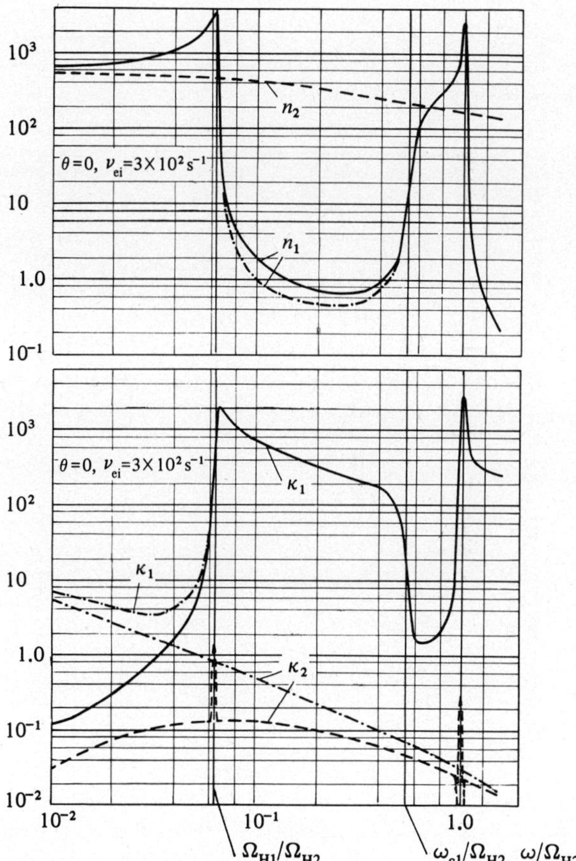

Fig. 4.10. Curves for the refractive indexes $n_{12}(\omega)$ and attenuation coefficients $\kappa_{12}(\omega)$ in a plasma consisting of two kinds of ions (O^+ and H^+) for $\theta = 0$ and $\nu_{ei} = 3 \times 10^2 s^{-1}$, calculated using various formulas (see text at beginning of Chapter 4; $\Omega_{H1} \simeq 2.57 \times 10^2 s^{-1}$, $\Omega_{H2} \simeq 4.1 \times 10^3 s^{-1}$, $\alpha_1 = \alpha_2 = 0.50$, $\omega_0 \simeq 3.75 \times 10^7 s^{-1}$). The dot-dash curves indicate the results of calculations using effective masses m_{eff} and M_{eff} (Al'pert, 1980a).

κ_2 and n_2 also exhibit a number of peculiarities. For instance, at some values of the plasma parameters, they go through sharp maxima.

The attenuation coefficients and refractive indexes behave in a complicated and unusual manner in the vicinity of the crossover frequencies ω_{12}, ω_{23}. In a certain interval of angles $\theta < \theta_{cr}$, the branches of the refractive-index curves $n_1(\omega, \theta)$ and $n_2(\omega, \theta)$ have, respectively, minima and maxima. For the specified plasma parameters, the relative depth of a minimum of $\kappa_1(\omega, \theta)$ and the height of a maximum $\kappa_2(\omega, \theta)$ reach their highest values at $\theta = \theta_{cr}$. At the point where $\omega = \omega_{cr}$ and $\theta = \theta_{cr}$, the

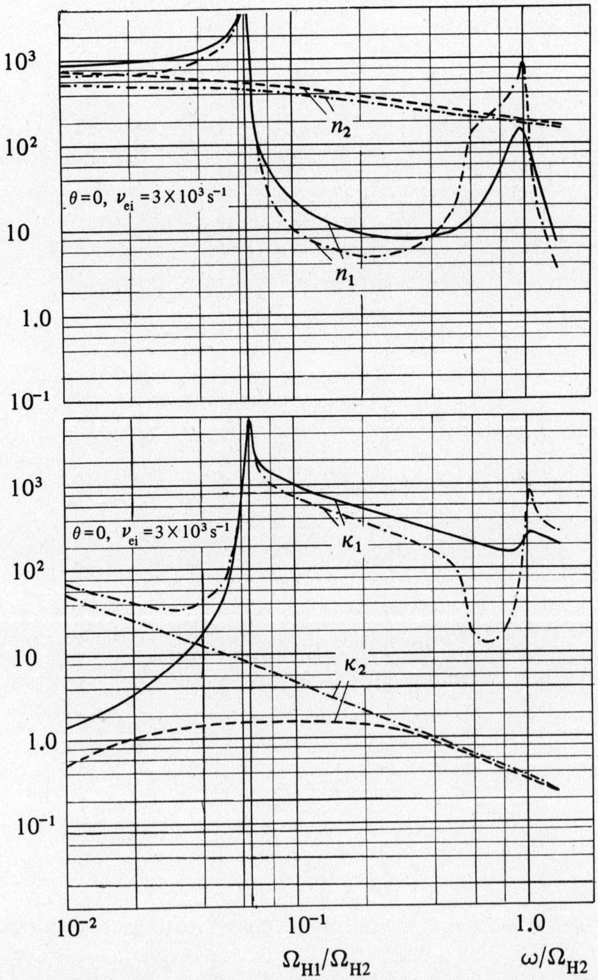

Fig. 4.11. As for Fig. 4.10, but with $\nu_{ei} = 3 \times 10^3\,s^{-1}$, $\alpha_1 = 0.85$, and $\alpha_2 = 0.15$.

equalities $n_1 = n_2$ and $\kappa_1 = \kappa_2$ are valid simultaneously. This is the *crossover region* for the complex refractive indexes, as described in Section 4.3.1. during our consideration of the properties of the polarization coefficient. For $\theta > \theta_{cr}$ the $\kappa_1(\omega, \theta)$ and $\kappa_2(\omega, \theta)$ branches intersect. When $\omega = \omega_{cr}$ and $\theta = \theta_{cr}$, both waves are linearly polarized; on either side of this point the waves are elliptically polarized in different senses. The properties of the refractive indexes near the crossover frequencies were first investigated by Jones, 1969, 1972.

The results of numerical calculations of the parameters of the outer ionosphere, observed in the daytime at altitudes $z = 500$ to 600 km and $z \simeq 300$ to 350 km (Figs. 4.10–4.21), illustrate the nature of the variation of $n_{12}(\omega, \theta)$ and $\kappa_{12}(\omega, \theta)$ in a collisional multicomponent plasma. Since the expressions for $\tilde{n}_{12}^2$ in a collisional, and especially multicomponent,

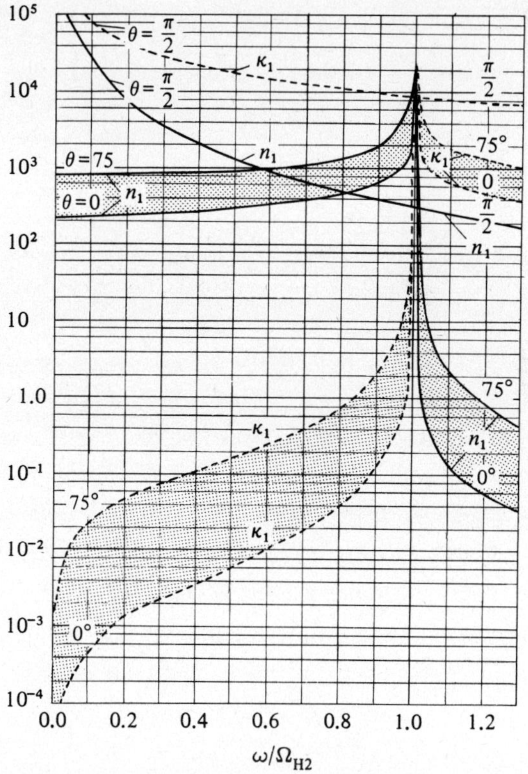

Fig. 4.12. Curves for the refractive indexes $n_1(\omega)$ and attenuation coefficients $\kappa_1(\omega)$ of the ordinary wave for various values of the angle ϕ in a plasma consisting of electrons and protons only, with $\nu_{ei} = 3 \times 10^2\,\mathrm{s}^{-1}$, $\Omega_{H2} = 4.1 \times 10^3\,\mathrm{s}^{-1}$, $\alpha_1 = \alpha_2 = 0.5$, $\omega_0 = 3.75 \times 10^7\,\mathrm{s}^{-1}$ (Al'pert, 1980a).

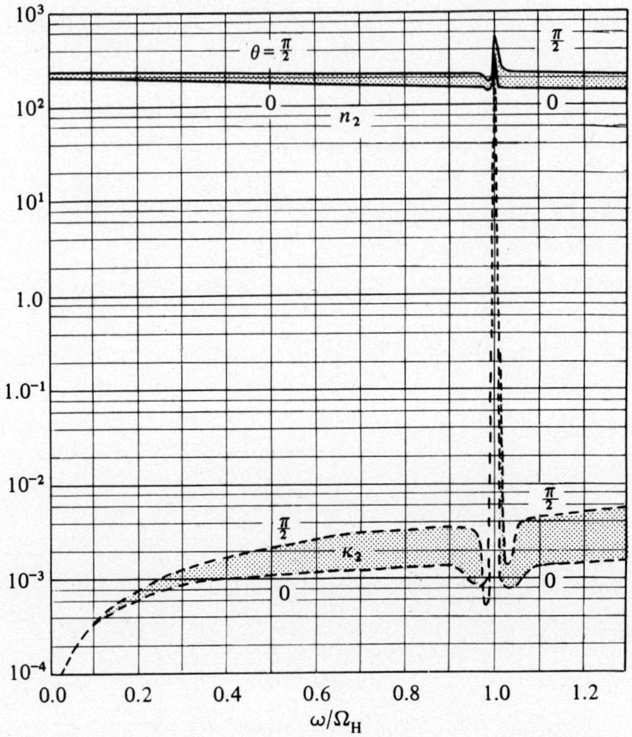

Fig. 4.13. As for Fig. 4.12, but for the extraordinary wave $\tilde{n}_2(\omega, \theta)$.

plasma are extremely complicated, a detailed study of their properties using analytical methods has not as yet been able to explain a number of important peculiarities to be described below.

The curves for n_{12} and κ_{12} given in Figs. 4.10 to 4.21 were obtained for a plasma containing protons (H^+) and two kinds of ions:

$$\text{atomic oxygen } O^+, \quad m/M_1 = 3.4256 \times 10^{-5},$$

$$\text{protons } H^+, \quad m/M_2 = 5.4810 \times 10^{-4}.$$

The following values of the characteristic frequencies were chosen:

$$\Omega_{H1} = 2.5688 \times 10^2\,\mathrm{s}^{-1}, \quad \Omega_{H2} = 4.1101 \times 10^3\,\mathrm{s}^{-1},$$

$$\omega_H = 7.5 \times 10^6\,\mathrm{s}^{-1}, \quad \omega_0 = 3.75 \times 10^7\,\mathrm{s}^{-1},$$

while the collision frequencies are, respectively,

$$\nu_{ei} = 3 \times 10^2\,\mathrm{s}^{-1}, \quad \nu_{ei} = 3 \times 10^3\,\mathrm{s}^{-1}.$$

The values of ν_{ei}, ν_{12}, and ν_{21} were determined using formulas (4.4) and (4.5). For altitudes $z \simeq 500$ to 600 km, the relative ion densities were taken

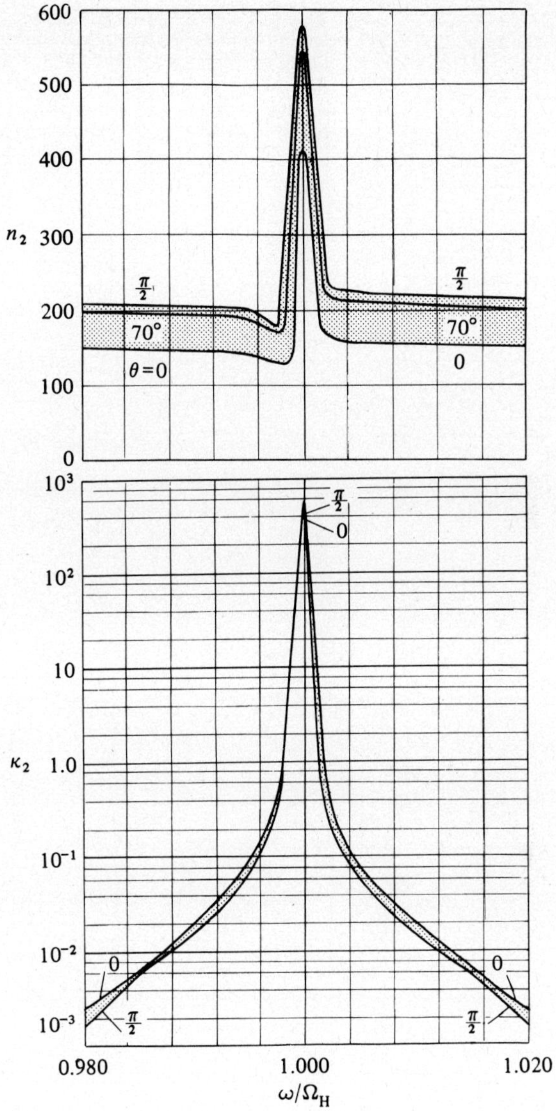

Fig. 4.14. Curves of $n_2(\omega)$ and $\kappa_2(\omega)$ for the extraordinary wave in a plasma consisting of electrons and protons in the vicinity of the gyroresonance $\omega/\Omega_H \simeq 1$ for various values of the angle θ; $\nu_{ei} = 3 \times 10^2\ \mathrm{s}^{-1}$, $\Omega_H = 4.1 \times 10^3\ \mathrm{s}^{-1}$, $\omega_0 = 3.75 \times 10^7\ \mathrm{s}^{-1}$.

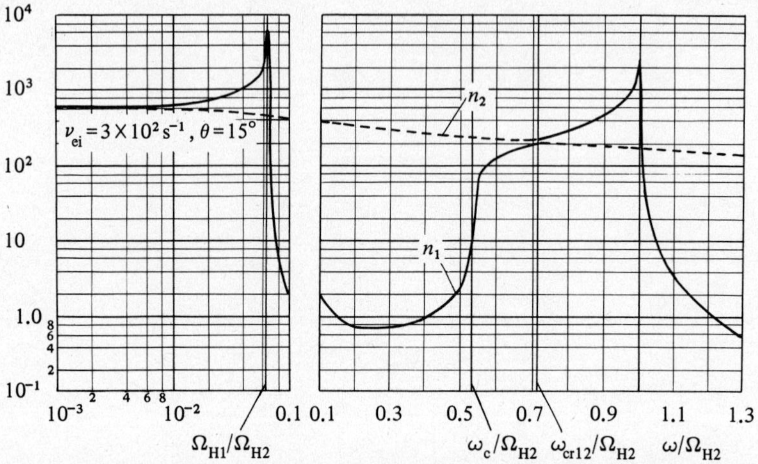

Fig. 4.15. Curves of $n_1(\omega)$ and of $n_2(\omega)$ in a two-component plasma (H^+ and O^+); $\nu_{ei} = 3 \times 10^2\ s^{-1}$, $\Omega_H \simeq 2.57 \times 10^2\ s^{-1}$, $\Omega_{H2} \simeq 4.1 \times 10^3\ s^{-1}$, $\omega_0 \simeq 3.75 \times 10^7\ s^{-1}$, $\alpha_1 = \alpha_2 = 0.5$ and $\theta = 15°$ (Al'pert, 1980a).

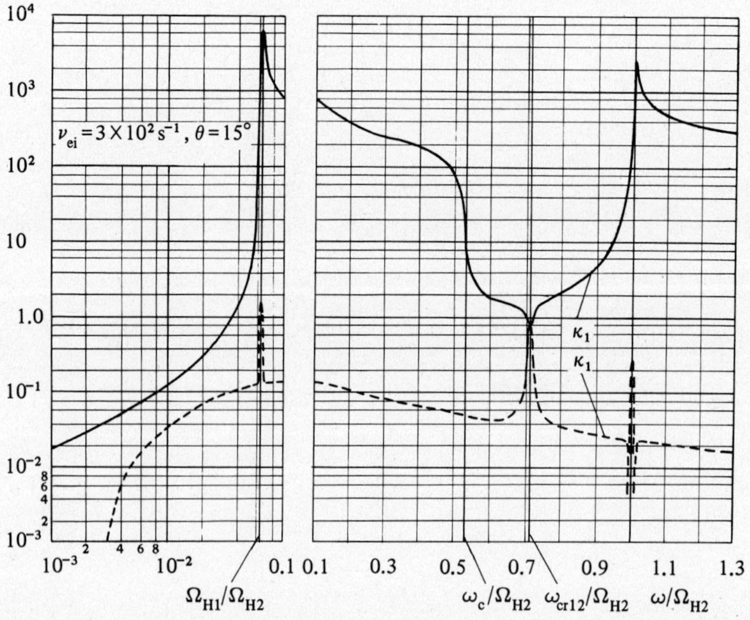

Fig. 4.16. Curves of $\kappa_1(\omega)$ and $\kappa_2(\omega)$ for the same values of the plasma parameters as in Fig. 4.15.

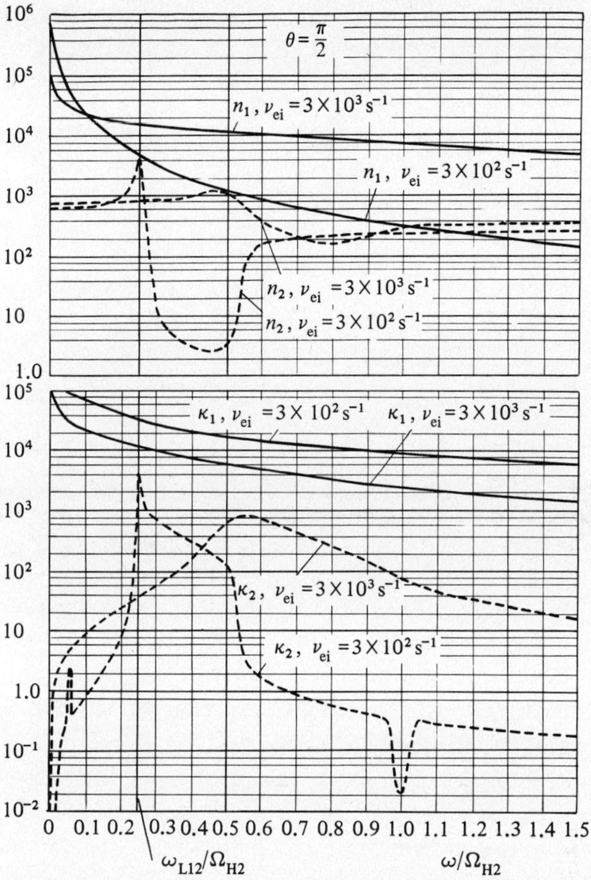

Fig. 4.17. Curves of $n_{12}(\omega)$ and $\kappa_{12}(\omega)$ for $\theta=\pi/2$ and the plasma-parameter values indicated in Figs. 4.10 and 4.11 (Al'pert, 1980a).

to be $\alpha_1=\alpha_2=0.5$, while for $z\simeq 300$ to 350 km it was assumed that $\alpha_1=0.85$ and $\alpha_2=0.15$. The calculations of n and κ were carried out with the aid of formulas (4.10).

It will also be of interest to compare here the curves for $n(\omega)$ and $\kappa(\omega)$ obtained using the tensor elements (4.10) with the curves obtained using the 'effective masses' M_{eff} and m_{eff}, which were discussed at the beginning of Section 4.1. The corresponding $n_{12}(\omega)$ and $\kappa_{12}(\omega)$ curves are shown in Figs. 4.10 and 4.11 (dot-dash curves), for an angle $\theta=0$ and various values of ν_{ei}; the curves pertain to a plasma with two kinds of ions. Inspection of these figures shows that the $n_{12}(\omega)$ and $\kappa_{12}(\omega)$ values found using the different formulas differ considerably in a number of cases; the formulas

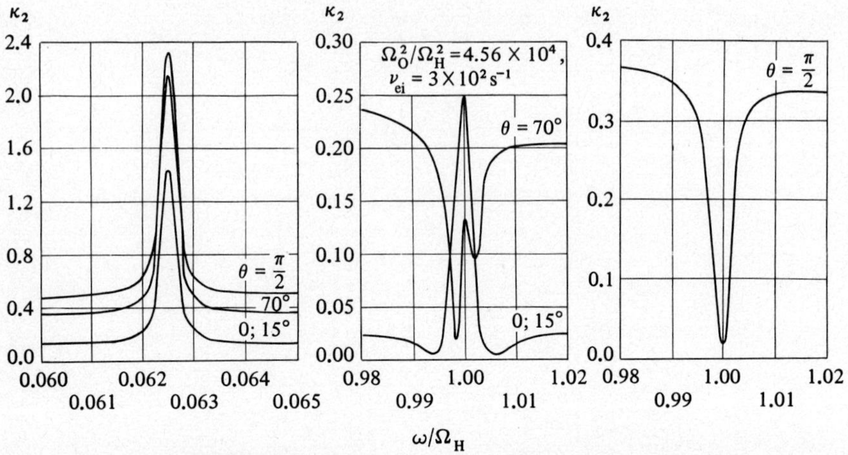

Fig. 4.18. Curves of the attenuation coefficient $\kappa_2(\omega)$ in a two-component plasma in the vicinity of the gyroresonances of O^+ and H^+ ions (the values of the parameters are the same as in Fig. 4.14) (Ashour-Abdalla, 1972.)

in which M_{eff} and m_{eff} were used do not reveal any singularities of κ_2 and n_2 in the vicinity of the gyroresonances (see Figs. 4.13 and 4.14). Moreover, in some frequency intervals they do not portray correctly the nature of the frequency dependences, especially as concerns $\kappa_2(\omega)$, the values of which are distinctly too low.

Figs. 4.12–4.14 illustrate the behavior of the refractive indexes and attenuation coefficients in a plasma containing just one kind of ion. It is important to note that in the case under review the attenuation coefficient $\kappa_2(\omega)$ of the extraordinary wave has a sharp narrow maximum at $\omega \simeq \Omega_H$ (Fig. 4.13). The peak of the refractive index $n_2(\omega)$ is less pronounced. Consequently, in a collisional plasma, in the ion gyroresonance region, both waves may be greatly attenuated. As far as is known to the author, these peculiarities in the propagation of the extraordinary wave in a plasma were not known previously.

Figs. 4.15–4.21 illustrate the behavior of $n_{12}(\omega)$ and $\kappa_{12}(\omega)$ in a plasma consisting of two kinds of ions. In this case the curves of the attenuation coefficient $\kappa_2(\omega, \theta)$ of the extraordinary wave are more diverse (see Figs. 4.16, 4.17, 4.18). For instance, for $\theta = \pi/2$ at the gyrofrequency of the second ion $\omega = \Omega_{H2} > \Omega_{H1}$, the $\kappa_2(\omega)$ curve has a minimum instead of a maximum (Fig. 4.17); the maximum begins to appear, for the plasma parameters selected, at $\theta \geqslant 30°$ to $50°$ (Fig. 4.18).

Figs. 4.16, 4.19, and 4.21 illustrate the $n_{12}(\omega, \theta)$ and $\kappa_{12}(\omega, \theta)$ relationships in the vicinity of the crossover frequency ω_{12}, where $\theta = \theta_{cr}$. On the

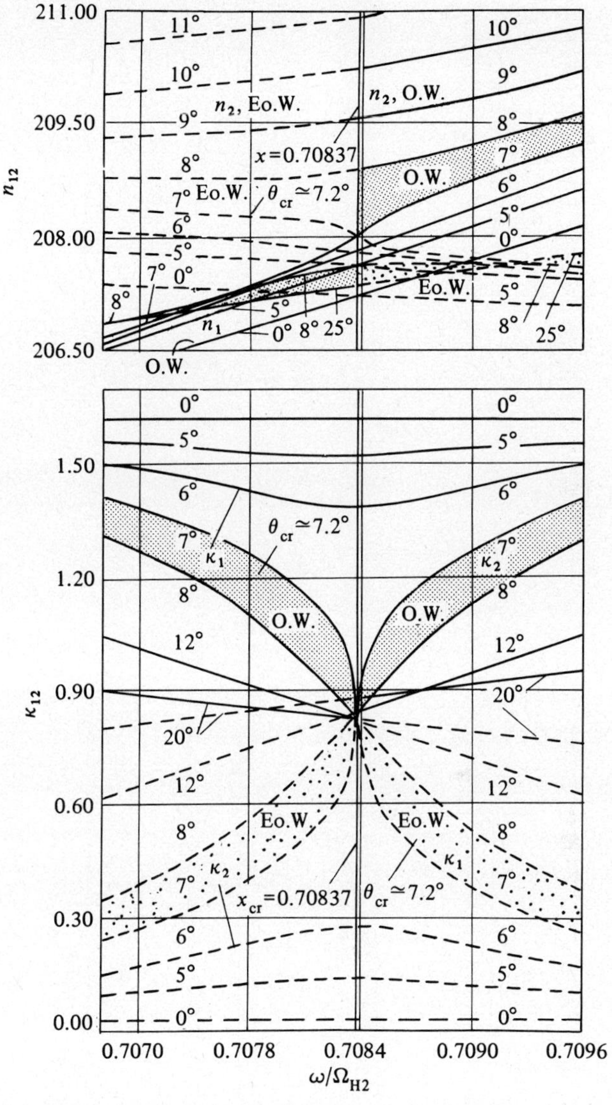

Fig. 4.19. Curves of $n_{12}(\omega)$ and $\kappa_{12}(\omega)$ in the vicinity of the crossover frequency ω_{12} in a two-component plasma for various values of the angle θ (numbers next to the curves). The values of the plasma parameters are the same as in Fig. 4.15 (Al'pert, 1980a).

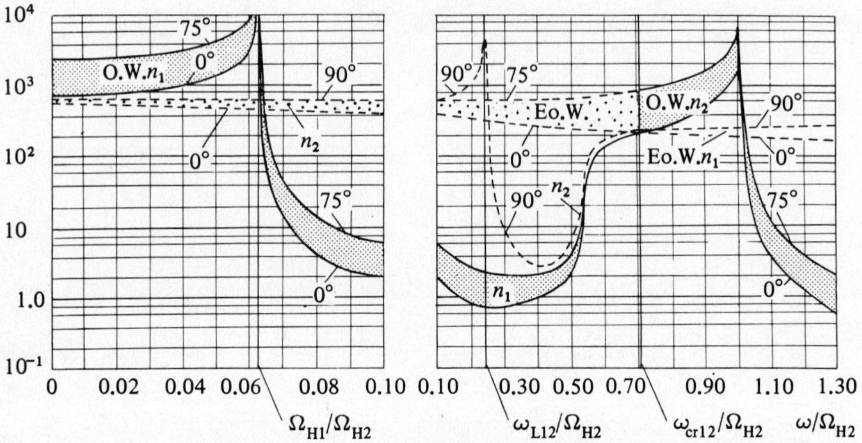

Fig. 4.20. General behaviour of the refractive indexes $n_{12}(\omega)$ for various values of θ (numbers next to the curves), in a plasma consisting of two kinds of ions (see Fig. 4.15).

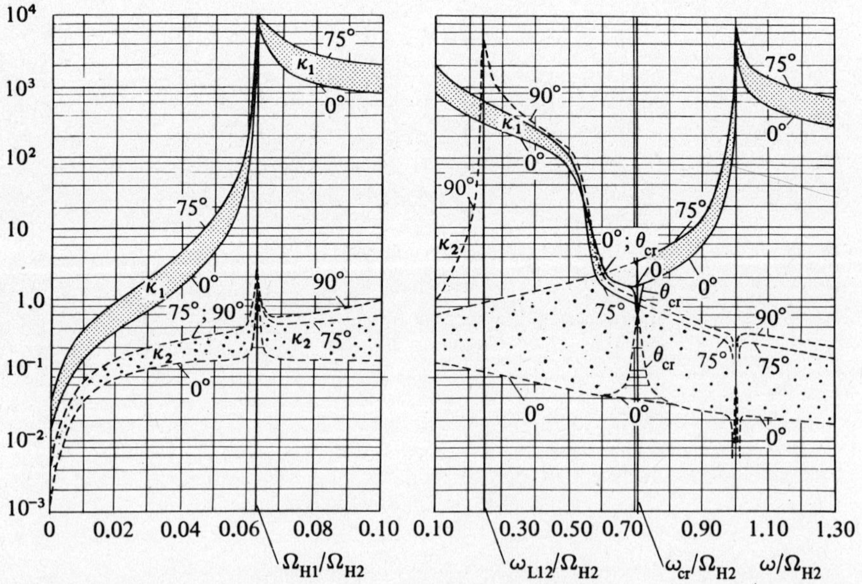

Fig. 4.21. General behaviour of the attenuation coefficients $\kappa_{12}(\omega)$ for various values of the angle θ (numbers next to the curves), in a plasma consisting of two kinds of ions (see Fig. 4.15).

figures the change in the polarization senses of the waves described by the refractive indexes n_1 and n_2, upon transition through the crossover frequencies, is indicated by a variation of the forms of the corresponding curves: a solid curve becomes a dashed curve or vice versa. Physically, a change in the polarization sense means that during the transition through the crossover region ($\omega = \omega_{cr}$, $\theta = \theta_{cr}$) *the extraordinary wave becomes an ordinary wave and vice versa.* The nature of the $n_{12}(\omega, \theta)$ and $\kappa_{12}(\omega, \theta)$ relationships in the vicinity of $\omega = \omega_{cr}$ and $\theta = \theta_{cr}$ can be traced out in detail in Fig. 4.19. This figure gives the families of $n_{12}(\omega, \theta = \text{constant})$ and $\kappa_{12}(\omega, \theta = \text{constant})$ curves.

Figs. 4.20 and 4.21 portray the general development of the $n_{12}(\omega, \theta)$ and $\kappa_{12}(\omega, \theta)$ curves over the entire frequency range of interest to us in this subsection, in the interval of angles $0 \leqslant \theta \leqslant \pi/2$. Despite the complexity of these figures and the cumbersomeness of some of the details in them, a careful examination of the $n(\omega, \theta)$ and $\kappa(\omega, \theta)$ curves gives us some idea of the behavior of the refractive indexes and absorption coefficients for ELF waves in a multicomponent plasma, and in addition it enables a quantitative evaluation of the variations of these parameters to be made under realistic conditions. These figures do not show the $n_1(\omega, \theta)$ and $\kappa_1(\omega, \theta)$ curves of the ordinary wave for $\theta = \pi/2$, since the values of these quantities increase rapidly for the plasma-parameter values used here, when $\theta > 75°$. Thus the corresponding curves would go beyond the limits of the figure.

A numerical analysis of the refractive indexes and of the tensor elements for a collisional plasma indicated that at the ion–ion hybrid frequencies ω_{L12}, ω_{L23} the real part of the tensor element $\mathrm{Re}\{\varepsilon_1\} = 0$, while at the crossover frequencies $\omega_{12} \simeq \omega_{23}$, the quantity $\mathrm{Re}\{\varepsilon_2\} = 0$.

5

Refractive indexes and attenuation factors in a warm plasma

When the thermal motion of the particles is taken into account, in other words when the spatial dispersion of the plasma is allowed for, the quantities A, B and C in the dispersion equation (3.1) depend on $\mathbf{K}$ and thus on n, that is, $A = A[\mathbf{K}(n)]$, etc. This means that (3.1) is a transcendental equation, and the number of roots, or branches of its solution, may be infinite. Moreover, only in a limited number of cases does the solution of (3.1) define branches on which the waves are only slightly attenuated; in a warm plasma the number of types of slightly attenuated branches of waves is quite limited.

When the magnetic field $\mathbf{H}_0 \neq 0$, the relations describing the refractive indexes $n^2(\omega, \mathbf{K})$ and the frequency $\omega(\mathbf{K})$ are difficult to use. Thus a number of the regularities and properties of these quantities are frequently revealed only during an analysis of particular cases, i.e., for certain ranges of values of the plasma parameters, ranges of frequency, etc. The corresponding cases will be considered below.

Kinetic theory, as we saw above in Chapter 3, reveals some basically new phenomena in a plasma. The most general of these are:

(1) Landau damping in a collisionless plasma, sometimes known as Čerenkov damping, and Doppler-shifted cyclotron (magnetobraking) damping;

(2) Čerenkov resonance (or Landau excitation) and Doppler-shifted cyclotron (magnetobraking) resonance.

Taking the thermal motion of particles into account usually leads only to minor corrections in the formulas for the refractive indexes $n_{1\,2}^2(\omega, \theta)$ of transverse electromagnetic waves in a cold plasma. However, at the same time, their singularities disappear in a collisionless plasma, which is an important consequence of the kinetic theory. The use of its results within the framework of the problems interesting us calls for a discriminating approach. While analysing a number of phenomena and processing the experimental data, it often suffices to use the results of the theory of a cold plasma. However, in many cases, as will be seen below, the phenomena in question can be explained only on the basis of a theory taking thermal motion of the particles into account, and to analyze these phenomena the appropriate formulas of kinetic theory have to be used.

5.1. Tensor elements and values of n_{12}^2 allowing for the thermal motion of the particles ($T_e \neq T_i \neq 0$).

In general the formulas defining the elements of the permittivity tensor in the kinetic theory are very complicated. However, if the conditions (4.3a) are satisfied, which is quite possible, with certain reservations, in the majority of cases of interest to us here, then the formulas are considerably simpler and have the following form:

$$
\left.
\begin{aligned}
\varepsilon_{11} &= \varepsilon_1 - \frac{1}{2}\frac{\omega_0^2 K^2 v_e^2}{(\omega^2-\omega_H^2)}\left[\frac{\omega^2+3\omega_H^2}{(\omega^2-\omega_H^2)^2}\cos^2\theta + \frac{3\sin^2\theta}{(\omega^2-4\omega_H^2)}\right] \\
&\quad - \frac{1}{2}\frac{\Omega_{01}^2 K^2 v_{i1}^2}{(\omega^2-\Omega_{H1}^2)}\left[\frac{\omega^2+3\Omega_{H1}^2}{(\omega^2-\Omega_{H1}^2)^2}\cos^2\theta + \frac{3\sin^2\theta}{(\omega^2-4\Omega^2)}\right] - \ldots, \\
\varepsilon_{12} &= \mathrm{i}\varepsilon_2 + \frac{\mathrm{i}}{2}\frac{\omega_H}{\omega}\frac{\omega_0^2 K^2 v_e^2}{(\omega^2-\omega_H^2)}\left[\frac{3\omega^2+\omega_H^2}{(\omega^2-\omega_H^2)^2}\cos^2\theta + \frac{6\sin^2\theta}{(\omega^2-4\omega_H^2)}\right] \\
&\quad - \frac{1}{2}\frac{\Omega_{H1}}{\omega}\frac{\Omega_{01}^2 K^2 v_{i1}^2}{(\omega^2-\Omega_{H1}^2)}\left[\frac{3\omega^2+\Omega_{H1}^2}{(\omega^2-\Omega_{H1}^2)^2}\cos^2\theta + \frac{6\sin^2\theta}{(\omega^2-4\Omega_{H1}^2)}\right] - \ldots, \\
\varepsilon_{13} &= -\sin\theta\cos\theta\left[\frac{\omega_0^2 K^2 v_e^2}{(\omega^2-\omega_H^2)^2} + \frac{\Omega_{01}^2 K^2 v_{i1}^2}{(\omega^2-\Omega_{H1}^2)^2} + \ldots\right], \\
\varepsilon_{22} &= \varepsilon_1 - \frac{1}{2}\frac{\omega_0^2 K^2 v_e^2}{(\omega^2-\omega_H^2)}\left[\frac{\omega^2+3\omega_H^2}{(\omega^2-\omega_H^2)^2}\cos^2\theta + \frac{\omega^2+8\omega_H^2}{\omega^2(\omega^2-4\omega_H^2)}\sin^2\theta\right] \\
&\quad - \frac{1}{2}\frac{\Omega_{01}^2 K^2 v_{i1}^2}{(\omega^2-\Omega_{H1}^2)}\left[\frac{\omega^2+3\Omega_{H1}^2}{(\omega^2-\Omega_{H1}^2)^2}\cos^2\theta + \frac{\omega^2+8\Omega_{H1}^2}{\omega^2(\omega^2-4\Omega_{H1}^2)}\sin^2\theta\right] - \ldots, \\
\varepsilon_{23} &= \frac{\mathrm{i}}{2}\sin\theta\cos\theta\left[-\frac{\omega_H\omega_0^2 K^2 v_e^2(3\omega^2-\omega_H^2)}{\omega^3(\omega^2-\omega_H^2)^2}\right. \\
&\quad \left. + \frac{\Omega_{H1}\Omega_{01}^2 K^2 v_{i1}^2(3\omega^2-\Omega_{H1}^2)}{\omega^3(\omega^2-\Omega_{H1}^2)^2} + \ldots\right], \\
\varepsilon_{33} &= \varepsilon_3 - \frac{1}{2}\frac{\omega_0^2 K^2 v_e^2}{\omega^4}\left(3\cos^2\theta + \frac{\omega^2\sin^2\theta}{(\omega^2-\omega_H^2)}\right) \\
&\quad - \frac{1}{2}\frac{\Omega_{01}^2 K^2 v_{i1}^2}{\omega^4}\left(3\cos^2\theta + \frac{\omega^2\sin^2\theta}{(\omega^2-\Omega_{H1}^2)}\right) - \ldots,
\end{aligned}
\right\}
\tag{5.1}
$$

where $\varepsilon_1, \varepsilon_2$, and ε_3 are the elements of the tensor (4.3) for a cold plasma (Sitenko & Stepanov, 1956; Akhiezer *et al.*, 1974).

Formulas (5.1) were written for electrons and ions of one kind. All the notation of formula (4.3) was used in these formulas. The dots... at the end of each formula signify that, if there are s ions of different kinds, then in the corresponding formula analogous terms with values Ω_{0s}, Ω_{Hs}, and ν_{is} have to be added. The infinities of the tensor elements (5.1) for $\omega = \omega_H, \omega = \Omega_{Hs}, \omega = 2\omega_H$, and $\omega = 2\Omega_H$ agree with the conditions (4.3*a*), being explained by the fact that at the gyrofrequencies for quasi-longitudinal propagation ($\theta \simeq 0$) the waves are greatly attenuated, while on the other hand, under the corresponding conditions, the growth rates for resonant excitation of a plasma at these frequencies are large. These cases require special consideration.

It should be noted in (5.1) that analogous effects do not appear at higher harmonics of the gyrofrequencies, because these formulas were derived without taking into account the terms $\beta^4 = (v/c)^4, \beta^6 = (v/c)^6$, etc., in the corresponding equations.

The limitations (4.3*a*) mean that the Larmor radii of electrons and ions of any kind should be considerably smaller than the wavelengths of the corresponding oscillations in a direction normal to the magnetic-field vector $\mathbf{H}_0$. Such conditions actually exist in the various wave processes observed in the near-Earth plasma.

5.2. High-frequency, Langmuir, and low frequency waves ($\omega \gtrsim \omega_0, \omega_H; \omega_L \ll \omega \leqslant \omega_H$)

In the high-frequency and low-frequency ranges it is enough just to take the motion of electrons into account; the effect of ions can be completely neglected. To sum up, in the approximation given above, if the tensor elements (5.1) are used, the dispersion equation reduces to the following:

$$(A - \tfrac{1}{2}\beta_e^2 n^2 A_1)n^4 + Bn^2 + C = 0 \tag{5.2}$$

where $\beta_e^2 = (v_e/c)^2$ and

$$A_1 = \frac{\omega_0^2}{\omega^2}\left[3\cos^4\theta + \frac{\omega^2(6\omega^4 + 3\omega^2\omega_H^2 + \omega_H^4)}{(\omega^2 - \omega_H^2)^3}\cos^2\theta\sin^2\theta + \frac{3\omega^4\sin^4\theta}{(\omega^2 - \omega_H^2)(\omega^2 - 4\omega_H^2)}\right] \tag{5.2a}$$

(Gershman, 1953a, b). Equation (5.2) is of the third degree in n^2 and thus, in contrast to (3.1), defines a third branch of the refractive index n_3^2 in addition to the two branches of n_{12}^2 given above (see formula (4.2)).

5.2.1. Refractive-index formulas

The branch n_3^2 describes the properties of potential *longitudinal* HF waves. Far away from resonance, it is defined by the condition

$$2A - \beta_e^2 n_3^2 A_1 = 0. \tag{5.2b}$$

Then, in (5.2*a*) the following conditions are imposed:

$$\left.\begin{aligned} &|\beta_e^2 n^2 \cos^2\theta| \ll 1, |\beta_e^2 n^2 (\omega/\omega_H)^2 \sin^2\theta| \ll 1, \\ &|1-(\omega_H/\omega)^2 + (\omega_0\omega_H/\omega_H^2)^2\cos^2\theta| \gtrsim \beta_e \ll 1. \end{aligned}\right\} \tag{5.3}$$

In this case longitudinal waves are only slightly attenuated and they are described by the refractive index

$$n_3^2 = \frac{\omega^2(\omega^2-\omega_U^2)+\omega^2\omega_H^2\cos^2\theta}{\frac{1}{2}\beta_e^2\omega^2\omega_0^2 F(\omega,\theta)} = \frac{(\omega^2-\omega_H^2)(\omega^2-\omega_0^2)-\omega_0^2\omega_H^2\sin^2\theta}{\frac{1}{2}\beta_e^2\omega^2\omega_0^2 F(\omega,\theta)}, \tag{5.4}$$

where $\omega_U = (\omega_H^2 + \omega_0^2)^{1/2}$ is the upper-hybrid frequency and

$$F(\omega,\theta) = \frac{3\omega^2\sin^4\theta}{(\omega^2-4\omega_H^2)} + \sin^2\theta\cos^2\theta\left[1+\frac{\omega^2(5\omega^2-\omega_H^2)}{(\omega^2-\omega_H^2)^2}\right] + 3\left(1-\frac{\omega_H^2}{\omega^2}\right)\cos^4\theta. \tag{5.4a}$$

An analysis of formula (5.4) shows that, when $\omega \to \omega_H$,

$$n_3^2 \simeq -\frac{(\omega^2-\omega_H^2)^2}{2\beta_e^2\omega_H^4\cos^2\theta} < 0. \tag{5.4b}$$

At the electron gyrofrequency a longitudinal wave is highly attenuated. However, in the vicinity of the second-harmonic gyroresonance $\omega \simeq 2\omega_H$, in a certain range of frequencies, conditions will exist for which $n_3^2 > 0$, i.e., the wave does not disappear. This is a consequence of the following formula, implied by (5.4),

$$n_3^2 \simeq \frac{(\omega^2-4\omega_H^2)[3\omega^2-\omega_0^2(4-\cos^2\theta)]}{12\beta_e^2\omega^2\omega_0^2\sin^4\theta}. \tag{5.4c}$$

For $\theta = 0$, the magnetic field does not influence the branch describing the resonant waves. In this case

$$n_3^2 = \tfrac{2}{3}(1/\beta_e^2)(\omega^2-\omega_0^2)/\omega_0^2. \tag{5.4d}$$

This conclusion is self-evident, since a magnetic field will not act upon particles moving in the direction of the vector $\mathbf{H}_0$. Thus it follows that the same value of n_3^2 is obtained when $H_0 = 0$, i.e., in an isotropic plasma. In the latter case the longitudinal *Langmuir waves* are described by the dispersion equation

$$\omega^2 = \omega_0^2(1 + 3K^2D_e^2) = \omega_0^2 + \tfrac{3}{2}(Kv_e)^2. \tag{5.5}$$

These waves are attenuated only slightly when their phase velocity is much higher than v_e, i.e., when

$$v_e/v_\phi = \beta_e n \ll 1, \tag{5.6}$$

or, equivalently, when their wavelength is much greater than the Debye length:

$$(KD_e)^2 = (2\pi D_e/\lambda)^2 \ll 1. \tag{5.6a}$$

In (5.2) to (5.6), the following notation is used:

$$\beta_e = \frac{v_e}{c}, \quad K = K_0 n_3 = \frac{\omega}{c} n_3, \quad D_e = \left(\frac{\varepsilon_0 k T_e}{Ne^2}\right)^{1/2} = \frac{1}{\sqrt{2}}\frac{v_e}{\omega_0}. \tag{5.7}$$

K is the wave number, while k is Boltzmann's constant.

It was during the study of the high-frequency resonances of an isotropic plasma that the collisionless attenuation mechanism known as Landau damping was first discovered (Landau, 1946). Collisionless Landau damping is explained by the absorption of longitudinal waves as they interact with electrons. In the case being considered, when $v_\phi \gg v_e$, only electrons situated in the *tail* of the velocity distribution play a role in this process. Consequently, the attenuation is also slight; it is characterized by the 'temporal' attenuation rate ($E \sim \exp(-\gamma_e t)$) and the 'spatial' coefficient ($E \sim \exp(-\kappa_e r)$) of attenuation (see (3.4)):

$$\gamma_e = e^{-3/2}\left(\frac{\pi}{8}\right)^{1/2} \frac{\omega_0}{(KD_e)^3} \exp\left(-\frac{1}{2K^2D_e^2}\right), \tag{5.8}$$

$$\kappa_e = \frac{e^{-3/2}}{3}\left(\frac{\pi}{8}\right)^{1/2} \frac{c}{\omega_0 D_e}\frac{1}{(KD_e)^4} \exp\left(-\frac{1}{2K^2D_e^2}\right). \tag{5.9}$$

In a magnetoplasma ($H_0 \neq 0$) the formula for the attenuation coefficient γ_{eH} is extremely cumbersome, so it will not be presented here. We note only that if the condition $(\beta_e n_3 \cos\theta)^2 = (v_e/v_\phi)^2 \cos^2\theta \ll 1$ is fulfilled (see formula (5.3)) the attenuation of longitudinal waves is slight ($\gamma_{eH} \ll \omega$), while for $\theta = \pi/2$ at frequencies far from gyroresonance the damping of longitudinal waves is completely absent ($\gamma_{eH} = 0$). Of course, as $\theta \to 0$, it follows that $\gamma_{eH} \to \gamma_e$ (see (5.8)). The frequency of the slightly attenuated plasma waves differs little from the Langmuir frequency ω_0, as is evident from formula (5.5) and the condition (5.6*a*). The phase velocity of the Langmuir waves is

$$v_\phi = (3/2)^{1/2} v_e (1 - \omega_0^2/\omega^2)^{-1/2}. \tag{5.10}$$

For $\theta = \pi/2$, sufficiently far away from the resonance frequencies, where

$$(\omega^2 - \omega_U^2)/\omega^2 \gg \beta_e^2, \quad (\omega^2 - 4\omega_H^2)/\omega^2 \gg \beta_e^2, \tag{5.11}$$

the refractive index for plasma waves is

$$n_3^2 = \frac{(\omega^2 - \omega_U^2)(\omega^2 - 4\omega_H^2)}{\frac{3}{2}\beta_e^2 \omega^2 \omega_0^2}. \tag{5.12}$$

The *high-frequency resonance branches* are defined, taking the thermal motion of electrons into account, by the formulas

$$\omega_1(\theta, \beta_e) \gtrsim \omega_1(\theta)(1 + \xi_1), \quad \omega_2(\theta, \beta_e) \gtrsim \omega_2(\theta)(1 + \xi_2),$$

where $\xi_{12} \ll 1$ (see (4.49)–(4.51)). The refractive indexes n_{123}^2 of the resonance branches $\omega_1(\theta)$ and $\omega_2(\theta)$ can be obtained as follows. To find n_1^2 of the ordinary wave, the first term in Eq. (5.2) is dropped. This gives a formula $n_{10}^2 = C/B$ (see (4.47)) which in the present approximation does not depend on β_e. The other two values of the refractive index are high when $\omega \simeq \omega_1, \omega_2$, so in (5.2) the term $C \simeq 1$ can be dropped. Thus, from (5.2) we obtain the final expression

$$n_{23}^2 = \frac{A \mp (A^2 + 2\beta_e^2 A_1 B)^{1/2}}{\beta_e^2 A_1}. \tag{5.13}$$

At the resonance frequencies, when $A \simeq 0$ ($A^2 \ll 2B\beta_e^2 A_1$), only one branch of the longitudinal plasma waves has meaning, its refractive index being

$$n_3^2 = \left(\frac{2B}{\beta_e^2 A_1}\right)^{1/2}, \quad n_3^2\left(\theta = \frac{\pi}{2}\right) = \frac{1}{\beta_e}\left(\frac{2}{3}\right)^{1/2} \frac{\omega_0 \omega_H}{\omega_U^3}(\omega_0^2 - 3\omega_H^2)^{1/2}. \tag{5.14}$$

It follows from (5.14) that the HF branch of the resonance waves $\omega_1(\theta)$ (see (4.49)) exists (is not damped out; $n_3^2 > 0$) provided that $\omega_0 > 3^{1/2}\omega_H$.

Far away from the resonance point, where $A \gg 2B\beta_e^2 A_1$, for the two waves we have

$$n_2^2 = -B/A, \quad n_3^2 = (2/\beta_e^2)(A/A_1), \tag{5.14a}$$

indicating that n_2^2 is equal to the value n_{20}^2 for a cold plasma.

In the nonresonance region the refractive indexes n_{12}^2, taking spatial dispersion into account, have, as noted earlier, negligibly small corrections. For instance, from formulas (3.6) and (4.2) for $\theta = 0$, with the aid of (5.1) we find that for the ordinary wave and the extraordinary wave, respectively,

$$n_1^2 = n_+^2 = (\varepsilon_{11} + \varepsilon_{12}) = n_{+0}^2\left[1 - \frac{\beta_e^2}{2}\frac{\omega\omega_0^2}{(\omega + \omega_H)^3}\right], \tag{5.15}$$

$$n_2^2 = n_-^2 = (\varepsilon_{11} - \varepsilon_{12}) = n_{-0}^2\left[1 - \frac{\beta_e^2}{2}\frac{\omega\omega_0^2}{(\omega - \omega_H)^3}\right],$$

where $n_{+0}^2 = (\varepsilon_1 + \varepsilon_2)$ (see (4.19)). If $\theta = \pi/2$, then

$$\left.\begin{aligned} n_2^2 = n_-^2 &= \frac{\varepsilon_{11}\varepsilon_{22} + \varepsilon_{12}^2}{\varepsilon_{11}} = n_{-0}^2\left[1 - \frac{\beta_e^2}{2}\frac{\omega_0^2(\omega^2 - \omega_0^2 + 2\omega_H^2)}{(\omega^2 - \omega_H^2)(\omega^2 - 4\omega_H^2)}\right] \\ n_1^2 = n_+^2 &= \varepsilon_{33} = n_{+0}^2\left[1 - \frac{\beta_e^2}{2}\frac{\omega_0^2}{(\omega^2 - \omega_H^2)}\right], \end{aligned}\right\} \tag{5.16}$$

while $n_{-0}^2 = (\varepsilon_1^2 - \varepsilon_2^2)/\varepsilon_1^2$ and $n_{+0}^2 = \varepsilon_3$ (see (4.20) and (4.21)). It is evident from (5.15) and (5.16) that in an isotropic plasma, without taking the effect of ions into account,

$$n^2 = \left(1 - \frac{\omega_0^2}{\omega^2}\right)\left(1 - \frac{\beta_e^2}{2}\frac{\omega_0^2}{\omega^2}\right). \tag{5.17}$$

Naturally, formula (5.5), like the kinetic correction in (5.15) and (5.16), requires satisfaction of the corresponding conditions

$$\frac{(\omega^2 - \omega_H^2)}{\omega_0^2}, \quad \frac{(\omega^2 - 4\omega_H^2)}{\omega^2}, \quad \frac{(\omega - \omega_H)^3}{\omega\omega_0^2} \gg \beta_e^2. \tag{5.18}$$

In the frequency range corresponding to the whistler mode for $\theta = 0$, the third condition (5.18) is equivalent to the limitation

$$\frac{(\omega - \omega_H)^2}{\omega^2} \gg \left(\frac{\beta_e^2}{2}\right) n_{-0}^2. \tag{5.18a}$$

Evaluations have shown that, in the near-Earth and interplanetary plasma, the condition (5.18*a*) is satisfied adequately even when $|\omega - \omega_H|/\omega_H \simeq (10^{-2}$ to $10^{-3})$, i.e., quite close to the electron gyroresonance.

At the gyrofrequency $\omega = \omega_H$, where $n_2^2 = n_{-\infty}^2 \to \infty$, taking thermal motion into account naturally reduces the infinity of the refractive index n_-^2 of the extraordinary wave to a simple maximum. Hence the wave is greatly attenuated, so

$$\left.\begin{aligned} \kappa_-(\omega = \omega_H) &= \frac{1}{2}\left(\frac{\pi^{1/2}}{\beta_e}\frac{\omega_0^2}{\omega_H^2}\right)^{1/3}, \\ n_-(\omega = \omega_H) = n_2 &= \left(\frac{3}{2}\right)^{1/2}\left(\frac{\pi^{1/2}}{\beta_e}\frac{\omega_0^2}{\omega_H^2}\right)^{1/3} \end{aligned}\right\} \tag{5.19}$$

These formulas cannot be obtained with the aid of tensor elements (5.1), which, as noted previously, do not apply for $\omega = \omega_H$, $\omega = 2\omega_H$, etc. For the second-harmonic gyroresonance

$$n_-(\omega = 2\omega_H) \simeq \kappa_-(\omega = 2\omega_H) \simeq (1/\beta_e)^{1/3}. \tag{5.20}$$

It should be noted that the damping of the resonances of the refractive indexes has in a number of cases only theoretical significance, and it plays a role only in plasma regions where the collision frequency is very low. The resonance values of the refractive indexes n^2_∞ in the near-Earth plasma often have to be determined taking the collision frequency ν into account, with the aid of formulas (4.76) and (4.18).

5.2.2. 'Kinetic' attenuation of LF and HF waves

In a collisionless plasma two kinds of Landau damping have an effect: Čerenkov damping and cyclotron damping. Waves are strongly absorbed in these cases only in the region where they interact effectively with fast electrons at the Čerenkov and cyclotron resonances. Thus, when calculating the attenuation factors, we can confine ourselves to the real values of the tensor elements in a cold plasma, since their imaginary parts are considerably less than $\varepsilon_1, \varepsilon_2$, and ε_3. To sum up, for the frequency range being considered here, namely $\omega^2 \gg \Omega_H \omega_H$, the refractive indexes n^2_{12} are determined by formula (4.17), except in the resonance regions near $\omega_1(\theta)$ and $\omega_2(\theta)$, for which the appropriate formulas were presented above.

In the frequency range under review, the spatial attenuation coefficient for transverse waves is

$$\kappa = \kappa_0 + \kappa_{H1} + \sum_{s \geqslant 2} \kappa_{Hs}. \tag{5.21}$$

Here the quantity

$$\kappa_0 = \pi^{1/2} \alpha_e \left(\frac{\omega_0}{\omega} \right)^2 \frac{1}{(\beta_e n_\pm \cos\theta)^3} \exp\left(- \frac{1}{\beta_e^2 n_\pm^2 \cos^2\theta} \right) \tag{5.21a}$$

defines the Čerenkov absorption. In the vicinity of gyroresonance, $\omega = \omega_H$ ($s = 1$), if the condition $(c/v_e)(\omega_0/\omega_H)^2 \gg 1$ is satisfied, which is practically always the case in the near-Earth plasma, the spatial attenuation coefficient is

$$\kappa_{H1} = \frac{1}{2\pi^{1/2}} \left(\frac{\omega_H}{\omega_0} \right)^2 \frac{\beta_e \cos\theta\, \omega_H^2 \phi_0(X_H) F(\omega_0^2/\omega^2, \theta)}{n_\pm [\omega_H^2 \sin^2\theta (n_\pm^2 - \frac{1}{2}) + (\omega_0^2 - \omega_H^2)]}. \tag{5.22}$$

At the multiples of the gyrofrequencies $s = \pm 2, \pm 3 \ldots$,

$$\kappa_{Hs} = \frac{\pi^{1/2}}{2} \alpha_H \frac{\omega_0^2}{\beta_e \omega^2 n_\pm |\cos\theta|} \frac{|s|^2}{2^{|s|}|s|!} \left(\frac{\beta_e \omega n_\pm \sin\theta}{(\surd 2)\omega_H} \right)^{2|s|-2}$$
$$\times \exp\left(- \frac{(\omega - s\omega_H)^2}{\beta_e^2 n_\pm^2 \omega^2 \cos^2\theta} \right) \tag{5.23}$$

(Stepanov, 1959, 1960; Gershman, 1960a). In formulas (5.21)–(5.23)

$$\left.\begin{aligned}\alpha_e &= \frac{n_\pm^4 \cos^2\theta - \varepsilon_1 n_\pm^2(1+\cos^2\theta) + (\varepsilon_1^2 - \varepsilon_2^2)}{2C + Bn_\pm^2},\\ \alpha_H &= \frac{n_\pm^4 \sin^2\theta - \varepsilon_3 n_\pm^2(1+\cos^2\theta) + 2(\varepsilon_1+\varepsilon_2)(\varepsilon_3 - n_\pm^2 \sin^2\theta)}{2C + Bn_\pm^2}\end{aligned}\right\} \tag{5.24}$$

The functions $F(\omega_0^2/\omega^2)$ and $\phi_0(X_H)$ are defined below, while $\varepsilon_1,\varepsilon_2,\varepsilon_3$, C, and B are found from (4.1) and (4.3). For longitudinal propagation of waves, $\theta = 0$, when $n_{-0}^2 = (\varepsilon_1 - \varepsilon_2)$ and $n_{+0}^2 = (\varepsilon_1 + \varepsilon_2)$ from (5.24) it follows that for both refractive indexes $\alpha_e = 0$, i.e., Čerenkov damping is absent for the propagation of all three types of waves: the electron-whistler (fast magnetoacoustic), together with the slow and fast extraordinary waves (see Figs. 4.2–4.5).

In formula (5.22)

$$\begin{aligned}F\left(\frac{\omega_0^2}{\omega^2},\theta\right) = {}& \left[1 - \left(\frac{\omega_0}{\omega_H}\right)^2\left(1 + \frac{\sin^4\theta}{4}\right)\right] n_\pm^4\\ & - \left[\left(1 - \frac{\omega_0^2}{\omega_H^2}\right)\left(1 - \tfrac{1}{4}\frac{\omega_0^2}{\omega_H^2}\right)(1+\cos^2\theta)\right.\\ & + \left(1 - \tfrac{1}{2}\frac{\omega_0^2}{\omega_H^2}\right)\left(1 + \frac{\omega_0^2}{\omega_H^2}\right)\sin^2\theta\\ & \left. - \tfrac{1}{4}\left(\frac{\omega_0}{\omega_H}\right)^2 \tan^2\theta(1+\cos^2\theta)\right] n_\pm^2\\ & + \left[\left(1 - \frac{\omega_0^2}{\omega^2}\right)\left(1 - \tfrac{1}{2}\frac{\omega_0^2}{\omega_H^2}\right)\right.\\ & \left. - \tfrac{1}{4}\left(\frac{\omega_0}{\omega}\right)^2\left(2 - \frac{\omega_0^2}{\omega_H^2}\right)\tan^2\theta\right],\end{aligned} \tag{5.25}$$

$$\left.\begin{aligned}\phi_0(X_H) &= \exp[-(X_H)^2 |W(X_H)|^{-2}],\\ W(X_H) &= \exp\left[-(X_H)^2\left(1 + 2i\pi^{-1/2}\int_0^{X_H} \exp t^2 dt\right)\right],\end{aligned}\right\} \tag{5.26}$$

where

$$X_H = \frac{\omega - \omega_H}{\beta_e \omega n_\pm \cos\theta} \tag{5.27}$$

and $W(\ldots)$ with the complex argument $Z = X + iY$, the so-called Kramp function, is the complex probability integral. This function has been tabulated (Fadeeva & Terent'ev, 1954; Fried & Conte, 1961), as has the integral

$$I(Z) = \exp(-Z^2)\int_0^Z \exp t^2 \mathrm{d}t. \tag{5.28}$$

The function $W(X_H)$ plays an important role in various problems in plasma physics and, in particular, will be encountered below in a number of formulas.

For $|Z| \gg 1$, the following asymptotic expansion of the integral (5.28) is useful:

$$I(Z) = \frac{1}{2Z}\left(1 + \frac{1}{2Z^2} + \frac{1\cdot 3}{4Z^4} + \frac{1\cdot 3\cdot 5}{8Z^6} + \ldots\right). \tag{5.29}$$

The general expansion of $I(Z)$, on the other hand, applicable for arbitrary values, but in particular for small values, of $|Z|$, has the form

$$I(Z) = Z - \frac{2Z^3}{3\cdot 1} + \frac{2\cdot 2Z^5}{5\cdot 3\cdot 1} - \frac{2\cdot 2\cdot 2Z^7}{7\cdot 5\cdot 3\cdot 1} + \frac{2\cdot 2\cdot 2\cdot 2Z^9}{9\cdot 7\cdot 5\cdot 3\cdot 1} - \ldots \tag{5.29a}$$

(Stix, 1962). For $|Z| \gg 1$ the following expansion of the Kramp function is generally used:

$$W(Z) = \frac{i}{\pi^{1/2}Z}\left(1 + \frac{1}{2Z^2} + \frac{3}{4Z^4} + \ldots\right) + \exp(-Z^2) \tag{5.30}$$

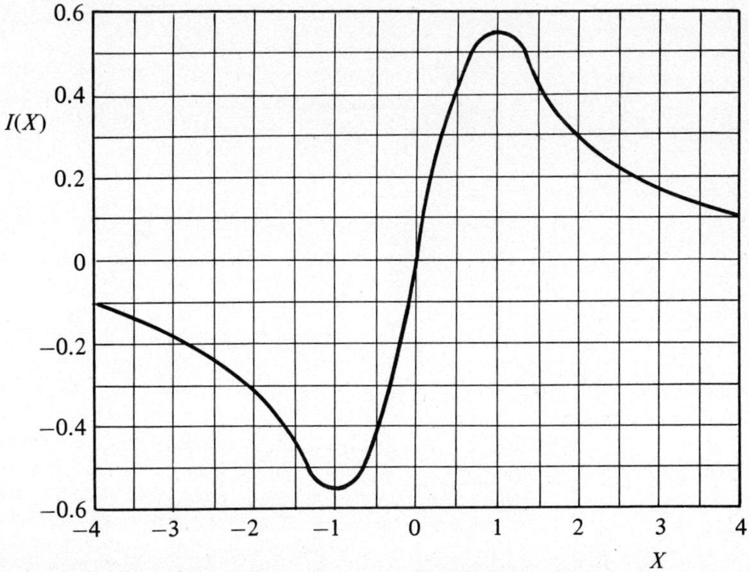

Fig. 5.1. The function $I(X)$ for real values of the argument $Z = X$ (formula (5.28)).

and for $|Z| \ll$ or $\lesssim 1$ the approximate formula

$$W(Z) \simeq 1 + 2iZ/\pi^{1/2}. \tag{5.30a}$$

Curves of the function $I(X)$ and $\phi_0(x)$ for real values of X are given in Figs. 5.1 and 5.2. With the aid of these curves it is possible, in particular, to find the values of κ_{H1} (see formula (5.22)) without using the corresponding tables.

An analysis of formula (5.22) shows that for arbitrary values of the angle θ the values of $\kappa_{\pm H1}$ are very low. Here $\kappa_{+H}/\kappa_{-H} \ll 1$. For $\theta = 0$, the attenuation of the slow extraordinary wave disappears ($\kappa_+ \to 0$), while the attenuation coefficients for the whistler wave and the fast extraordinary wave increase. At the same time, far enough away from the resonance region, where $X_- \gg 1$,

$$\kappa_{H-}(s=1) = \sqrt{\pi}\frac{\omega_0^2}{2\beta_e\omega^2 n_-}\exp\left[-\left(\frac{\omega - \omega_H}{\beta_e\omega n_-}\right)^2\right] \tag{5.31}$$

(see (5.27)), whereas for $\theta = 0$ the temporal attenuation rate of these waves is

$$\gamma_{H-}(s=1) = \pi^{1/2}\frac{(\omega_H - \omega)^2}{\beta_e\omega n_-}\exp\left[-\left(\frac{\omega - \omega_H}{\beta_e\omega n_-}\right)^2\right]. \tag{5.32}$$

It can be concluded from the above formulas that both the cyclotron damping and the Čerenkov damping of HF transverse waves are exponen-

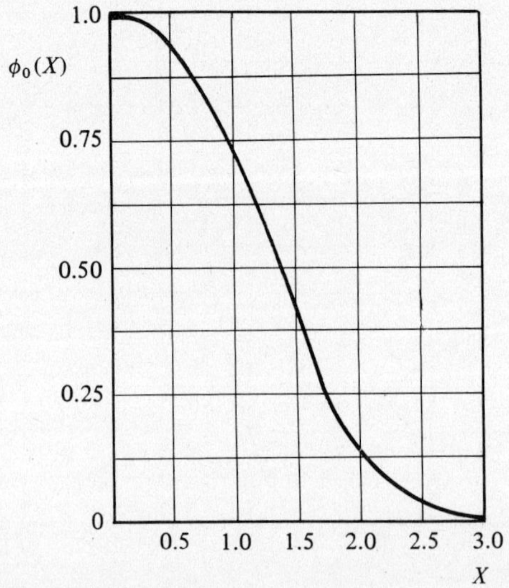

Fig. 5.2. The function $\phi_0(X)$ (formulas (5.26) and (5.28)).

tially small over the entire frequency range. An exception is the resonance region for a whistler wave around $\omega = \omega_H$, where the attenuation coefficient $\kappa_{H-}(s=1)$ is large, being given by the formula (5.19). However, gyro-resonance absorption of higher harmonic order is absent for $\theta = 0$ is (see 5.23)).

5.3. Very-low-frequency (VLF) and extremely-low-frequency (ELF) waves ($0 \leqslant \omega \ll \omega_H$)

In this frequency range one of the branches of the refractive index $n_1^2 = n_+^2$ describes ELF ion waves (Alfvén waves, hydromagnetic and ion whistlers). The second branch $n_2^2 = n_-^2$ describes VLF, ELF, and LF 'electron' waves, the so-called modified Alfvén, fast magnetoacoustic, and whistler-mode waves. In nonresonance regions the refractive indexes of these waves, which taking the thermal motion of the particles into account can be described by (4.2) and the tensor elements (5.1), have only negligibly small corrections for HF waves as well. For instance, for $\theta = 0$, we have

$$n_+^2 = n_{+0}^2\left[1 - \frac{\beta_i^2}{2}\frac{\omega\Omega_0^2}{(\omega - \Omega_H)^3}\right],$$

$$n_-^2 = n_{-0}^2\left[1 - \frac{\beta_i^2}{2}\frac{\omega\Omega_0^2}{(\omega + \Omega_H)^3}\right], \tag{5.33}$$

where the expressions for a cold plasma $n_{\pm 0}^2 = (\varepsilon_1 \pm \varepsilon_2)$ are as defined above in Chapter 4. For $\theta = \pi/2$, the branch $n_+^2 = \varepsilon_3$ is absent, and

$$n_-^2 = n_{-0}^2\left[1 - \frac{\beta_i^2}{2}\frac{\Omega_0^2(\omega^2 - \Omega_0^2 + 2\Omega_H^2)}{(\omega^2 - \Omega_H^2)(\omega^2 - 4\Omega_H^2)}\right], \tag{5.34}$$

where $\beta_i = v_i/c, n_{-0}^2 = n_{20}^2 = (\varepsilon_1^2 - \varepsilon_2^2)/\varepsilon_1$ (see (4.21)). In the resonance regions the values of refractive indexes (4.14) and (4.15) for both waves no longer become infinite, of course. Thus, for an ion wave, when the resonance frequency $\omega_3(\theta) \to \Omega_H$,

$$n_+ = \left(\frac{c^2}{V_A^2\beta_i}\frac{1 + \cos^2\theta}{\cos^3\theta}\right)^{1/3} \tag{5.35}$$

(Doyle & Neufeld, 1959; Stepanov, 1960). The quantity V_A in (5.35) is the Alfvén velocity. For a fast magnetoacoustic wave as $\theta \to \pi/2$, when the resonance frequency $\omega_2(\theta) \to \omega_L$ (where ω_L is the lower-hybrid frequency, see (4.15)), formula (5.1) can be used to find the refractive index

$$n_2^2(\theta = \pi/2) = n_-^2 = \frac{\omega_0\omega_H}{(\frac{3}{2})^{1/2}\beta_e(\omega_L^4 + 4\omega^2\Omega_H^2)^{1/2}}. \tag{5.36}$$

'Kinetic' attenuation of VLF and ELF waves. In these frequency ranges Čerenkov absorption on electrons plays a role, as does cyclotron absorption for $\omega = s\Omega_H (s = 1, 2, 3, \ldots)$. Under the conditions often present in the near-Earth plasma, i.e., when the gas-kinetic pressure of the charged particles NkT is considerably lower than the magnetic pressure $\mu_0 H_0^2/2$, in different parts of the frequency range being considered the attenuation coefficients will be given by the following formulas.

For $\omega \ll \Omega_H$ the Čerenkov attenuation coefficients for the two magnetohydrodynamic waves are, respectively,

$$\left.\begin{aligned}\kappa_{-e} &= \frac{\pi^{1/2}}{4}\frac{m}{M}\frac{v_e}{V_A}\frac{\sin^2\theta}{\cos\theta}\exp(-X_{-e}^2),\\ \kappa_{+e} &= \frac{\pi^{1/2}}{4}\frac{m}{M}\frac{v_e}{V_A}\frac{\omega^2}{\Omega_H^2}n_+\{\cot^2\theta\exp(-X_{+e}^2) + \tan^2\theta X_{+e}\phi_1(X_{+e})\},\end{aligned}\right\}$$

where

$$\phi_1(X) = \frac{\exp(-X^2)}{X|1 + i\pi^{1/2}XW(X)|^2}, \quad X_{\pm e} = \frac{c}{n_\pm v_e \cos\theta} \tag{5.38}$$

(Stepanov, 1958). To determine κ_{+e}, the curve of the function

$$\phi_2(X) = \frac{1}{2}\left[\phi_1(X) + \frac{\exp(-X^2)}{X}\right] \tag{5.39}$$

in Fig. 5.3 can be used; this will also be employed below (see (5.45)). From (5.37) it follows that $\kappa_{+e}/\kappa_{-e} \simeq \omega^2/\Omega_H^2 \ll 1$. Formulas (5.37) are, however, inapplicable for $\theta^2 < \omega/\Omega_H$. In this range of angles $\kappa_{\pm e} \simeq (m/M)(v_e/V_A)\theta^2$; as $\theta \to 0$, we have $\kappa_{\pm e} \to 0$.

In the gyroresonance region, $\omega \lesssim \Omega_H$, the Čerenkov attenuation coefficient of an Alfvén (ion-whistler) wave increases, while

$$\kappa_{+e} = \frac{\pi^{1/2}}{8}\frac{m}{M}\tan^2\theta\frac{\omega n_\pm}{(\Omega_H - \omega)}\phi_1(X_{+e}), \tag{5.40}$$

(Akheizer *et al.*, 1964). For $|X| \ll 1$ and $|X| \gg 1$, respectively, we have $\phi_1(X) = 1/X$ and $\phi_1(X) = 4X^3\exp(-X^2)$.

The cyclotron attenuation coefficient of an ion-whistler wave for $\omega \lesssim \Omega_H$ is

$$\kappa_+ = \frac{\pi^{1/2}}{2}\frac{\Omega_0^2}{\beta_i\omega^2 n_+\cos\theta}\exp\left[-\left(\frac{\Omega_H - \omega}{\beta_i\omega n_+\cos\theta}\right)^2\right] \tag{5.41}$$

For $\omega = \Omega_H$ the resonance attenuation coefficient of an ion wave becomes

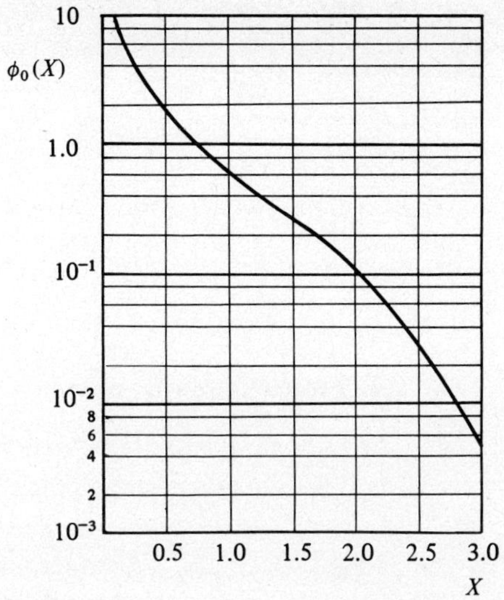

Fig. 5.3. The function $\phi_2(X)$ (formulas (5.38) and (5.39)).

very large. The propagation of these waves is then practically impossible:

$$\kappa_+(\omega = \Omega_H) = \left[\left(\frac{c}{V_A}\right)^2 \frac{1}{\beta_i} \frac{1 + \cos^2\theta}{\cos\theta}\right]^{1/3}. \tag{5.42}$$

In the gyroresonance regions $\omega = s\Omega_H$, $s \pm 1, 2, 3, \ldots,$ the cyclotron damping of a fast magnetoacoustic wave for $\theta = \pi/2$ is given by the formula

$$\kappa_2 = \kappa_-(\omega = s\omega_H) = \alpha_H \frac{\pi^{1/2}}{2} \frac{|s|^2}{2^{|s|}|s|!} \frac{\Omega_0^2}{\omega^2 \beta_i n_- \cos\theta}$$

$$\times \left(\frac{\omega\beta_i n_- \sin\theta}{(\sqrt{2})\Omega_{Hi}}\right)^{2|s|-2} \exp\left[-\left(\frac{\omega - s\Omega_H}{\omega\beta_i n_- \cos\theta}\right)^2\right] \tag{5.43}$$

where $\beta_i = v_i/e$ and

$$\alpha_H = \frac{n_-^2(1 + \cos^2\theta) - 2(\varepsilon_1 - \varepsilon_2)}{2n_-^3 \cos^2\theta - \varepsilon_1 n_-(1 + \cos^2\theta)}. \tag{5.44}$$

In the VLF range, $\Omega_H < \omega \ll \omega_H$, *the Čerenkov attenuation coefficient* of a fast magnetoacoustic wave (VLF tail of electron-whistler wave) for $\theta \neq \pi/2$ is

$$\kappa_{-e} = \frac{\pi^{1/2}}{4} \frac{\sin^2\theta}{\cos\theta} \frac{\omega}{\omega_H} \phi_2(X_{-e}), \tag{5.45}$$

where the expression for the function $\phi_2(X)$ is given above (see (5.38) and (5.39), and also Fig. 5.3). For $|X_e| \ll 1$ and $|X_e| \gg 1$, respectively,

$$\left.\begin{aligned} \kappa_{-e} &= \tfrac{1}{4}\pi^{1/2}(\omega/\omega_H)\beta_e n_-^2 \sin^2\theta, \\ \kappa_{-e} &= \tfrac{1}{2}\pi^{1/2}\frac{\sin^2\theta}{\cos\theta}\frac{\omega}{\omega_H\beta_e^3 n_-^2\cos^3\theta}\exp\left[-\left(\frac{1}{\beta_e n_-\cos\theta}\right)^2\right] \end{aligned}\right\} \tag{5.46}$$

The damping of this wave for $(\Omega_H\omega_H)^{1/2} \ll \omega \lesssim \omega_H$ was described in the previous section.

5.4. Electron-acoustic waves

Langmuir waves, as we saw above, exist only in a very narrow frequency range: $\Delta\omega = \omega - \omega_0 \ll \omega_0$. Within this range, they are only slightly attenuated, because the thermal velocity of the electrons v_e is low in comparison with the phase velocity v_ϕ. A wider spectrum of highly attenuated, longitudinal, high-frequency plasma waves can be excited in a plasma if

$$\omega \gg Kv_i, \quad v_\phi \gg v_i, \tag{5.47}$$

i.e., on condition that the thermal velocity of the ions v_i is low in comparison with the phase velocity of the waves v_ϕ. These HF waves are associated with the motion of electrons, and in a certain frequency interval with the motion of ions as well.

The dispersion equation for these waves, sometimes known as *electron-acoustic waves*, can, in an isotropic plasma, be written as follows for a Maxwellian distribution function (see Stix, 1962):

$$\omega^2 = \Omega_0^2 + 2\omega^2\alpha_e^2[2\alpha_e I(\alpha_e) - 1], \tag{5.48}$$

where

$$\alpha_e = \omega/Kv_e = v_\phi/v_e \tag{5.48a}$$

and the function $I(\alpha_e)$ was defined above ((5.28)–(5.30), (5.48), and Fig. 5.1).

The dispersion relation for these waves in an isotropic plasma has two branches, which come together at a frequency $\omega > \omega_0$, or more precisely at

$$\omega_M = (1.29\omega_0^2 + \Omega_0^2)^{1/2}. \tag{5.49}$$

At this point

$$v_\phi \simeq 1.5v_e. \tag{5.50}$$

For frequencies $\omega < \omega_M$, i.e., in the frequency range

$$\Omega_0 \leqslant \omega \leqslant \omega_M \tag{5.51}$$

there is a branch of *slow* electron-acoustic waves. It is cut off at

$$v_\phi = 0.924 v_e \tag{5.52}$$

for a frequency Ω_0. Therefore, this branch crosses two frequency regions: LF and HF waves. The branch describing *fast* electron-acoustic waves lies in the range

$$\omega_M < \omega \leqslant (\Omega_0^2 + 1.647\omega_0^2)^{1/2}, \tag{5.53}$$

while its phase velocities vary from $v_\phi \simeq 1.5 v_e$ (see (5.50)) to $v_\phi \gg v_e$. So far, however, no detailed theoretical analysis of the properties of these waves has apparently been carried out, in particular not for a magneto-plasma.

Electron sound in a nonisothermal plasma at frequencies $\omega \gg \Omega_H$ constitutes another type of high-frequency longitudinal oscillation. It appears in a highly nonisothermal plasma with $T_i \gg T_e$ if the following conditions are satisfied:

$$\omega \ll \omega_H, \quad K v_e \cos\theta \ll \omega \ll K v_i. \tag{5.54}$$

The second of these conditions will evidently be satisfied only if $\theta \simeq \pi/2$, since $v_e \simeq (10^2\text{–}10^3) v_i \gg v_i$ even for $T_i \gg T_e$. The oscillations being considered are of short wavelength. They are possible in a natural plasma provided that special conditions are created in it, namely the injection of hot ions. These waves are described by the dispersion equation

$$\omega(K, \theta) = \frac{K v_{se} \cos\theta}{(1 + K^2 \rho_s^2 + K^2 D_i^2)^{1/2}} \tag{5.55}$$

where $v_{se} = (kT_i/m)^{1/2}$, $\rho_s = v_{sa}/\omega_H$, and $D_i = v_i/(\sqrt{2})\Omega_0$ (Mikhailovskii, 1965).

The total Čerenkov attenuation rate for these oscillations, due to absorption by both electrons and ions, is

$$\gamma = \gamma_e + \gamma_i = \frac{\pi^{1/2}}{8}\left\{\frac{\omega^4}{(K v_e \cos\theta)^3}\exp\left[-\left(\frac{\omega}{K v_e \cos\theta}\right)^2\right] + \left(\frac{m}{M}\right)^{1/2}\omega(K,\theta)\right\}. \tag{5.56}$$

It is easy to see that, when the wavelength of these oscillations is sufficiently long

$$K\rho_s, K D_i \ll 1, \tag{5.56a}$$

then

$$\omega(K, \theta) = K v_{se} \cos\theta, \tag{5.57}$$

i.e., these waves are described by an equation analogous to the sound equation. The quantity v_{se} can thus be called the velocity of *high-frequency electron sound*. In a plasma with hot ions, for $\theta = \pi/2$ there may also exist *low-frequency electron sound*, $\omega \ll \Omega_H$, if $K\rho_{Hi} \gg 1$ and $KD_i \ll 1$ (Mikhailovskii, 1967). Low-frequency electron-acoustic oscillations are an extension of the Alfvén branch into the short-wavelength region.

5.5. Ion-acoustic waves

Now let us examine slightly attenuated waves in a nonisothermal plasma with $T_e \gg T_i$, the phase velocities of which lie in an intermediate region between the thermal velocities of the ions and of the electrons. The dispersion relation for these waves has been well-studied theoretically, both in an isotropic plasma and in a magnetized plasma, when

$$v_i \ll v_\phi/\cos\theta \ll v_e. \tag{5.58}$$

In the isotropic case (when $H_0 = 0$ and $\cos\theta = 1$) the dispersion equation of these waves, which are often called electrostatic waves, or Langmuir–Tonks waves, has the form

$$\omega^2(K) = \omega_{10}^2 = \frac{(Kv_s)^2}{1+(KD_e)^2} = \frac{\Omega_0^2}{1+(KD_e)^{-2}}, \tag{5.59}$$

where the subscript '10' has been introduced in order to distinguish these waves from the corresponding two branches of the waves in a magnetoplasma. While deriving (5.59), it was assumed that $(KD_i)^2 \ll 1$. In the limiting case, i.e., when the condition

$$(KD_e)^2 = \left(\frac{2\pi D_e}{\Lambda}\right)^2 \ll 1 \tag{5.60}$$

is satisfied, which means that the waves are quite long, we have

$$\omega(K) = Kv_s(1 + \tfrac{3}{2}T_i/T_e), \tag{5.61}$$

where $v_s = \left(\dfrac{kT_e}{M}\right)^{1/2}$ is the velocity of nonisothermal sound. Since equation (5.61) is similar to the sound-wave equation, the waves described by equations (5.59) and (5.61) are called ion-acoustic waves. For $(KD_e)^2 \gg 1$,

$$\omega(K) = \Omega_0[1 + 3(KD_i)^2], \tag{5.62}$$

an equation which describes low-frequency Langmuir ion oscillations.

In a magnetized plasma the following approximate dispersion equation can be used to describe the waves in question:

$$\left.\begin{aligned} \frac{1}{\omega^2} + \frac{\tan^2\theta}{\omega^2 - \Omega_H^2} &= \frac{1}{\cos^2\theta}\left(\frac{1}{(Kv_s)^2} + \frac{1}{\Omega_0^2}\right) = \frac{1}{\cos^2\theta}\left(\frac{1+(KD_e)^2}{(Kv_s)^2}\right) \\ &= \frac{1}{\cos^2\theta}\left(\frac{1+(KD_e)^{-2}}{\Omega_0^2}\right). \end{aligned}\right\} \tag{5.63}$$

This equation defines two branches $\omega_1(K,\theta)$ and $\omega_2(K,\theta)$, representing the fast and slow ion-acoustic waves:

$$\omega_{12}^2(K,\theta)=\tfrac{1}{2}\{(\omega_{10}^2+\Omega_H^2)\pm[(\omega_{10}^2+\Omega_H^2)^2-4\omega_{10}^2\Omega_H^2\cos^2\theta]^{1/2}\}. \quad (5.64)$$

It follows directly from (5.64) that for $H_0=0$ (or $\cos\theta=1$) the frequency $\omega_1(K,\theta)=\omega_{10}$. It is also evident from condition (5.58) that formulas (5.63) and (5.64) should be applied with caution as θ approaches $\pi/2$ and as $\omega\to\Omega_H$. The corresponding criteria for the applicability of these formulas have the form

$$\theta^2 \gg \frac{2Kv_i}{\Omega_H}\frac{|\Omega_H^2-\omega_{10}^2|}{\omega_{10}^2}, \quad \frac{|\Omega_H-\omega_{10}|}{Kv_i} \gg \cos\theta. \quad (5.65)$$

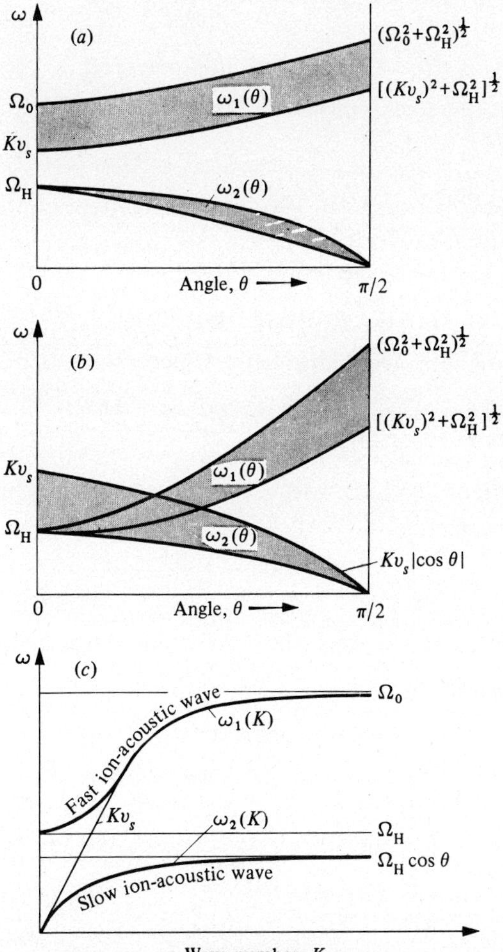

Fig. 5.4. Branches describing the fast $\omega_1(K,\theta)$ and slow $\omega_2(K,\theta)$ ion-acoustic waves in a nonisothermal magnetoplasma.

The dispersion curves for the fast and slow ion-acoustic waves, i.e. the curves of $\omega_{12}(\theta)$ and $\omega_{12}(K)$, are shown schematically in Fig. 5.4. It is seen that the range of fast and slow ion-acoustic waves varies in different cases within the following limits:

$$\Omega_H < \omega_1 < (\Omega_0^2 + \Omega_H^2)^{1/2}, \quad 0 < \omega_2 < \Omega_H. \tag{5.66}$$

The refractive indexes of the ion-acoustic waves are

$$n_{12}^2 = \left(\frac{c}{v_s}\right)^2 \frac{\Omega_0^2(\omega_{12}^2 - \Omega_H^2)}{\Omega_0^2(\omega_{12}^2 - \Omega_H^2 \cos^2\theta) - \omega_{12}^2(\omega_{12}^2 - \Omega_H^2)}. \tag{5.67}$$

If $(KD_e) \ll 1$, then

$$n_{12}^2 = \frac{\omega_{12}^2 - \Omega_H^2}{\omega_{12}^2 - \Omega_H^2 \cos^2\theta}. \tag{5.68}$$

Note, too, that for $H_0 = 0$, taking the ion temperature into account,

$$n_{10}^2 = \left(\frac{c}{v_s}\right)^2 \Omega_0^2 \left[(\Omega_0^2 - \omega_{10}^2) + 3\frac{T_i}{T_e}\frac{\Omega_0^2}{(\Omega_0^2 - \omega_{10}^2)}\right]^{-1} \tag{5.69}$$

(Ginzburg & Rukhadze, 1970). The spatial coefficient of attenuation and the temporal attenuation rate of these waves, which are attenuated mainly due to absorption by electrons, are respectively

$$\left.\begin{aligned} \kappa_{ei} &= \frac{\pi^{1/2}}{2}\frac{c}{v_e\cos\theta}\frac{(\omega_1^2 - \Omega_H^2)[(\omega_1^2 - \Omega_H^2) + (\omega_1^2 - \omega_{10}^2)]}{(\omega_1^2 - \Omega_H^2\cos^2\theta)[(\omega_1^2 - \Omega_H^2)\cos^2\theta + \omega_1^2\sin^2\theta]}, \\ \kappa_{e2} &= \frac{\pi^{1/2}}{2}\frac{c}{v_e\cos\theta}[(\omega_2^2 - \Omega_H^2)^2 + (\omega_2^2 - \Omega_H^2)(\omega_2^2 - \omega_{10}^2)] \\ &\quad \{[\omega_2^2 + \Omega_H^2\cos^2\theta) - (\omega_{10}^2 + \Omega_H^2)][(\omega_2^2 - \Omega_H^2)\cos^2\theta + \omega_2^2\sin^2\theta]\}^{-1}, \end{aligned}\right\} \tag{5.70}$$

and

$$\gamma_{e12} = \left(\frac{\pi}{8}\right)^{1/2}\left(\frac{m}{M}\right)^{1/2}\frac{\omega_{12}^4}{(Kv_s)^3\cos\theta}\left(\cos^2\theta + \frac{\omega_{12}^4\sin^2\theta}{(\omega_{12}^2 - \Omega_H^2)^2}\right)^{-1} \tag{5.70a}$$

For $\omega_{12} \simeq Kv_s \simeq \Omega_H$, we have

$$\gamma_{e12} \simeq (m/M)^{1/2}\Omega_H. \tag{5.71}$$

If the frequency of a fast ion-acoustic wave $\omega_1(K,\theta) = s\Omega_H$, where $s = 2, 3, \ldots$, then absorption becomes perceptible, due to interaction of the wave with ions. This absorption is described by the attenuation rate

$$\begin{aligned} \gamma_{i1} &= \frac{\pi}{2^s s!}\frac{\omega^4}{(Kv_i)^3} = \left(\frac{Kv_i\sin\theta}{\sqrt{2}\Omega_H}\right)^{2s}\frac{(\omega^2 - \Omega_H^2)^2\cos^2\theta}{[\omega^4\sin^2\theta + (\omega^2 - \Omega_H^2)^2\cos^2\theta]} \\ &\quad \times \exp\left[-\left(\frac{\omega - s\Omega_H}{Kv_i\cos\theta}\right)^2\right] \end{aligned} \tag{5.72}$$

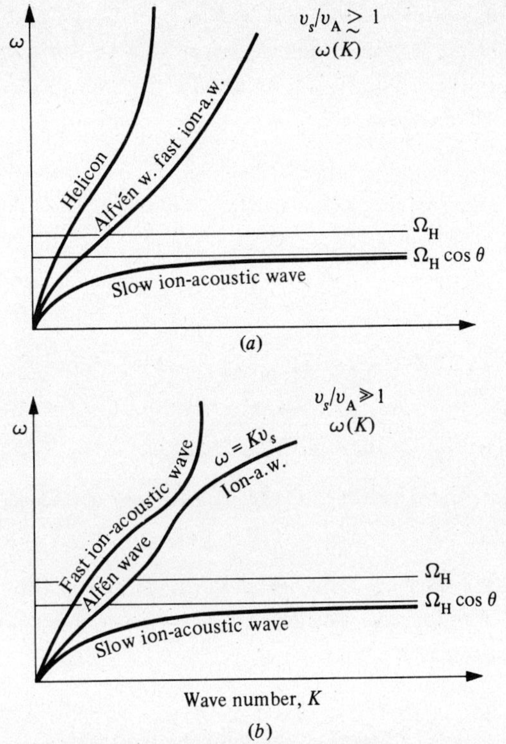

Fig. 5.5. Branches ω(K) describing ion-acoustic waves for various ratios of the nonisothermal velocity v_s to the Alfvén velocity.

(See Akhiezer & Fainberg, 1951b and Rytov, 1947). It should be noted that in the limiting case, when $Kv_s \ll \Omega_H$ and $KD_e \ll 1$, an ion-acoustic wave changes into a slow magnetoacoustic wave (see below):

$$\omega_2 = Kv_s \cos\theta, \quad \gamma_{e2} = (\pi/8)^{1/2}(m/M)^{1/2}\omega^2. \tag{5.73}$$

The properties of ion-acoustic waves differ considerably for different ratios v_s/V_A. Figure 5.4 shows the curves of $\omega_{12}(K)$ for $(v_s/V_A) \ll 1$. For $v_s/V_A \simeq 1$ or $\gg 1$, the branch describing a fast ion-acoustic wave ω_1 divides into two. The three branches of ion-acoustic waves which appear in these cases are plotted schematically in Fig. 5.5.

5.6. Magnetoacoustic waves

The ion-acoustic waves considered above are only slightly attenuated, as was seen earlier, when condition (5.58) is satisfied. This condition signifies that these waves are excited in a slightly ionized plasma, in which the pressure of the electrons is low in comparison with the magnetic pressure.

Actually, the phase velocity of these waves $v_\phi \simeq v_s \gg v_i$, and hence $V_A \gg v_i$ and $NkT \ll \mu_0 H_0^2/2$. In the frequency range $\omega \ll \Omega_H$, in a nonisothermal plasma there are three slightly attenuated ELF waves instead of two. One of these is an Alfvén wave

$$\omega = KV_A \cos\theta. \tag{5.74}$$

The other two are a fast magnetoacoustic wave and a slow magnetoacoustic wave:

$$\omega = Kv_{12} \tag{5.75}$$

where

$$v_{12} = \tfrac{1}{2}\{(V_A^2 + v_s^2) \pm [(V_A^2 + v_s^2)^2 - 4V_A^2 v_s^2 \cos^2\theta]^{1/2}\}. \tag{5.76}$$

The coefficients and rates of Čerenkov attenuation for these waves are defined, respectively, by the following formula:

$$\frac{\kappa_{12}}{n} = \frac{\gamma_{12}}{\omega} = \left(\frac{\pi}{8}\right)^{1/2}\left(\frac{m}{M}\right)^{1/2} \frac{v_{12}^4 + 2v_s^2(v_s^2\cos^2\theta - v_{12}^2)\cos^2\theta}{(v_{12}^2 - v_s^2\cos^2\theta)(2v_{12}^2 - V_A^2 - v_s^2)}. \tag{5.77}$$

If $V_A \simeq v_s$, then $\gamma/\omega \simeq (m/M)^{1/2}$. The attenuation of the Alfvén wave, $\gamma/\omega \simeq (m/M)^{1/2}(\omega/\Omega_H)^2$, is considerably less than the attenuation of the magnetoacoustic waves.

5.7. Electron and ion cyclotron waves

At angles $\theta \simeq \pi/2$ the Landau damping of longitudinal gyroresonance waves excited in a plasma is negligible. Consequently, gyroresonance waves of very high harmonic order, known in the literature as Bernstein modes, are readily generated in the near-Earth plasma (see Volume 2). These waves are described, for electrons and ions respectively, by the following dispersion equations:

$$K_e^2(\omega) = 4\frac{\omega_0^2}{v_{e\perp}^2}\exp(-p_e)\sum_{s=1}^{\infty} I_s(p_e)\frac{s^2\omega_H^2}{\omega^2 - s^2\omega_H^2}, \tag{5.78}$$

$$K_i^2(\omega) = 4\frac{\Omega_0^2}{v_{i\perp}^2}\exp(-p_i)\sum_{s=1}^{\infty} I_s(p_i)\frac{s^2\Omega_H^2}{\omega^2 - s^2\Omega_H^2}, \tag{5.79}$$

where I_s is a Bessel function with an imaginary argument, the subscript $\perp$ indicates the component normal to the direction of the vector $\mathbf{H}_0$, and

$$p_e = \frac{(Kv_{e\perp})^2}{2\omega_H^2}, \quad p_i = \frac{(Kv_{i\perp})^2}{2\Omega_H^2}, \quad v_{e\perp}^2 = 2kT_{e\perp}/m, \quad v_{i\perp}^2 = \frac{2kT_{i\perp}}{M} \tag{5.80}$$

(Bernstein, 1958).

Equations (5.78) and (5.79) were obtained from the dispersion equation for longitudinal waves, i.e. the condition $A=0$, leading to the condition $\varepsilon_{11}=0$. Their solution gives the resonance frequencies of the oscillations in question. For electron HF waves the first resonance frequency $\omega_1(K)$, corresponding to $s=1$, is naturally related to the upper-hybrid frequency (4.22). Taking spatial dispersion into account, for instance for $Kv_e/\omega_H \ll 1$, it follows from (5.78) that

$$\omega_1(K)=\omega_U\left[1+\frac{3}{4}\frac{(Kv_e\omega_0)^2}{(\omega_0^2-3\omega_H^2)}\right]=\omega_U\left[1+\left(\frac{3}{8}\right)^{1/2}\frac{\beta_e\omega_0^2\omega_H}{\omega_U^2(\omega_0^2-3\omega_H^2)^{1/2}}\right], \tag{5.81}$$

where $\omega_U=(\omega_H^2+\omega_0^2)^{1/2}$ is the upper-hybrid frequency in a cold plasma, and where $\omega_1(K)\to\omega_U$ as $K\to 0$. The resonance value n_∞^2 is used in (5.81) (see formula (5.13)).

A general analysis of $\omega_1(K)$ shows that, as the wave number K varies, the frequency $\omega_1(K)$ at first increases monotonically until it reaches a maximum, after which with a further rise in K the value of $\omega_1(K)$ decreases to ω_H. If $Kv_e/\omega_H \ll 1$, the resonance frequencies of high harmonic order, $s=2,3,\ldots$, are given by the formula

$$\omega_s(K)=s\omega_H\left\{1+\frac{(s^2-1)\omega_0^2(Kv_e)^{s-2}}{2^s[(s^2-1)\omega_H^2-\omega_0^2]}\right\}. \tag{5.82}$$

With an increase in K, so that $Kv_e/\omega_H \gg 1$, the frequency $\omega_s(K)\to s\omega_H$. However, the electron resonance frequencies never become exactly equal to the values $s\omega_H$, but rather differ from them by finite intervals, the so-called Gross 'gaps' (Gross, 1951).

For the ion waves described by equation (5.79), when $Kv_i/\Omega_H \ll 1$, the resonance frequency $\omega_1(K)$ is close to the lower-hybrid frequency for $s=1$, namely

$$\omega_1(K)=\omega_L(1+\xi_1) \tag{5.83}$$

For values $s=2,3,\ldots$, the resonance branches are also described by formulas of the form

$$\omega_s(K)=s\Omega_H(1+\xi_s) \tag{5.84}$$

where ξ_1 and ξ_s are function of K. The relations giving the ion-cyclotron resonance frequencies as functions of K are similar in nature to the corresponding relations for the electron resonance frequencies.

The cyclotron oscillations and waves being considered are, on the other hand, described by different relations giving the resonance frequencies as a function of the wave number K. Under certain conditions, the resonance frequencies $\omega_{es}(K)$ and $\omega_{is}(K)$ will, as $K \to 0$, approach the values $s\omega_H$ and $s\Omega_H$, while with an increase in K as $K \to \infty$ we have $\omega_{es}(K) \simeq (s-1)\omega_H$ and $\omega_{is} \simeq (s-1)\Omega_H$ (see Fig. 5.6). Conditions are possible under which, with an increase in K, the frequencies $\omega_{is}(K)$ and $\omega_{es}(K)$ both increase initially, approaching the values $\omega \simeq (s+1)\Omega_H$, i.e., the $\omega_s(K)$ curves go through maxima and then decrease, with the frequencies $\omega_s(K)$

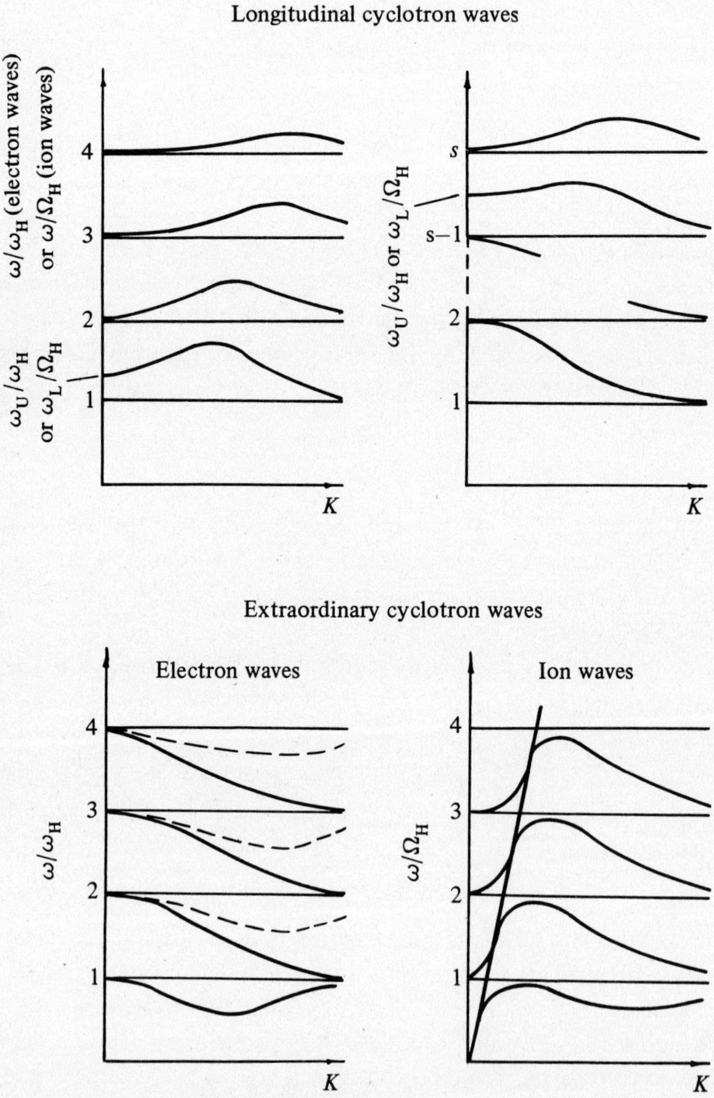

Fig. 5.6. The branches $\omega(K)$ describing the cyclotron waves.

approaching the values $s\Omega_H$. Consequently, when examining the experimental data, a precise theoretical analysis is necessary, in order to ascertain correctly the true branches describing the observed resonance oscillations. A detailed theoretical analysis of some different relations for the resonance frequencies of cyclotron oscillations is given in Akhiezer *et al.*, 1974. The reader is thus referred to that monograph in order to become better acquainted with this subject, as well as with a number of other properties of a plasma, taking spatial dispersion into account, properties which were only touched upon briefly in this section.

6

Growth rates for the different oscillation branches

As noted above in Chapter 3, for the definition of a harmonic wave selected here, namely $\exp[i(\omega + i\gamma)t]$, the amplitude of the plasma oscillations may increase with time, provided the following conditions for 'kinetic instability' are satisfied:

$$\partial f/\partial v > 0, \quad \gamma < 0, \tag{6.1}$$

where $f(v)$ is the distribution function for the charged particles. When the quantity $\gamma < 0$, it is called the *growth rate of the oscillations.* Under the given conditions the state of the plasma is, so to speak, unstable. This means that the branch of the waves for which $\gamma < 0$ has more energy imparted to it by the particles moving in the plasma than is imparted to the particles by the waves. This may result in the excitation of waves of the corresponding type, i.e., to a growth in their amplitude.

Below we will consider briefly a number of cases in which conditions (6.1) are satisfied and oscillations and waves of various kinds are excited. The growth rates for the oscillations of these waves will be obtained. As in the foregoing, different frequency ranges will be considered. When evaluating the values of γ, it should be kept in mind that growth of the oscillations is possible only if the effective frequency of collisions between particles $\nu_{\mathrm{eff}} < 0$, where the quantity ν_{eff} has to be specially defined.

A plasma can become unstable only when a state of thermodynamic disequilibrium exists in it, i.e., when the velocity distribution of the charged particles in the plasma is not Maxwellian. With regard to the phenomena of interest to us in this monograph, the three most important kinds of instabilities leading to the spontaneous excitation of oscillations in a plasma are:

(1) *beam instability*, caused by the presence in the plasma of directed particle fluxes, when some of the beam energy can be converted into energy of plasma oscillations;

(2) instability appearing due to an *anisotropic distribution* of particle velocities in the plasma – an anisotropic distribution of the plasma

temperature, for instance, only of the *hot tail* of the particle distribution; (3) instability due to *spatial nonuniformity* of the plasma.

In this chapter growth rates of plasma oscillations for only the first two kinds of instability will be considered.

6.1. Generation of high-frequency (HF) waves ($\omega \gg \Omega_0$) in an isotropic plasma

6.1.1. Beam instability

In an isotropic ($\mathbf{H}_0 = 0$) plasma there may be excitation of the HF branch of *longitudinal Langmuir waves* at the frequency

$$\omega(\mathbf{K}) \simeq \omega_0 = (N_0 e^2/\varepsilon_0 m)^{1/2} \tag{6.2}$$

under the influence of a quasi-neutral beam of charged particles moving in the plasma with a velocity $\mathbf{V}_b$. A beam with a *displaced Maxwellian velocity distribution*

$$f_{eb} = N_{eb}\left(\frac{m}{2\pi k T_{eb}}\right)^{3/2} \exp\left[-\frac{m(\mathbf{v} - V_b)^2}{2kT_{eb}}\right] \tag{6.3}$$

(where, as everywhere below, the subscripts b, eb, and ib refer to a beam, electron beam, or ion beam, respectively), leads to the excitation of Langmuir potential oscillations of the plasma. The growth of these oscillations can be attributed to resonant electrons in the beam which are in phase with the wave, i.e., which satisfy the condition

$$\omega(\mathbf{K}) = \mathbf{K}\cdot\mathbf{v}_e. \tag{6.4}$$

If the number density of the electrons in the beam $N_{eb} \ll N_0$ and if their temperature is high enough, then such a beam is usually called 'hot' – it has a wide velocity distribution. The solution of the dispersion equation for the plasma permeated by the hot beam, when the electrons of the plasma (N_0, T_{e0}) also have a Maxwellian distribution, leads to the following results.

The frequency of the excited plasma oscillations $\omega(\mathbf{K}) \simeq \omega_0$; it differs from the Langmuir frequency only by a negligibly small amount. The imaginary part of the complex frequency is given by the formula

$$\gamma_{e0}(\mathbf{K}) = \pi^{1/2}\omega(\mathbf{K})\left\{\left[\frac{\omega^2\omega(\mathbf{K})}{(Kv_e)^3}\exp-\left(\frac{\omega(\mathbf{K})}{Kv_e}\right)^2\right] + \left(\frac{\omega_{0b}}{Kv_{eb}}\right)^2\right.$$

$$\left.\times\frac{(\omega(\mathbf{K}) - K_\parallel V_b)}{Kv_{eb}}\exp-\left(\frac{\omega(\mathbf{K}) - K_\parallel V_b}{Kv_{eb}}\right)^2\right\}. \tag{6.5}$$

Here it is assumed that the following conditions are satisfied:

$$\omega_{0b} \ll K v_{eb} \quad \text{and} \quad (\gamma_{e0})_{max} \ll K v_{eb} \tag{6.5a}$$

(Bohm & Gross, 1949).

In formula (6.5) $v_{eb} = (2kT_{eb}/m)^{1/2}$, $\omega_{0b} = (N_{eb}e^2/\varepsilon_0 m)^{1/2}$, and $K_\parallel$ is the longitudinal component of the wave vector parallel to the vector $\mathbf{V}_b$. In the derivation of (6.5) it was assumed that the inherent Langmuir oscillations in the plasma are only slightly damped, namely that $\omega(\mathbf{K})/Kv_e = v_\phi/v_e \gg 1$ (see formula (5.7)).

Formula (6.5) implies that $\gamma_{e0} < 0$, i.e., there may be a growth in the amplitude of the oscillations $\omega(\mathbf{K})$, provided that the second term in (6.5) is greater than the first. This leads to the following condition for the excitation of Langmuir waves:

$$K_\parallel V_b > \omega(\mathbf{K}) \simeq \omega_0, \quad V_b > \omega_0/K_\parallel = v_{\phi\parallel}, \tag{6.5b}$$

i.e., the beam velocity V_b must be higher than the longitudinal component $v_{\phi\parallel}$ of their phase velocity. Hence, in the case under review, in an isotropic plasma *the excitation of transverse* (nonpotential) *electromagnetic waves* is impossible, since their phase velocity would always be greater than c, the velocity of light.

The resonance excitation of plasma oscillations being considered is, so to speak, of the *Čerenkov type* (see (3.18) and (6.4)). The minimum value $V_{b,min}$ leading to a growth in the oscillations is determined from the condition $\gamma_e(\mathbf{K}) = 0$, i.e.,

$$V_{b,min} = \frac{\omega_0}{K}\left\{1 + \frac{N_0}{N_{eb}}\left(\frac{T_{eb}}{T_0}\right)^{3/2} \exp\left[-\left(\frac{\omega(\mathbf{K})}{Kv_e}\right)^2\right]\right\}. \tag{6.6}$$

The value $V_{b,min}$ is often called the *critical velocity*. The maximum value of the growth rate, as can be seen from formula (6.5), corresponds to the condition.

$$\frac{K_\parallel V_b - \omega(\mathbf{K})}{Kv_{eb}} = \frac{1}{\sqrt{2}}, \quad \omega(\mathbf{K}) = \omega_0 = K_\parallel V_b - \frac{Kv_{eb}}{\sqrt{2}} \tag{6.7}$$

and is determined by the formula

$$(\gamma_{e0})_{max} = -\left(\frac{\pi}{2e}\right)^{1/2} \omega_0 \left(\frac{\omega_{0b}}{K_\parallel v_{eb}}\right)^2 = -\left(\frac{\pi}{2e}\right)^{1/2}\left(\frac{N_{eb}}{N_0}\right)\left(\frac{V_b}{v_{eb}}\right)^2 \omega_0 \tag{6.8}$$

(here e is the base of natural logarithms), since $v_{\phi\parallel} = V_b$.

Another case of beam instability is possible if a cold beam with a *narrow velocity spread* passes through the plasma, if its directed velocity

$$V_b < v_e \tag{6.9}$$

and $(\gamma_{e0})_{max} \ll K v_{eb}$ (see Bohm & Gross, 1949; Akhiezer & Fainberg, 1949, 1951a, b). When the density of the beam $N_{eb} \ll N_0$ and the condition $(KD_e)^2 \ll 1$ $(v_\phi^2 \gg v_e^2)$ is satisfied, which implies that the Langmuir oscillations are only slightly attenuated in the absence of the beam (see formula (5.7)), the growth rate in the presence of the beam is

$$(\gamma_{e0}) \simeq -(\pi/8)^{1/2}(N_{eb}/N_0)^{1/2}\frac{V_b}{v_\phi}\omega_0 \tag{6.10}$$

(Mikhailovskii, 1975, 1977). Here the critical velocity may be quite low:

$$V_{b,\min} \simeq (N_{eb}/N_0)^{1/2} v_e/[1+(KD_e)^2]^{1/2}. \tag{6.11}$$

A buildup of the Langmuir oscillations by a cold beam is also possible for $(KD_e)^2 \gtrsim 1$. In this case

$$(\gamma_{e0}) \simeq \left(\frac{N_{eb}}{N_0}\right)^{1/2}\left(\frac{V_b}{v_e}\right)\omega_0. \tag{6.12}$$

Under the resonance conditions $\omega \simeq K_\parallel V_b$, $(v_\phi \simeq V_b)$, we have

$$(\gamma_{e0})_{max} = \frac{3^{1/2}}{2(2)^{1/3}}\left(\frac{N_{eb}}{N_0}\right)^{1/3}\omega_0, \tag{6.13}$$

while the beam velocity

$$V_b \gg \left(\frac{N_{eb}}{N_0}\right)^{1/3} v_{eb}. \tag{6.14}$$

Now let us cite some particular cases in which a directed beam can lead to the excitation in an isotropic plasma of nonpotential oscillations of *transverse electromagnetic waves*. However, the conditions required for the excitation of these waves are more exacting, and the rates of growth of their amplitude are considerably smaller than in the cases considered above (Fried, 1959, Neufeld & Doyle, 1961). A charge-neutral beam of hot electrons which is directed through a cold plasma excites in it transverse waves with a phase velocity

$$v_e \ll (\omega/K = v_\phi) \ll v_{eb} \tag{6.15}$$

if

$$V_b^2 > [(K^2c^2 + \omega_0^2)/\omega_{0b}^2]v_{eb}^2. \tag{6.16}$$

Consequently, for $\omega = \omega_0$, when $K^2c^2 = \omega_0^2 c^2/v_\phi^2$, the requirements imposed on the beam velocity will be quite strict. For instance, it has to satisfy the condition

$$V_b^2 > (N_0/N_{eb})(c/v_\phi)^2 v_{eb}^2. \tag{6.17}$$

If the electrons and ions of the plasma are hot enough, and the beam is cold, then transverse electromagnetic waves may be excited in the plasma, with

$$v_i, v_{ib} \ll (\omega/K = v_\phi) \ll v_e, v_{eb} \tag{6.18}$$

at the Langmuir frequency $\omega \simeq \omega_0$' when

$$V_b^2 > \frac{1+(N_{eb}/N_0)(c/v_\phi)^2}{1+(N_{eb}/N_0)(v_e/v_i)^2} v_e^2. \tag{6.19}$$

Condition (6.19) evidently allows quite wide ranges of variations in the beam velocity and density. When $T_e \gg T_i$ and condition (6.18) is satisfied, while in addition

$$N_{eb} \simeq N_0, \quad V_b \simeq v_e \simeq v_{eb}, \tag{6.20}$$

the growth rate of these transverse waves will be

$$\gamma = -(m/M)^{1/2}(v_e/v_\phi)\omega_0. \tag{6.21}$$

6.1.2. 'Anisotropic' instability

An unstable state of the plasma can also appear, as mentioned above, in the abșence of directed fluxes, when the distribution of particle velocities is anisotropic. This can happen, for instance, if the velocity distribution of the electrons is anisotropic-Maxwellian, i.e., if it has the form

$$f_e = \frac{N_0 m^{3/2}}{(2\pi k)^{3/2} T_{e\|}^{1/2} T_\perp} \exp\left[-\left(\frac{mv_{e\|}^2}{2kT_{e\|}} + \frac{mv_{e\perp}^2}{2kT_{e\perp}}\right)\right], \tag{6.22}$$

where $T_\|$, $T_\perp$, $v_\|$ and $v_\perp$ are, respectively, the longitudinal (along the specified direction) and transverse (normal to this direction) values of the temperature and of the electron velocity. In this case, when $T_\| > T_\perp$, the Čerenkov growth rate of the longitudinal Langmuir oscillations is

$$(\gamma_{e0}) = -\pi^{-1/2}(v_{e\perp}/v_\phi)(c/v_\phi)^2(T_\perp/T_\|)\omega_0 \tag{6.23}$$

where $v_{e\perp} = (2kT_\perp/m)^{1/2}$ and it is assumed that

$$(c/v_\phi)^2 \gg (T_\|/T_\perp) - 1. \tag{6.23a}$$

The direction of propagation of these waves is parallel to the axis of rotational symmetry of the velocity distribution. If, however, $T_\perp > T_\|$, the instability of the plasma leads to the excitation of oscillations that are propagated perpendicularly to the symmetry axis. These oscillations may grow quite rapidly.

6.2. Excitation of HF and LF waves ($\omega \gg \omega_L$) in a magnetoplasma

The passage of a compensated quasi-neutral beam of charged particles with a directed velocity $\mathbf{V}_e$ through a magnetoplasma causes the latter to become unstable due to the effects of both the electrons and the ions of the beam. Below we will consider various instances of excitation of the corresponding branches describing longitudinal (potential) and transverse (nonpotential) oscillations, assuming that the densities of the charged particles in the beam N_{eb} and N_{ib} are much less than N_0.

6.2.1. Beam instability of longitudinal waves

In an anisotropic ($\mathbf{H}_0 \neq 0$) plasma, beam instability manifests itself especially strongly in the resonance regions, i.e., when the following conditions are satisfied:

$$\omega(\mathbf{K}) = \omega_{12}(\theta) = \mathbf{K}_{\parallel} V_b, \quad \omega(\mathbf{K}) = \omega_{12}(\theta) = K_{\parallel} V_b - s\omega_H, \tag{6.24}$$

where $\omega_1(\theta)$ and $\omega_2(\theta)$ are the HF branches describing the inherent resonance oscillations of the plasma (see formula (4.49) and the text following it.) The first of formulas (6.24) describes the Čerenkov excitation of plasma oscillations, or *Čerenkov resonance*, while the second describes the magnetic bremsstrahlung of the plasma, or *cyclotron resonance*, which exists in the case of beam instability only when $s = +1, +2, \ldots$, i.e., under conditions of the *anomalous Doppler effect* (see Chapter 3). When the velocities of the electrons in the plasma and in the beam moving in it both have a Maxwellian distribution, when the beam density is so low that $N_{eb} \ll N_0$, and when the directed velocity of the beam $\mathbf{V}_e$ is parallel to $\mathbf{H}_0$ (in which case γ is a maximum), then the plasma oscillations grow if

$$V_b > \frac{\omega_{12}(\theta)}{K_{\parallel}} = v_{\phi\parallel}, \quad \cos\theta \neq 0. \tag{6.25}$$

This means that the beam velocity must be higher than the longitudinal component of the phase velocity $v_{\phi\parallel}$ of the waves along the direction of the magnetic field $\mathbf{H}_0$. The condition opposite to (6.25) leads to increased attenuation of the waves.

Let us consider the growth rates of plasma oscillations for various cases of interaction of a beam with a magnetoplasma (Kitsenko & Stepanov, 1961; Kovner, 1960; Rapoport, 1960; Akhiezer *et al.*, 1974; Mikhailovskii, 1975).

In a system consisting of a *cold plasma and a hot beam*, when the velocities of the beam particles are widely spread and

$$V_b \gg v_{eb}, \tag{6.26}$$

the quantity γ in the frequency ranges of the resonance branches $\omega_1 = \omega_1(\theta)$ and $\omega_2 = \omega_2(\theta)$ for θ values sufficiently far away from $\pi/2$ is given by the formula

$$(\gamma_e)_{12} = \pi^{1/2}\frac{\omega_{12}}{2^{2s}s!}\frac{\omega_{0b}^2}{(Kv_{eb})^2}\left(\frac{K_\perp v_{eb}}{\omega_H}\right)^{2s}\left[\left(\frac{\omega_0}{\omega_{12}}\right)^2\cos^2\theta + \left(\frac{\omega_0\omega_{12}\sin\theta}{\omega_{12}^2-\omega_H^2}\right)^2\right]^{-1}$$
$$\times\frac{(\omega_{12}-K_\parallel V_b)}{K_\parallel v_{eb}}\exp\left[-\left(\frac{\omega_{12}+s\omega_H-K_\parallel V_b}{K_\parallel v_{eb}}\right)^2\right], \tag{6.27}$$

where $s=0,1,2,\ldots$ Here, in (6.27), it is taken into account that $(K_\perp v_{eb}/\omega_H)^2 \ll 1$ (see formula (4.38)). It is seen from (6.27) that $(\gamma_e)<0$, i.e., growth of the plasma oscillations may take place, provided that the condition (6.25) is satisfied. With an increase in the order s of the cyclotron resonance, the growth rate of the oscillations decreases rapidly.

For resonances of the first order $s=1$ where $(\omega_{12}-K_\parallel V_b) = -\omega_H$, the following growth-rate formulas are obtained for, respectively, a *dense* (slightly magnetized, $\omega_0^2 \gg \omega_H^2$) plasma and a *tenuous* (highly magnetized, $\omega_0^2 \ll \omega_H^2$) plasma:

$$\left.\begin{aligned}(\gamma_{eH})_1 &\simeq -\frac{\pi^{1/2}}{4}\left(\frac{N_{eb}}{N_0}\frac{V_b}{v_{eb}}\right)\left(\frac{\omega_0}{\omega_H}\right)\omega_0\sin^2\theta,\\ (\gamma_{eH})_2 &\simeq -\frac{\pi^{1/2}}{4}\left(\frac{N_{eb}}{N_0}\frac{V_b}{v_{eb}}\right)\frac{\sin^4\theta\cos\theta}{(1+\cos\theta)}\omega_H,\end{aligned}\right\} \tag{6.28a}$$

$$\left.\begin{aligned}(\gamma_{eH})_1 &\simeq -\frac{\pi^{1/2}}{4}\left(\frac{N_{eb}}{N_0}\frac{V_b}{v_{eb}}\right)\left(\frac{\omega_0}{\omega_H}\right)^3\omega_0 cos^4\theta,\\ (\gamma_{eH})_2 &\simeq -\frac{\pi^{1/2}}{4}\left(\frac{N_{eb}}{N_0}\frac{V_b}{v_{eb}}\right)\left(\frac{\omega_0}{\omega_H}\right)^3\omega_H\sin^2\theta\cos\theta,\end{aligned}\right\} \tag{6.28b}$$

where in (6.28*a*) the equations $\omega_1^2(\theta) = (\omega_0^2 - \omega_H^2\cos^2\theta)$, $\omega_2^2(\theta) = \omega_H^2\cos^2\theta$ are used, and in (6.28*b*) the equations $\omega_1^2(\theta) = (\omega_H^2 - \omega_0^2\cos^2\theta)$, $\omega_2^2(\theta) = \omega_0^2\cos^2\theta$, obtained from (4.49), are used. If $\omega_0 = \omega_H$ (for $s=1$),

$$(\gamma_{eH})_{12} = -\frac{\pi^{1/2}}{8}\left(\frac{N_{eb}}{N_0}\frac{V_b}{v_{eb}}\right)\frac{\sin^2\theta(1\pm\sin^2\theta)^{1/2}}{1+(1\pm\sin^2\theta)^{1/2}}\omega_H, \tag{6.28c}$$

where it is taken into account that $\omega_{12}^2(\theta) = \omega_H^2(1\pm\sin^2\theta)$.

The maximum growth rates for Čerenkov resonance correspond to the condition $(\omega_{12}-K_\parallel V_b) = (K_\parallel v_{eb})/\sqrt{2}$, and for $\omega_0^2 \gg \omega_H^2$, $\omega_0^2 \ll \omega_H^2$, and $\omega_0 \simeq \omega_H$, taking the condition $V_b \gg v_{eb}$ into account, we obtain from (6.27):

$$\left.\begin{aligned}(\gamma_{e0})_1 &= -\left(\frac{\pi}{2e}\right)^{1/2}\left(\frac{N_{eb}}{N_0}\right)\left(\frac{V_b}{v_{eb}}\right)^2\omega_0\cos^2\theta,\\ (\gamma_{e0})_2 &= -\left(\frac{\pi}{2e}\right)^{1/2}\left(\frac{N_{eb}}{N_0}\right)\left(\frac{V_b}{v_{eb}}\right)^2\omega_H\sin^2\theta\cos\theta,\end{aligned}\right\} \tag{6.29a}$$

$$\left.\begin{aligned}(\gamma_{e0})_1 &= -\left(\frac{\pi}{2e}\right)^{1/2}\left(\frac{N_{eb}}{N_0}\right)\left(\frac{V_b}{v_{eb}}\right)^2\left(\frac{\omega_0}{\omega_H}\right)^3\frac{\cos^6\theta}{\sin^2\theta}\omega_0,\\ (\gamma_{e0})_2 &= -\left(\frac{\pi}{2e}\right)^{1/2}\left(\frac{N_{eb}}{N_0}\right)\left(\frac{V_b}{v_{eb}}\right)^2\omega_0\cos\theta,\end{aligned}\right\}\qquad (6.29b)$$

$$(\gamma_{e0})_{12} = -\tfrac{1}{2}\left(\frac{\pi}{2e}\right)^{1/2}\left(\frac{N_{eb}}{N_0}\right)\left(\frac{V_b}{v_{eb}}\right)^2\frac{\cos^2\theta}{(1\pm\sin^2\theta)^{1/2}}\omega_H. \qquad (6.29c)$$

Note that formulas (6.28) and (6.29) require the satisfaction of the conditions $(\omega_0/\omega_H)^2 < (N_0/N_{eb})^{1/3}$ and $\theta \neq 0$ (or $\sin^2\theta \gg (\omega_0^2/\omega_H^2)\cos^2\theta$, where $\omega_0^2/\omega_H^2 \gg 1$). An examination of these formulas indicates that Čerenkov resonances generally build up more rapidly than cyclotron resonances. Favourable for cyclotron resonances are cases in which the wave vector $\mathbf{K}$ is close to the direction perpendicular to the vector $\mathbf{H}_0$.

In a system consisting of a *cold beam and a cold plasma*, for Čerenkov resonance $\omega_{12} = K_\parallel V_b$ it is assumed that $K_\parallel v_{eb} \ll \gamma \ll \omega_H$ (properties of a cold beam):

$$(\gamma_{e0})_{12} = -\frac{3^{1/2}}{2(2)^{1/3}}\left\{\frac{\omega_{0b}^2\cos^2\theta[\omega_1^2(\theta)-\omega_2^2(\theta)]\omega_{12}}{|\omega_{12}^2-\omega_H^2|}\right\}^{1/3}. \qquad (6.30)$$

If $\omega_0 \simeq \omega_H$, then

$$(\gamma_{e0})_{max} \simeq -\frac{3^{1/2}}{2(2)^{1/3}}\left(\frac{N_{eb}}{N_0}\right)^{1/3}\omega_0 \qquad (6.31a)$$

and this is the same as the value of γ_{e0} in an isotropic plasma as given by (6.13). When $\omega_0^2 \gg \omega_H^2$ and $\omega_0^2 \ll \omega_H^2$, respectively,

$$\left.\begin{aligned}(\gamma_{e0})_1 &\simeq -\frac{3^{1/2}}{2(2)^{1/3}}\left(\frac{N_{eb}}{N_0}\right)^{1/3}(\cos\theta)^{2/3}\omega_0,\\ (\gamma_{e0})_2 &\simeq -\frac{3^{1/2}}{2(2)^{1/3}}\left(\frac{N_{eb}}{N_0}\right)^{1/3}\left(\frac{\omega_H}{\omega_0}\sin\theta\right)^{-2/3}\omega_0\cos\theta,\end{aligned}\right\}\qquad (6.31b)$$

$$\left.\begin{aligned}(\gamma_{e0})_1 &\simeq -\frac{3^{1/2}}{2(2)^{1/3}}\left(\frac{N_{eb}}{N_0}\right)^{1/3}\omega_H,\\ (\gamma_{e0})_2 &\simeq -\frac{3^{1/2}}{2(2)^{1/3}}\left(\frac{N_{eb}}{N_0}\right)^{1/3}\omega_0\cos\theta.\end{aligned}\right\}\qquad (6.31c)$$

Cyclotron resonances of the first order, where $\omega_{12} = K_\parallel V_b - \omega_H$, are characterized in the given case by the following growth rates of the oscillations:

$$(\gamma_{eH})_{12} \simeq -\tfrac{1}{2}\left[\frac{\omega_{0b}^2\sin^2\theta|\omega_{12}^2-\omega_H^2|\omega_{12}}{\omega_H(\omega_1^2-\omega_2^2)}\right]^{1/2}. \qquad (6.32)$$

If $\omega_0 \simeq \omega_H$, then

$$(\gamma_{eH})_{12} \simeq -\frac{1}{2\sqrt{2}}\left(\frac{N_{eb}}{N_0}\right)^{1/2}\omega_0(1 \pm \sin\theta), \tag{6.33a}$$

and if $\omega_0^2 \gg \omega_H^2$ or $\omega_0^2 \ll \omega_H^2$, respectively, then

$$\left.\begin{aligned}(\gamma_{eH})_1 &\simeq -\tfrac{1}{2}\left(\frac{N_{eb}}{N_0}\right)^{1/2}\left(\frac{\omega_0}{\omega_H}\right)^{1/2}\omega_0\sin\theta,\\ (\gamma_{eH})_2 &\simeq -\tfrac{1}{2}\left(\frac{N_{eb}}{N_0}\right)^{1/2}(\cos\theta)^{1/2}\omega_H\sin^2\theta,\end{aligned}\right\} \tag{6.33b}$$

$$\left.\begin{aligned}(\gamma_{eH})_1 &\simeq -\tfrac{1}{2}\left(\frac{N_{eb}}{N_0}\right)^{1/2}\left(\frac{\omega_0}{\omega_H}\right)\omega_0\sin\theta\cos\theta,\\ (\gamma_{eH})_2 &\simeq -\tfrac{1}{2}\left(\frac{N_{eb}}{N_0}\right)^{1/2}\left(\frac{\omega_0}{\omega_H}\right)^2(\cos\theta)^{1/2}\omega_0\sin\theta.\end{aligned}\right\} \tag{6.33c}$$

Note that for motion of the electrons relative to the ions with a velocity $V_{eb} \gg v_e$, caused by an electric field applied parallel to the magnetic field, an instability will appear with a growth rate for HF Langmuir oscillations $\gamma \simeq -(m/M)^{1/3}\omega_0$.

For the passage of a *cold beam through a hot plasma*, the conditions for the excitation of plasma oscillations are more rigid than in the cases considered above. If $V_b \simeq v_e$, then the oscillations caused by Čerenkov resonance, $\omega = K_\parallel V_b$, and cyclotron resonance, $\omega = (K_\parallel V_b \pm \omega_H)$, will be unstable. In the first case, if $KD_e \lesssim 1$, then $(\gamma_{e0}) \simeq -(N_{eb}/N_0)^{1/2}Kv_e$; in the second case, $(\gamma_{eH}) \simeq -(N_{eb}/N_0)Kv_e$ when $(KD_e, K\rho_{He}) \ll 1$, $\omega_H \simeq \omega_0$. It should be noted that in the given case the oscillations $\omega = (K_\parallel V_b + \omega_H)$ are also unstable.

The instability situation in a *cold beam–hot plasma system* can bc illustrated by the following examples (see Mikhailovskii, 1975). In a highly magnetized plasma, $\omega_0^2 \ll \omega_H^2$, Čerenkov excitation is characterized by the growth rate

$$(\gamma_{e0}) \simeq -\frac{\pi}{4}\left(\frac{N_{eb}}{N_0}\right)^{1/2}\left(\frac{\omega_0}{\omega_H}\right)^2\left(\frac{V_b}{v_e}\right)\omega_0\sin^2\theta\cos\theta \tag{6.34a}$$

if $v_e/V_b > (\omega_0/\omega_H)^2$. However, if this condition is satisfied for cyclotron resonance, then we have

$$(\gamma_{eH}) \simeq -\frac{\pi^{1/2}}{8}\left(\frac{N_{eb}}{N_0}\right)^{1/2}\left(\frac{\omega_0}{\omega}\right)^3\left(\frac{V_b}{v_e}\right)\omega_0\sin^4\theta. \tag{6.34b}$$

Finally, if $v_e/V_b > \omega_0/\omega_H$, then

$$(\gamma_{e0}) \simeq -\frac{\pi^{1/2}}{e}\left(\frac{N_{eb}}{N_0}\right)\left(\frac{V_b}{v_e}\right)^2\left(\frac{\omega_0}{\omega_H}\right)^3 \omega_0 \sin^2\theta \cos^2\theta. \qquad (6.34c)$$

In all the foregoing cases the oscillations build up slowly.

6.2.2. Beam instability of transverse electromagnetic waves

Beam instability may lead to the excitation of an *electron-whistler* wave $\omega_L \ll \omega \lesssim \omega_H$, described by a refractive index n_2^2 and a frequency $\omega_2(K,\theta)$ (see formula (4.60)).

The dispersion equation of this wave branch can be written simply, as follows:

$$\omega^2(K,\theta) = \omega_H \cos\theta/[1 + (\omega_0/Kc)^2]. \qquad (6.35)$$

For this branch, provided the resonance conditions for Čerenkov radiation and magnetic bremsstrahlung are satisfied:

$$\omega_2(K,\theta) = \omega_2 = K_{\parallel} V_b, \quad \omega_2 = K_{\parallel} V_b - \omega_H, \qquad (6.36)$$

in a *cold plasma–cold beam* system the growth rates will be, respectively,

$$(\gamma_{e0})_2 = -\frac{3^{1/2}}{2(2)^{1/3}}\left[\left(\frac{N_{eb}}{N_0}\right)\frac{\sin^2\theta}{1 + (\omega_0/\omega_2)^2 n_2^2}\right]^{1/3} \omega_2(K,\theta) \qquad (6.37)$$

and

$$(\gamma_{eH})_2 = -\tfrac{1}{2}\left(\frac{N_{eb}}{N_0}\right)^{1/2}\left[1 + \left(\frac{\omega_0}{n_2\omega_2}\right)^2 + |\cos\theta|\right]\left|\left[|\cos\theta| - \left(\frac{\omega_0}{\omega_2 n_2}\right)^2\right]\right|$$
$$\times (1 - |\cos\theta|)\left\{\left[1 + \left(\frac{\omega_0}{n_2\omega_2}\right)^2\right]|\cos\theta|^{3/2}\right\}^{-1}, \qquad (6.38)$$

where the angle range is $0 < \theta < \pi/2$ (Kitsento & Stepanov, 1961; Kovner, 1960). The growth rate of the oscillations of these waves increases with the frequency. When $\omega_2 \simeq \omega_H$, the condition $\cos\theta > \omega_2(K,\theta)/\omega_H$ must be satisfied, and from (6.37) and (6.38) we obtain

$$(\gamma_{e0})_2 \simeq -\frac{3^{1/2}}{2(2)^{1/3}}\left(\frac{N_{eb}}{N_0}\right)^{1/3}\left(\frac{\sin^2\theta}{\cos\theta}\right)^{1/3} \omega_H, \qquad (6.37a)$$

$$(\gamma_{eH})_2 \simeq -\left(\frac{N_{eb}}{N_0}\right)^{1/2}\frac{\omega_H}{(\cos\theta)^{3/2}}. \qquad (6.38a)$$

In a *cold plasma–hot beam* system

$$(\gamma_{e0})_2 = -\pi^{1/2}\frac{\omega_2^3(K,\theta)}{(K_{\parallel}v_{eb})^2}\frac{\sin^2\theta}{[1 + (\omega_0/n_2\omega_2)^2]}\left(\frac{N_{eb}}{N_0}\right)\left(\frac{K_{\parallel}V_b - \omega_2}{K_{\parallel}v_{eb}}\right)$$
$$\times \exp\left[-\left(\frac{K_{\parallel}V_b - \omega_2}{K_{\parallel}v_{eb}}\right)^2\right], \qquad (6.39)$$

$$
\left.
\begin{aligned}
(\gamma_{eH})_2 = &-\left(\frac{\pi}{2}\right)^{1/2}\frac{s}{2^s s!}\frac{\omega_H\omega_2(K,\theta)}{Kv_{eb}}\left(\frac{K_\perp v_{eb}}{\omega_H}\right)^{2s-2}\left(\frac{N_{eb}}{N_0}\right)\\
&\times\frac{(1-\cos\theta)^2\left\{s\left(\dfrac{\omega_0}{n_2\omega_2}\right)^2[1+(\omega_0/n_2\omega_2)^2]-\cos\theta(1+\cos\theta)\right\}}{\cos^3\theta[1+(\omega_0/\omega_2 n_2)^2]^3}\\
&\times\exp\left[-\left(\frac{K_\| V_b-\omega_2(K,\theta)+s\omega_H}{K_\| v_{eb}}\right)^2\right],
\end{aligned}
\right\}
\tag{6.40}
$$

where $s=1,2,\ldots$, and in (6.40), as everywhere above, it is assumed that $(Kv_{eb})^2 \ll \omega_H^2$. The maximum value of γ_{e0} corresponds to the value $K_\| V_b-\omega_2(K,\theta)=K_\| v_{eb}/\sqrt{2}$, and, since $V_b \gg v_{eb}$, for the first-order resonance at $\omega=\omega_H$, we get:

$$
(\gamma_{e0})_{max} = -\left(\frac{\pi}{2e}\right)^{1/2}\left(\frac{N_{eb}}{N_0}\right)\left(\frac{V_b}{v_{eb}}\right)^2\frac{\sin^2\theta}{\cos\theta}\omega_H \tag{6.39a}
$$

$$
\begin{aligned}
(\gamma_{eH})_{max} = &-2(2\pi)^{1/2}\left(\frac{N_{eb}}{N_0}\right)\left(\frac{c}{v_{eb}}\right)\frac{\omega_H^2}{\omega_0}\exp\left[-\left(\frac{V_b}{v_{eb}}\right)^2\right]\\
&(\cos\theta-\omega_2/\omega_H)^{1/2}(1-\cos^2\theta)\cos^{-4}\theta.
\end{aligned}
\tag{6.40a}
$$

During the derivation of formulas (6.39*a*) and (6.40*a*), the relations $\omega_0/n_2\omega_2=\cos\theta-\omega_2/\omega_H$, $\cos\theta>\omega_2/\omega_H$ were used.

6.2.3. Cyclotron instability for an anisotropic distribution of electron velocities

In the various processes observed in the near-Earth plasma, a major role is apparently played by the presence of an additional population of hot electrons with an anisotropic velocity distribution, the density N_{eb} of which is considerably lower than the total electron density N_0. The temperature of these particles satisfies the condition $T_\perp > T_\| \gg T_0$, where T_0 is the plasma temperature. Under these conditions instability of the plasma will appear in the vicinity of the electron gyroresonance frequency $\omega \simeq \omega_H$.

In the frequency band corresponding to the electron-whistler branch of the oscillations n_2 (see (4.61), and Fig. 4.2), when the gyroresonance condition

$$
Kv_{eH} = \omega - \omega_H \tag{6.41}
$$

is satisfied, where v_{eH} is the velocity of the resonant electrons, the growth rate of the whistler waves is

$$
(\gamma_{eH})_2 = -\pi\omega_H\left(\frac{\omega_H-\omega}{\omega_H}\right)^2\left[A(v_{eH})-\frac{\omega}{\omega_H-\omega}\right]F_1(v_{eH}) \tag{6.42}
$$

(Kennel & Petschek, 1966; Liemohn, 1967). Formula (6.42) was obtained for an anisotropic velocity distribution of the additional population of hot electrons $F_e(v_\perp, v_\parallel)$ and a Maxwellian distribution of the plasma electrons f_{e0}. However, the functions $F_1(v_{eH})$ and $A(v_{eH})$ defining $(\gamma_{eH})_2$ are

$$F_1(v_{eH}) = 2\pi \frac{\omega_H - \omega}{K} \frac{1}{N_0} \int v_{e\perp} F_e(v_{e\perp}, v_{e\parallel}) dv_{e\perp} \Big|_{v_{e\parallel} = v_{eH}}. \tag{6.43}$$

$$A(v_{eH}) = \frac{\int_0^\infty \left\{ v_{e\perp} \left[v_{e\parallel} \left(\frac{\partial F_e}{\partial v_\parallel} \right) - v_{e\perp} \left(\frac{\partial F_e}{\partial v_\parallel} \right) \right]_{v_{e\parallel} = v_{eH}.} \right\} dv_{e\perp}}{2 \int \{ v_{e\perp} [F_e(v_{e\perp}, v_{e\parallel})]_{v_{e\parallel} = v_{eH}.} \} dv_{e\perp}}. \tag{6.44}$$

It follows directly from (6.42) that growth of the amplitude of the electron-whistler waves ($\gamma < 0$) is possible if

$$A(v_{eH}) > \omega/(\omega_H - \omega), \tag{6.45}$$

since the frequency ω of the waves in question is always lower than ω_H. Note that formula (6.43) describes an *anisotropic velocity distribution* of the hot electrons or, so to speak, it determines the *measure of their anisotropy with respect to the pitch angles* $\alpha = \tan^{-1}(-v_\perp/v_\parallel)$. If the distribution function $F_e(v_{e\perp}, v_{eH})$ is anisotropic-Maxwellian (see (6.22); it is sometimes called a bi-Maxwellian function), then it follows from (6.44) that

$$A(v_{eH}) = (T_\perp/T_\parallel) - 1 \tag{6.46}$$

and

$$(\gamma_{eH})_2 = -\pi^{1/2} \frac{\omega}{\omega_H} \left(\frac{\omega_H - \omega}{K v_{e\parallel}} \right) \left(\frac{N_{eb}}{N_0} \right) \left(\frac{T_\perp - T_\parallel}{T_\parallel} - \frac{\omega}{\omega_H - \omega} \right)$$
$$\times \exp\left[-\left(\frac{\omega_H - \omega}{K v_{e\parallel}} \right)^2 \right], \tag{6.47}$$

where $v_{e\parallel} = (2kT_{e\parallel}/m)^{1/2}$. Note that (6.43) can be rewritten as

$$F_1(v_{eH}) = \frac{\frac{1}{N_0} \int_0^\infty \left[v_{e\perp} \frac{\partial F_e}{\partial \alpha} \tan\alpha \right] dv_{e\perp}}{2 \int_0^\infty v_{e\perp} F_e dv_\perp}. \tag{6.48}$$

Now the meaning of the function $F_1(v_{eH})$ becomes clear. It defines the *proportion of the additional population of hot electrons* contained in the plasma, the velocities of which exceed the resonance velocity $|v_{eH}|$. Formula (6.48) also implies that $\partial F_e/\partial\alpha > 0$. This is a sufficient condition for instability, when the magnetic energy per resonant electron is less than the kinetic energy of a resonant electron, i.e., $W_H = \mu_0 H_0^2/2M_{eb} < W = \frac{1}{2}mv_{eH}^2$. The latter inequality follows from the resonance condition (6.41) and the expression (4.60) for n_2^2.

6.3. Excitation of the very-low-frequency (VLF) and extremely-low-frequency (ELF) oscillation branches ($0 < \omega \ll \omega_H$)

Various types of waves, connected either wholly or partially with the motion of ions, are in a number of cases excited in a plasma under conditions analogous to those for the excitation of LF and HF waves $\omega \gg \omega_L$. In some cases they are described by growth rates for the oscillation amplitudes in which it is just necessary to replace v_{eb} by v_{ib}, ω_H by Ω_{Hb}, and ω_0 by Ω_{0b}, and to use the refractive indexes characterizing the corresponding oscillation branches.

6.3.1. Beam instability of ion-acoustic waves ($\omega \ll \omega_0$) in an isotropic plasma

A charge-neutral beam of hot electrons with a Maxwellian distribution (see (6.3) and the text following it) may in an isotropic plasma ($\mathbf{H}_0 = 0$) excite *ion-acoustic waves* (for $T_e \gg T_i$) having a frequency $\omega(\mathbf{K}) = \omega_{10}$ (see (5.59)). The growth rate of these waves is

$$\gamma_{e0} = -\pi^{1/2}\frac{\omega^3(\mathbf{K})}{\Omega_0^2}\left\{\frac{\omega_0^2\omega(\mathbf{K})}{(Kv_e)^3} + \frac{\omega_{0b}^2}{(Kv_{eb})^2}\left(\frac{K_\| V_b - \omega(\mathbf{K})}{Kv_{eb}}\right)\right. \\ \left. \times \exp\left[-\left(\frac{K_\| V_b - \omega(\mathbf{K})}{Kv_{eb}}\right)^2\right]\right\} \tag{6.49}$$

Here it is sufficient that the condition

$$V_b > \omega(\mathbf{K})/K = (\omega_{10}/K)v_s \tag{6.50}$$

be satisfied, where $v_s = (kT_e/M)^{1/2}$ is the velocity of nonisothermal sound. However, lest these oscillations be quenched by Landau damping on the electrons, the following condition has to be satisfied:

$$v_{eb} \gg V_b > \frac{N_0}{N_{eb}}\left(\frac{T_{eb}}{T_e}\right)^{3/2}\frac{v_s}{(1+K^2D_e^2)^{1/2}} \tag{6.51}$$

(see Akhiezer *et al.*, 1974). The growth rates for the ion-acoustic and Langmuir oscillations excited by *resonant ions of the beam* are defined by formulas (6.4) and (6.49), but with ω_{0b} and v_{eb} replaced in them by Ω_{0b} and v_{ib}. The maximum growth rate for ion Langmuir oscillations (keeping in mind that $T_{eb} \simeq T_{ib}$) is

$$(\gamma_{i0})_{max} = -(\pi/2)^{1/2}(\Omega_{0b}/Kv_{ib})^2\omega_0. \tag{6.52}$$

On the other hand, the interaction between the beam ions and the Langmuir oscillations is bound to be less effective, since the range of phase velocities of the unstable oscillations is considerably narrower.

In the case of instability caused by a *beam with a narrow spread of velocities* ($V_b < v_e$), the maximum growth rate for ion-acoustic waves under conditions of resonance

$$\omega(\mathbf{K})_s = Kv_s/(1+K^2D_e^2)^{1/2} = K_\parallel V_b, \tag{6.53}$$

caused by the electrons in the beam, is

$$(\gamma_{e0})_{max} = -\frac{3^{1/2}}{2(2)^{1/3}}\left(\frac{N_{eb}}{N_0}\right)^{2/3}\left(\frac{M_e}{m}\right)^{2/3}\omega(\mathbf{K})_s. \tag{6.54}$$

Here the following conditions have to be satisfied:

$$V_b \gg v_{eb}\left(\frac{N_0}{N_{eb}}\frac{m}{M_b}\right)^{1/3}, \quad v_i \ll v_{eb} \ll v_e, \tag{6.55}$$

which will indeed be the case for a sufficiently low beam velocity V_b.

The excitation of ion-acoustic oscillations under resonance conditions

$$\omega_1(K)_s = \frac{\Omega_{0b}}{[1+(KD_e)^{-2}]^{1/2}} = K_\parallel V_b, \tag{6.56}$$

when

$$(N_0/N_{eb})(m/M_b) \ll 1, \tag{6.57}$$

is greatly enhanced. The growth rate of these oscillations has a maximum value

$$(\gamma_{e0})_{max} = -\frac{3^{1/2}}{2(2)^{1/3}}\left(\frac{N_0}{N_{eb}}\right)^{1/3}\left(\frac{m}{M_b}\right)^{1/3}\omega_1(\mathbf{K})_s. \tag{6.58}$$

6.3.2. Instability of ion-acoustic waves caused by motion of the electrons relative to the ions in an isotropic plasma

The excitation of ion-acoustic oscillations is possible if there is motion of electrons relative to ions, for instance when a constant electric field $\mathbf{E}_0$ is present in the plasma. In this case, if the energy imparted to an electron under the influence of the field over the mean free path $W_E = eE_0\Lambda > kT_e$, then the velocity of the electron $\dot{r}_e \simeq (e/m)E_0t$, and in principle it can increase without limit. This is the so-called 'electron run-away' condition (see below, Section 7.1; Dreicer, 1959; Gurevich, 1960). However, this process will in fact be limited by the scattering of electrons by longitudinal plasma oscillations, due to the plasma instability involving ion-acoustic waves. The latter ($T_e \gg T_i$) are excited in the given case if $v_e > (V_e)_{min}$; the *critical electron velocity is*

$$(V_e)_{min} = e^{-3/2}\frac{\omega_{10}}{K}\left\{1+\left(\frac{T_e}{T_i}\right)^{3/2}\left(\frac{M}{m}\right)^{1/2}\exp\left(-\frac{kT_e}{2kT_i(1+K^2D_e^2)}\right)\right\}, \tag{6.59}$$

while the growth rate of the oscillations is

$$\gamma_e = (m/M)\Omega_0 \tag{6.60}$$

(see Gordeev, 1954). Ion-acoustic waves of frequency $\omega_{10} = Kv_s$ (see (5.61)) are excited for not particularly high values of T_e/T_i if $(KD_e)^2 \ll 1$, while ion-Langmuir waves $\omega_{10} = \Omega_0$ (see (5.62)) are excited if $KD_e \gtrsim 1$ and $T_e/T_i > \ln[(T_e/T_i)^3(M/m)]$. It should be noted that ion-acoustic waves can also be excited under the influence of an electron beam $V_b > v_e$ in a quasi-isothermal ($T_e \simeq T_i$) plasma (see, for example, Gurevich, 1964).

6.3.3. Beam instability of longitudinal waves in a magnetoplasma

Passage of a beam through the plasma excites longitudinal, pure ion or ion–electron, oscillations with $\omega^2 \ll \omega_H^2$ most markedly in the vicinity of the VLF and ELF resonance branches $\omega_2(\theta)$ and $\omega_3(\theta)$ (see (4.51), (4.52), and (4.53)). If the following resonance conditions are satisfied:

$$\omega(\mathbf{K}) = \omega_{23}(\theta) = K_\parallel V_b, \quad \omega_{23}(\theta) = K_\parallel V_b - \Omega_{H1} \tag{6.61}$$

in a *cold beam–cold plasma* system, then the growth rates of the oscillations will be

$$(\gamma_{e0})_{23} = -\frac{3^{1/2}}{2(2)^{1/3}}\left(\frac{N_{eb}}{N_0}\right)^{1/3}\left[\frac{\omega_{23}(\theta)|\omega_{23}^2-\Omega_H^2|\cos^2\theta}{\omega_H(1+\omega_H^2/\omega_0^2)(\omega_2^2-\omega_3^2)}\right]^{1/3}\omega_H, \tag{6.62}$$

$$(\gamma_{iH})_{23} = -\tfrac{1}{2}\left(\frac{N_{ib}}{N_0}\right)^{1/2}\left[\frac{\Omega_0\omega_{23}(\theta)|\omega_{23}^2-\Omega_H^2|}{(\omega_H^2+\omega_0^2)(\omega_2^2-\omega_3^2)}\right]^{1/2}\omega_H. \tag{6.63}$$

Here Čerenkov excitation is effective because of the interaction of the plasma with the beam electrons, and cyclotron excitation is effective because of the interaction with the beam ions.

It follows from (6.62) and (6.63) that, if $\omega_0 \simeq \omega_H$ and $\theta \simeq \pi/2$ (or $\cos\theta \simeq (m/M)^{1/2}$), then at frequencies $\omega_2 = \omega_L = (\Omega_H/\omega_H)^{1/2}$ (where ω_L is the lower-hybrid frequency) and $\omega_3 \simeq \Omega_H/\sqrt{2}$, respectively, the growth rates will be

$$\left.\begin{aligned}(\gamma_{e0})_2 &= -\frac{3^{1/2}}{2^{5/3}}\left(\frac{N_{eb}}{N_0}\right)^{1/3}\left(\frac{M}{m}\right)^{1/3}\omega_L,\\ (\gamma_{e0})_3 &= -\frac{3^{1/2}}{2^{13/6}}\left(\frac{N_{eb}}{N_0}\right)^{1/3}\omega_L,\end{aligned}\right\} \tag{6.62a}$$

$$\left.\begin{aligned}(\gamma_{iH})_2 &= -\frac{1}{2(2)^{1/2}}\left(\frac{N_{ib}}{N_0}\right)^{1/2}\omega_L,\\ (\gamma_{iH})_3 &= -\frac{1}{2^{9/4}}\left(\frac{N_{ib}}{N_0}\right)^{1/2}\left(\frac{m}{M}\right)^{1/4}\Omega_H.\end{aligned}\right\} \tag{6.63a}$$

Here the beam density must satisfy the condition $N_{eb}/N_0 \ll m/M$. Obviously, the amplitude of the oscillations of the wave branch $\omega_2(\theta)$ in the vicinity of the lower-hybrid frequency will grow more rapidly than for the branch of the ion wave $\omega_3(\theta)$. If $\omega_0^2 \gg \omega_H^2$ or $\omega_0^2 \ll \omega_H^2$ and $\cos\theta \simeq (m/M)^{1/2}$, then we have

$$\left.\begin{aligned}(\gamma_{e0})_2 &= -\frac{3^{1/2}}{2^{7/6}}\left(\frac{N_{eb}}{N_0}\right)^{1/3}\omega_L,\\ (\gamma_{e0})_3 &= -\frac{3^{1/2}}{2^{13/6}}\left(\frac{N_{eb}}{N_0}\right)^{1/3}\Omega_H,\end{aligned}\right\} \tag{6.62b}$$

or

$$\left.\begin{aligned}(\gamma_{e0})_2 &= -\frac{3^{1/2}}{2^{7/6}}\left(\frac{N_{eb}}{N_0}\right)^{1/3}(\Omega_0\omega_0)^{1/2},\\ (\gamma_{e0})_3 &= -\frac{3^{1/2}}{2^{13/6}}\left(\frac{N_{eb}}{N_0}\right)^{1/3}\Omega_H,\end{aligned}\right\} \tag{6.62c}$$

and

$$\left.\begin{aligned}(\gamma_{iH})_2 &= -\frac{1}{2^{3/4}}\left(\frac{N_{ib}}{N_0}\right)^{1/2}\left(\frac{\omega_H}{\omega_0}\right)^{1/2}\omega_L,\\ (\gamma_{iH})_3 &= -\frac{1}{2^{9/4}}\left(\frac{N_{ib}}{N_0}\right)^{1/2}\left(\frac{\Omega_H}{\Omega_0}\right)^{1/2}\Omega_H,\end{aligned}\right\} \tag{6.63b}$$

or

$$\left.\begin{aligned}(\gamma_{iH})_2 &= -\frac{1}{2^{3/4}}\left(\frac{N_{ib}}{N_0}\right)^{1/2}\left(\frac{m}{M}\right)^{1/2}(\Omega_0\omega_0)^{1/2}\\ (\gamma_{iH})_3 &= -\frac{1}{2^{9/4}}\left(\frac{N_{ib}}{N_0}\right)^{1/2}\left(\frac{\Omega_H}{\omega_0}\right)^{1/2}\Omega_H.\end{aligned}\right\} \tag{6.63c}$$

In the sets of equations (6.62) and (6.63) the values $\omega_2^2(\theta) = 2\Omega_H\omega_H$, $2\omega_0\Omega_0, \Omega_H\omega_H$, and $\omega_3^2(\theta) = \Omega_H^2/2$ were used, respectively, for $\omega_0^2 \gg \omega_H^2$, $\omega_0^2 \ll \omega_H^2$, $\omega_0^2 \simeq \omega_H^2$, and $\cos^2\theta \simeq m/M$. In the formulas (6.62) and (6.63) for the growth rates, it was assumed, as everywhere above, that $\gamma < \omega$. For the wave branches being examined here, this means that both γ_{e0} and γ_{eH} are less than Ω_{Hb}. If $\gamma \gg \Omega_{Hb}$, which is possible only for the branch $\omega_2(\theta)$ under certain conditions, the growth rate for Čerenkov resonance $\omega_2(\theta) = K_\parallel V_b$ is

$$(\gamma_{e0})_2 = -\frac{3^{1/2}}{2(2)^{1/3}}\left[\frac{(\Omega_0^2 + \omega_{0b}^2\cos\theta)\omega_2(\theta)}{1 + \omega_0^2/\omega_H^2}\right]^{1/3}. \tag{6.62d}$$

When a *hot beam* passes through a *cold plasma* with a velocity $\mathbf{V}_e$ parallel to the magnetic field $\mathbf{H}_0$ under resonance conditions

$$\omega_{23}(\mathbf{K}) = \mathbf{K}_\parallel V_b - s\omega_H, \tag{6.64}$$

and

$$\omega_{23}(\mathbf{K}) = K_\parallel V_b - s\Omega_{Hb}, \tag{6.65}$$

where $s = 0, 1, 2, \ldots$, the growth rates due to the interaction of the plasma with the beam electrons will be

$$(\gamma_e)_s = -\pi^{1/2}\omega_{23}(\theta)\frac{1}{2^{2s}s!}\left(\frac{\omega_{0b}}{Kv_{eb}}\right)^2\left(\frac{K_\perp v_{eb}}{\omega_H}\right)^{2s}\frac{|\omega_{23}^2 - \Omega_H^2|\omega_H^2}{(\omega_0^2 + \omega_H^2)(\omega_2^2 - \omega_3^2)}$$
$$\times\left(\frac{K_\parallel V_b - \omega_{23}(\theta)}{K_\parallel v_{eb}}\right)\exp\left[-\left(\frac{K_\parallel V_b - \omega_{23} - s\omega_H}{K_\parallel v_{eb}}\right)^2\right]. \tag{6.66}$$

The growth rates $(\gamma_{iH})_s$ due to interaction with the beam ions are defined by the same formula, but with ω_{0b}^2 replaced by Ω_{0b}^2, v_{eb} by v_{ib}, and ω_H by Ω_{Hb} in the multiplying factors $(K_\perp v_{eb}/\omega_H)^{2s}$, and in the exponential. In (6.66), $(K_\perp v_{eb})^2 \ll \omega_H^2$ and $(K_\perp v_{ib})^2 \ll \Omega_{Hb}^2$, while everywhere above $N_{eb} = N_{ib}$.

It follows from (6.66) that, close to the lower-hybrid resonance for $\cos^2\theta \simeq m/M$, the maximum values of the growth rates for Čerenkov excitation ($s = 0$) of the wave branches $\omega_2(\theta)$, when $\omega_0^2 \gg \omega_H^2$, $\omega_0^2 \ll \omega_H^2$ or $\omega_0 \simeq \omega_H$ and $\omega_2^2 \simeq 2\omega_H\Omega_H, 2\omega_0\Omega_0$, or $\omega_H\Omega_H$, respectively, are,

$$\left.\begin{aligned}(\gamma_{e0})_2 &= -\tfrac{1}{2}\left(\frac{N_{eb}}{N_0}\right)\left(\frac{V_b}{v_{eb}}\right)^2\omega_L,\\ (\gamma_{e0})_2 &= -\tfrac{1}{2}\left(\frac{N_{eb}}{N_0}\right)\left(\frac{V_b}{v_{eb}}\right)^2\left(\frac{m}{M}\right)^{1/2}(\Omega_0\omega_0)^{1/2},\\ (\gamma_{e0})_2 &= -\tfrac{1}{2}\left(\frac{\pi}{2e}\right)^{1/2}\left(\frac{N_{eb}}{N_0}\right)\left(\frac{V_b}{v_{eb}}\right)^2\omega_L.\end{aligned}\right\} \tag{6.66a}$$

The interaction with the ions of the beam is less effective. From (6.66), by analogy with (6.66*a*), the following formulas can be obtained for the maximum values of the growth rates:

$$\left.\begin{aligned}(\gamma_{\text{io}})_2 &\simeq -\tfrac{1}{2}\left(\frac{\pi}{\text{e}}\right)^{1/2}\left(\frac{N_{\text{ib}}}{N_0}\right)\left(\frac{V_{\text{b}}}{v_{\text{ib}}}\right)^2\left(\frac{m}{M}\right)^{1/2}\Omega_{\text{H}},\\ (\gamma_{\text{io}})_2 &\simeq -\tfrac{1}{2}\left(\frac{\pi}{\text{e}}\right)^{1/2}\left(\frac{N_{\text{ib}}}{N_0}\right)\left(\frac{V_{\text{b}}}{v_{\text{ib}}}\right)^2\left(\frac{m}{M}\right)^{3/4}\Omega_0,\\ (\gamma_{\text{io}})_2 &\simeq -\tfrac{1}{2}\left(\frac{\pi}{2\text{e}}\right)^{1/2}\left(\frac{N_{\text{ib}}}{N_0}\right)\left(\frac{V_{\text{b}}}{v_{\text{ib}}}\right)^2\left(\frac{m}{M}\right)^{1/2}\Omega_{\text{H}}.\end{aligned}\right\} \tag{6.66b}$$

For the branch $\omega_3(\theta)$ for $\cos^2\theta \simeq m/M$, when $\omega_3 = \Omega_{\text{H}}/\sqrt{2}$,

$$\left.\begin{aligned}(\gamma_{\text{eo}})_3 &\simeq -\tfrac{1}{4}\left(\frac{\pi}{\text{e}}\right)^{1/2}\left(\frac{N_{\text{eb}}}{N_0}\right)\left(\frac{V_{\text{b}}}{v_{\text{eb}}}\right)^2\Omega_{\text{H}},\\ (\gamma_{\text{io}})_3 &\simeq -\tfrac{1}{4}\left(\frac{\pi}{\text{e}}\right)^{1/2}\left(\frac{N_{\text{ib}}}{N_0}\right)\left(\frac{V_{\text{b}}}{v_{\text{ib}}}\right)^2\frac{m}{M}\Omega_{\text{H}}.\end{aligned}\right\} \tag{6.66c}$$

At the first-order ($s = 1$) cyclotron resonance, when $\omega_2(\theta = \pi/2) = \omega_{\text{L}}$ and $\omega_3(\theta \simeq \pi/2) = \Omega_{\text{H}}/\sqrt{2}$, we get:

$$\left.\begin{aligned}(\gamma_{\text{eH}})_2 &= -\frac{\pi^{1/2}}{4}\left(\frac{N_{\text{eb}}}{N_0}\right)\left(\frac{V_{\text{b}}}{v_{\text{eb}}}\right)\frac{\omega_{\text{L}}}{(1+\omega_{\text{H}}^2/\omega_0^2)},\\ (\gamma_{\text{iH}})_2 &= -\frac{\pi^{1/2}}{4}\left(\frac{N_{\text{eb}}}{N_0}\right)\left(\frac{V_{\text{b}}}{v_{\text{eb}}}\right)\frac{\omega_{\text{H}}}{(1+\omega_{\text{H}}^2/\omega_0^2)},\end{aligned}\right\} \tag{6.66d}$$

and

$$\left.\begin{aligned}(\gamma_{\text{eH}})_3 &= -\frac{1}{16}\left(\frac{\pi}{2}\right)^{1/2}\left(\frac{N_{\text{eb}}}{N_0}\right)\left(\frac{V_{\text{b}}}{v_{\text{eb}}}\right)\left(\frac{m}{M}\right)\Omega_{\text{H}},\\ (\gamma_{\text{iH}})_3 &= -\frac{1}{8(2)^{1/2}+1}\left(\frac{\pi}{2}\right)^{1/2}\left(\frac{N_{\text{ib}}}{N_0}\right)\left(\frac{V_{\text{b}}}{v_{\text{ib}}}\right)\Omega_{\text{H}}.\end{aligned}\right\} \tag{6.66e}$$

6.3.4. Beam instability of transverse electromagnetic waves in a magnetoplasma

The excitation of transverse *Alfvén* ($0 < \omega_1 < \Omega_{\text{H}}$) *and fast magnetoacoustic* ($0 < \omega_2 \gtrless \omega_{\text{H}}$) *waves* by a beam instability is quite effective close to the cyclotron resonances

$$\omega_{12} = K_{\parallel}V_{\text{b}} - \omega_{\text{H}}, \quad \omega_{12}(\mathbf{K}) = K_{\parallel}V_{\text{b}} - \Omega_{\text{Hb}}. \tag{6.67}$$

The frequencies $\omega_{12}(\mathbf{K})$ and the refractive indexes n_{12}^2 of these branches are described by the formulas (4.58) and (4.59).

In a *hot plasma–cold beam* system, of which the beam velocity is parallel to $\mathbf{H}_0$, the growth rates of these waves, due to their interaction with the beam electrons, are

$$(\gamma_{eH})_{12} = -\frac{\omega_{0b}}{2}\left\{\frac{\omega_H}{\omega_{12}} \times \frac{[(1+\cos^2\theta)n_{12}^2 - 2(\Omega_0/\Omega_H)^2(1 \mp \omega_{12}/\Omega_H)^{-1}](\omega_{12}^2 - \Omega_H^2)^2}{\Omega_0^4[(1+\cos^2\theta)(n_{12}\Omega_H/\Omega_0)^2 + (\omega_{12}/\Omega_H)^2 - 2]}\right\}^{1/2}, \tag{6.68}$$

while the quantity $(\gamma_{iH})_{12}$, associated with the interaction between the plasma and the beam ions, is described by the same formula, but with ω_H replaced by Ω_{Hb} and ω_{0b} by Ω_{0b}, and with a minus sign in the corresponding term of the numerator. An analysis of formula (6.68) shows that the excitation of an ion-acoustic wave, with $\omega_1(\mathbf{K}) \lesssim \Omega_H$, is ineffective over the entire range of angles θ. The branch of the fast magnetoacoustic wave $\omega_2(\mathbf{K})$, on the other hand, is readily excited. In the vicinity of the frequency $\omega_2(\mathbf{K}) \leqslant \omega_L$, when $\cos^2\theta \simeq m/M$, i.e., close to resonance $\theta \simeq \pi/2$, at the lower-hybrid frequency ω_L the refractive index is

$$n_2^2 \simeq \omega_0^2/(\omega_L - \omega^2)(1 + \omega_H^2/\omega_0^2) \tag{6.69}$$

(see (4.21) and (4.25)). Now, with the aid of (6.69), from (6.68) we obtain the following growth rates:

$$\left.\begin{aligned} (\gamma_{eH})_2 &= -\tfrac{1}{2}\left(\frac{N_{eb}}{N_0}\right)^{1/2}\left(\frac{M}{m}\right)^{5/4}\omega_L, \\ (\gamma_{iH})_2 &= -\tfrac{1}{2}\left(\frac{N_{ib}}{N_0}\right)^{1/2}\left(\frac{\omega_L}{\Omega_H}\right)^{1/2}\omega_L. \end{aligned}\right\} \tag{6.70}$$

In the frequency range $\omega \ll \Omega_H$, both branches of the waves in question, i.e., the Alfvén and modified Alfvén waves, described by the refractive index (4.37), are easily excited. Their growth rates are

$$\left.\begin{aligned} (\gamma_{eH})_1 \simeq (\gamma_{eH})_2 &\simeq -\tfrac{1}{2}\left(\frac{N_{eb}}{N_0}\right)^{1/2}\left(\frac{M}{m}\right)^{1/2}\left(\frac{\omega_H}{\omega_{12}}\right)^{1/2}\Omega_H, \\ (\gamma_{iH})_1 \simeq (\gamma_{iH})_2 &\simeq -\tfrac{1}{2}\left(\frac{N_{eb}}{N_0}\right)^{1/2}\left(\frac{\Omega_H}{\omega_{12}}\right)^{1/2}\Omega_H. \end{aligned}\right\} \tag{6.71}$$

When calculating the growth rates γ_i in (6.70) and (6.71), it is assumed that the masses of the ions in the plasma and in the beam are equal: $M_0 = M_b$.

It follows from (6.70) and (6.71) that the interaction between the plasma and the beam electrons is more effective than the interaction with the ions. Note that as $\omega \to 0$, the formulas (6.71) become meaningless. Taking the limit requires a special consideration of formula (6.68), and in addition it must be made more precise. Beam instability of the branches in question appears when

$$V_b^2 > V_A^2 + V_{Ab}^2 \tag{6.72}$$

where V_A and V_{Ab} are respectively, the Alfvén velocities of the plasma and the beam.

In a *hot beam–cold plasma* system, when $N_{eb} \ll N_0$, we have, for the Čerenkov and cyclotron resonances respectively, the following growth rates:

$$\omega_{12} = K_{\parallel} V_b, \quad \omega_{12} = K_{\parallel} V_b - s\omega_H, \quad \omega_{12} = K_{\parallel} V_b - s\Omega_{Hb}, \tag{6.73}$$

$$\left.\begin{aligned}(\gamma_{e0})_{12} = &-\frac{\pi^{1/2}}{2}\frac{\omega_{0b}^2}{\omega_{12}}\left(\frac{K_\perp v_{eb}}{\omega_H}\right)^2 \\ &\times \frac{[n_{12}^2 \cos^2\theta(\omega_{12}^2 - \Omega_H^2) + \Omega_0^2](\omega_{12}^2 - \Omega_H^2)}{[(1+\cos^2\theta)(n_{12}\Omega_H/\Omega_0)^2 + (\omega_{12}/\Omega_H)^2 - 2]\Omega_0^4} \\ &\times \left(\frac{K_\parallel V_b - \omega_{12}}{K_\parallel v_{eb}}\right)\exp\left[-\left(\frac{K_\parallel V_b - \omega_{12}}{K_\parallel v_{eb}}\right)^2\right],\end{aligned}\right\} \tag{6.74}$$

$$\left.\begin{aligned}(\gamma_{eH})_{12} = &-\frac{\pi^{1/2}s^2}{2^s s!}\frac{\omega_{0b}^2}{\omega_{12}}\left(\frac{K_\perp v_{eb}}{\omega_H}\right)^{2s-2} \\ &\times \frac{[(1+\cos^2\theta)n_{12}^2 - 2(\Omega_0/\Omega_H)^2(1 \mp \omega_{12}/\Omega_H)^{-1}](\omega_{12}^2 - \Omega_H^2)^2}{[(1+\cos^2\theta)(n_{12}\Omega_H/\Omega_0)^2 + (\omega_{12}/\Omega_H)^2 - 2]\Omega_0^4} \\ &\times \left(\frac{K_\parallel V_b - \omega_{12}}{K_\parallel v_{eb}}\right)\exp\left[-\left(\frac{K_\parallel V_b - \omega_{12} - s\omega_H}{K_\parallel v_{eb}}\right)^2\right],\end{aligned}\right\} \tag{6.75}$$

where $s = 1, 2, \ldots$ Here the growth rates $(\gamma_{i0})_{12}$ and $(\gamma_{iH})_{12}$ caused by the interaction between the plasma and the ions of the beam are described by the same formulas, but with ω_{0b} replaced by Ω_{0b}, ω_H by ω_{0b} and v_{eb} by v_{ib}, and with a minus sign in the corresponding multiplying factor in the numerator. As everywhere above, the conditions $(K_\perp v_{eb})^2 \ll \omega_H^2$ and $(K_\perp v_{ib})^2 \ll \Omega_{Hb}^2$ have to be satisfied.

An analysis of formulas (6.74) and (6.75) indicates that Čerenkov excitation under the influence of both the electrons of the beam and the ions of the beam is less effective than cyclotron excitation for $s = 1$. For the branch $\omega_2(\theta = \pi/2)$, (6.74) and (6.75) imply that in the vicinity of the lower-hybrid frequency $\omega_2 = \omega_L$ the maximum values of the growth rates for $\cos^2\theta \simeq m/M$ are:

$$\left.\begin{aligned}(\gamma_{e0})_1 &\simeq -\tfrac{1}{2}\left(\frac{\pi}{2e}\right)^{1/2}\left(\frac{N_{eb}}{N_0}\right)\left(\frac{v_{eb}}{V_b}\right)^2\left(\frac{m}{M}\right)\left(1-\frac{\omega^2}{\Omega_H^2}\right)\omega_L\cos^2\theta,\\(\gamma_{e0})_2 &\simeq -\tfrac{1}{2}\left(\frac{\pi}{2e}\right)^{1/2}\left(\frac{N_{eb}}{N_0}\right)\left(\frac{v_{eb}}{V_b}\right)^2\omega_L.\end{aligned}\right\}\quad(6.75a)$$

Under the influence of the beam ions, at frequencies such that $0<\omega\lesssim\Omega_H$, the branches $\omega_1(\theta)$ and $\omega_2(\theta)$ may be excited if the following conditions, respectively, are satisfied: $\cos^2\theta>1-\omega_1^2/\Omega_H^2$ and $\theta\neq 0$. In both cases the growth rates are very small.

For first-order cyclotron resonance it follows from (6.75) that

$$\left.\begin{aligned}(\gamma_{eH})_2 &= -\frac{\pi^{1/2}}{2}\left(\frac{N_{eb}}{N_0}\right)\left(\frac{V_b}{v_{eb}}\right)\left(\frac{M}{m}\right)^2\omega_L,\\(\gamma_{iH})_2 &= -\left(\frac{\pi}{2}\right)^{1/2}\left(\frac{N_{eb}}{N_0}\right)\left(\frac{V_b}{v_{eb}}\right)\Omega_H.\end{aligned}\right\}\quad(6.75b)$$

Excitation of the branch $\omega_1(\theta)$ is, however, less effective.

6.3.5. Cyclotron instability for an anisotropic distribution of ion velocities

Just as for the electron-whistler branch of the oscillations, near the cyclotron resonances

$$K_{\parallel}v_{i\parallel}=\omega-\Omega_H \tag{6.76}$$

there may be excitation of an ion-whistler branch of the oscillations, described by the refractive index n_1^2 (see (4.58)), in which case an *anisotropic distribution of ion velocities* is observed. The appropriate conditions are realized, for instance, in a Maxwellian plasma (N_0, T_0) which contains an additional *population of hot ions* $N_{ib}\ll N_0$, $T_{i\perp}>T_{i\parallel}\gg T_0$. The growth rate of these oscillations is

$$(\gamma_i)=-\pi\frac{\Omega_H(\Omega_H-\omega)^2}{\omega\,(2\Omega_H-\omega)}\left[A(v_{i\parallel})-\frac{\omega}{\Omega_H-\omega}\right]F_1(v_{i\parallel}), \tag{6.77}$$

where $F_1(v_{i\parallel})$ and $A(v_{i\parallel})$ are defined by formulas (6.43) and (6.44), but with ω_H replaced by Ω_H, and $T_{e\perp}$ by $T_{i\perp}$, and $T_{e\parallel}$ by $T_{i\parallel}$, and $v_{e\parallel}$ by $v_{i\parallel}$. For an anisotropic-Maxwellian distribution function, (6.22), of the additional population of hot particles, it follows from (6.77) that

$$(\gamma_i) = -\pi^{1/2}\frac{\Omega_H}{K_{\parallel}v_{i\parallel}}\frac{(\Omega_H-\omega)^2}{(2\Omega_H-\omega)}\left(\frac{N_{ib}}{N_0}\right)\left(\frac{T_{i\perp}}{T_{i\parallel}}-1-\frac{\omega}{\Omega_H-\omega}\right)$$
$$\times\exp\left[-\left(\frac{\omega-\Omega_H}{K_{\parallel}v_{i\parallel}}\right)^2\right], \tag{6.78}$$

where $v_{i\parallel} = (2kT_{i\parallel}/M)^{1/2}$ and $(T_{i\perp}/T_{i\parallel}) - 1 > \omega/(\Omega_H - \omega)$. At the limit, for $\omega \ll \Omega_H$, the *growth rate of the amplitude of the Alfvén waves* will be

$$(\gamma_i) = -\frac{\pi^{1/2}}{2}\frac{\Omega_H^2}{K_{\parallel}v_{i\parallel}}\left(\frac{N_{ib}}{N_0}\right)\left(\frac{T_{i\perp}}{T_{i\parallel}}-1\right)\exp\left[-\left(\frac{\Omega_H}{K_{\parallel}v_{\parallel}}\right)^2\right]. \tag{6.79}$$

The branch of the modified Alfvén wave, or *magnetoacoustic wave* $\omega \ll \Omega_H$, will be excited if $T_{i\parallel}/T_{i\perp} > 1$. Similarly, a fast magnetoacoustic wave may be excited in the vicinity of the resonance $K_{\parallel}v_{i\parallel} = \omega + \Omega_H$, when $T_{i\parallel}/T_{i\perp} > 1$, i.e., when the condition $1 - T_{i\perp}/T_{i\parallel} > \omega/(\omega + \Omega_H)$ is satisfied.

6.4. Some remarks about growth rates

So far in this chapter we have presented various formulas for the growth rates characterizing the buildup of oscillations and waves in a plasma, caused by their interaction with resonant particles. The *instability* of the plasma leading to these processes is thus of a *kinetic nature*, although some of the formulas for the growth rates can be obtained with sufficient precision from the hydrodynamic approximation as well. An important feature of this class of instabilities, especially within the framework of the topics of interest to us, is that the interaction between the waves and the particles is of a *local nature*. It may well take place within a limited region of the plasma, i.e., only a very select group of the plasma particles are able to participate in it.

On the other hand, there may be instability of the plasma due to, for example, a drifting motion of large-scale inhomogeneous plasma formations in a steady (but nonuniform) field, such as the magnetic field of the Earth. In this case the buildup of the plasma oscillations takes place at the expense of the energy associated with the pressure of the inhomogeneity, or of the energy of the nonuniform pressure of the magnetic field. Gravitational forces, for instance the attractive force of the Earth, also play a role in these processes. Obviously, the large-scale oscillations excited in this case have very low frequencies, $\omega \ll \Omega_H$, and thus long wavelengths, commensurable with the scale of an inhomogeneity, for instance, as thick as the F2 layer of the ionosphere several hundred kilometres or more in size. The large-scale magnetohydrodynamic and internal gravity waves which appear during this process are observed, for instance, in the ionosphere; numerous studies of these have been carried out (see, for instance, Yeh & Liu, 1972). However, since the description of these goes beyond the framework of the topics being examined here, we will not discuss the conditions for the excitation of these waves at all.

Instabilities of the *nonlinear type* form a large and important class of plasma instability. They lead to an effective buildup of electron, ion–electron, or purely ion plasma oscillations and to the excitation of waves of various kinds. In a number of cases the corresponding waves can also be described quite adequately in the hydrodynamic approximation. This type of instability appears most frequently under the influence of the field set up by electromagnetic waves, and it becomes particularly effective if the plasma is influenced by, so to speak, *strong pump waves.* Such waves alter the properties of the plasma, making its structure, parameters, and medium considerably inhomogeneous. The linear scales of the inhomogeneities depend on the field intensity, and they may be substantially greater than, or of the order of, the lengths of the 'pump waves'. Instabilities of the *parametric type* are an example of this type of plasma instability; they appear in a collisionless plasma. On the other hand, under the influence of powerful electromagnetic waves, instabilities of the *heating type* may also be observed in a collisional plasma. This effect involves a variation of the temperature of the medium, which can also give rise to inhomogeneities of the plasma but of a considerably larger size. These are actually instabilities of the type: *wave–particle–plasma inhomogeneity.* They are described by nonlinear field equations, and they will be considered briefly in the next section, which gives a general description of certain nonlinear effects in a plasma that are related to the things we are interested in here. Note, too, that this type of instability is closely related to the excitation of plasma oscillations when isolated inhomogeneities are present in the plasma, for instance, in the vicinity of a body moving through the plasma at a supersonic speed or when a beam of particles is injected into the plasma.

It goes without saying that the linear *growth rates* presented above only describe the initial stage in the development of the buildup of plasma oscillations. Consequently, they may prove to be very useful for the analysis of the results of various experiments. If we know certain parameters characterizing a group of experimental data, including recordings of plasma oscillations, then the growth-rate formulas may well make it possible to ascertain the type of waves observed, the kind of instability, etc. Experimental data containing information about the *development of the growth* of the oscillation amplitude also make it possible to evaluate the parameters of a beam of oscillating particles, to determine the degree of anisotropy of the temperature, or to obtain other data on the mechanism of plasma excitation. It appears that such a detailed analysis of plasma instability will be greatly aided by, for instance, the results of experimental studies of so-called hydromagnetic (ion) whistlers and electron whistlers (whistling atmospherics; see Volume 2). However, any buildup of the oscillations will naturally be limited. The growth of the oscillation amplitude gradually ceases, the process becomes quasi-steady-state, and a quasi-constant amplitude is established, or, alternatively, the amplitude begins to decrease, if damping (dispersal) of the energy of the excited waves prevails over growth. A process for which, in particular, there occurs a gradual decrease of the growth rate will, of course, be purely nonlinear in nature. Its mechanism will depend on the specific conditions and on the nature of the phenomena taking place in the plasma. Theoretical investigations of such problems often run up against serious difficulties and can be carried out only for a very limited number of cases. Some other nonlinear phenomena observed in the near-Earth plasma will be described in the next chapter.

7

Nonlinear effects in a plasma

7.1. General remarks

The theoretical results considered in the previous chapters were obtained for a system of linearized kinetic equations plus Maxwell's (Poisson's) equations, (3.7) and (3.8). This means that the dependence of the plasma parameters on the electric field of the waves propagating in, and generated in, the plasma were not taken into account. Thus the formulas presented above describe, so to speak, the behavior of a plasma in the linear approximation. However, in order to arrive at a complete picture of the phenomena taking place in a plasma, it is in principle necessary to solve the corresponding nonlinear problems. This becomes obvious from the following simple considerations, which illustrate to some degree under what conditions it is possible to confine ourselves to the linear theory.

Let us assume, for instance, that in an isotropic collisionless plasma there is a constant electric field $\mathbf{E}_0$. Then, from the equation of motion of an electron $m\mathrm{d}\dot{\mathbf{r}}_e/\mathrm{d}t = -e\mathbf{E}_0$, it follows that the velocity imparted to it will be $\dot{\mathbf{r}}_e = (e\mathbf{E}_0/m)t$. Obviously, the velocity $\dot{\mathbf{r}}_e$ increases linearly with time. After some time, according to this formula, it may even exceed the thermal velocity of the electrons v_e and then increase without limit. If, to be more explicit, it is assumed that $v_e \simeq 5 \times 10^5\,\mathrm{m\,s^{-1}}$ and $\mathbf{E}_0 \simeq 1\,\mathrm{V\,m^{-1}}$, then it is easy to show that $\dot{\mathbf{r}}_e \simeq v_e$ after only a very short time $\tau_0 \simeq 3 \times 10^{-6}\,\mathrm{s}$ (!), and that $\dot{\mathbf{r}}_e$ continues to increase steadily. (Note that such conditions are realized, for instance, in the outer ionosphere, where the collision frequency $\nu_{ei} \simeq 10\,\mathrm{s}^{-1}$ and the mean free time for an electron $\tau_\nu \equiv \nu_e^{-1} \simeq 10^{-1}\,\mathrm{s}$). Clearly, however, such a situation is impossible. Consequently, some other description of the process being considered is necessary. It can evidently proceed *linearly* only during a time interval $\Delta t \lesssim$ or $\ll \tau_0$, and the actual development to be expected is one that limits the continual increase of the electron velocity $\dot{\mathbf{r}}_e$ with time.

But what actually takes place? It was mentioned above (see Section

6.3.2) that, under the influence of a constant field, instability *vis-à-vis* ion-acoustic waves appears in a plasma. Thus, ultimately, $\dot{\mathbf{r}}_e$ is limited by the scattering of electrons by longitudinal oscillations of the waves excited in the plasma, and the linear relationship between $\dot{\mathbf{r}}_e$ and t disappears. Naturally, the velocity $\dot{\mathbf{r}}_e$ will also be limited by collisions ν_e between electrons and other particles, if the mean free path is not especially large. In the given case $\tau_\nu \gg \tau_0$ and collisions play a relatively minor role. On the other hand, the growth rate γ_i of the ion-acoustic waves excited by the electric field $\mathbf{E}_0$ will also be time-dependent and will decrease. Otherwise a new contradiction arises: the amplitude of the excited waves $A \sim \exp(\gamma_i t)$ increases without limit and rapidly. This contradiction is eliminated by assuming a nonlinear relationship for the growth rate $\gamma_i(E_0, t)$. Therefore, to describe the entire process under review, the corresponding nonlinear problem has to be solved. A linear description is possible only within a limited time interval Δt, which depends on the field strength $\mathbf{E}_0$. In the case at hand, the time interval Δt is very small, being of the order of, or less than, 10^{-6} s.

Now let us consider another example: when an isotropic homogeneous collisionless plasma is subjected to the influence of an alternating harmonic field $\mathbf{E}_0 \exp(i\omega t)$, and the electrons have imparted to them an ordered velocity $\dot{\mathbf{r}}_e \simeq |e|\mathbf{E}/m\omega$, oscillating periodically in the direction of $\mathbf{E}_0$. The amplitude of the displacements of the electrons in the field of the wave will be $\mathbf{r}_e = e\mathbf{E}/m\omega^2$, while the distribution of their number density will become spatially inhomogeneous in the direction of the vector $\mathbf{E}_0$, oscillating in time. The mean energy acquired by each electron from the field during a period $T = 2\pi/\omega$ will be $W_E \simeq e^2 E_0^2 \nu/4\pi m\omega^2$. Now let us assume that, while the field is acting, the energy W_E is additive, (i.e.), that *the course of the process is linear*. Then, during a time interval $\tau_0 = (4\pi m\omega^2/\nu e^2 E_0^2)(\frac{3}{2}kT_e)$, an electron acquires an energy comparable with its thermal energy $W_{Te} = \frac{3}{2}kT_e$.

Let this process take place, for example, in the ionosphere at a height $z \simeq 300\,\mathrm{km}$, where the thermal energy of an electron $W_{Te} = \frac{3}{2}kT_e \simeq 0.2\,\mathrm{eV} \simeq 3 \times 10^{-19}\,\mathrm{J}$ and $\nu = 10^3\,\mathrm{s}^{-1}$. We further assume that the source of an alternating electric field E_0 is located at the Earth's surface, its frequency being $\omega \simeq 10^8\,\mathrm{s}^{-1}$ and its power being $W \simeq 100\,\mathrm{kW}$. Then $E_0(z \simeq 300\,\mathrm{km}) \simeq 10\,\mathrm{mV\,m}^{-1}$. For the specified values of W_{Te}, ω, E_0, and ν, we obtain a time $\tau_0 \simeq 1.5 \times 10^6\,\mathrm{s}$ (!). Consequently, only after about 400 hours of continuous action of an alternating field will the energy W_E acquired by an electron become equal, according to this evaluation, to its gas-kinetic energy W_{Te}. However, this process cannot take place in the described manner. As W_E increases, there will be a gradual change in the state of the plasma. The velocity of an electron $\dot{\mathbf{r}}_e$ becomes a nonlinear function of time. The time t_E after which the nonlinear description begins to apply is considerably less than τ_0. In a plasma, phenomena appear which limit an increase in the energy of the oscillating electrons. The plasma becomes unstable, waves are generated in it, and the function $\dot{\mathbf{r}}_E(E_0, t)$ becomes nonlinear. However, the evaluation carried out nevertheless shows that in the given case the time of action of the linear process can apparently be quite long, since the amplitude of the alternating field $\mathbf{E}_0$ is not especially great, while an external source scarcely affects the state of the plasma and the phenomena taking place in it.

Another aspect of this statement becomes clear if we estimate the amplitude of the displacement of an electron $\dot{\mathbf{r}}_E$ in the plasma due to the field. For the values of $\mathbf{E}_0$ and ω used above, $r_E \simeq 1.8 \times 10^{-9}\,\mathrm{m} \approx 10^{-10}\lambda$. This means that the linear scale of the spatial inhomogeneity, appearing in the plasma due to the field, is considerably less than the wavelength of the source $\lambda = 2\pi c/\omega$. The field of the wave alters the plasma structure only very slightly, so that a linear description of its interaction with the medium is possible. It is obvious, however, that this situation must change radically when $r_E \simeq \lambda$. But, for this to be the case, the corresponding value of the field strength would have to be $E_\lambda \simeq 2\pi c\omega/e \simeq 10^{10} E_0$. Thus the field amplitude assumed above is still very low, compared to the amplitude that is required if the field is to have a significant effect on the plasma structure and if is to engender nonlinear phenomena in the plasma.

The general *qualitative conclusion* ensuing from the foregoing elementary considerationsis as follows: as long as $\dot{\mathbf{r}}_e \ll v_e$, it can be assumed that $r_e \ll \lambda$ and the time of action of the field $\Delta t \ll \tau_0$, i.e., it is considerably shorter than the typical time required for the indicated quantities to become commensurable, and the processes in the plasma are only slightly nonlinear. It can also be shown that the condition for $\dot{\mathbf{r}}_E$ to be small in comparison with the phase velocity v_ϕ of the waves is analogous.

The appearance of nonlinear effects in a collisionless plasma can also be illustrated effectively by means of an important example associated with the so-called *nonlinear Landau damping*. As in the foregoing, the considerations are purely qualitative, not to say elementary and non-rigorous. Let us consider the conditions for Čerenkov resonance $\omega = \mathbf{K}\cdot\mathbf{v}_{eR}$ (see (3.18)) in the field of a one-dimensional periodic travelling wave, the potential of which $\phi(z, t) = \phi_0(t)\exp[i(\omega t - Kz)]$, where the amplitude $\phi_0(t)$ is only slightly attenuated, so that $\phi_0(t) = \text{const}$. This case describes the interaction of Langmuir waves (5.6), the phase velocity of which is v_ϕ, with resonant electrons, having a mean velocity $\bar{v}_{eR}$. The electrons interact strongly with a wave in a certain velocity range $(v_{eR}+\Delta v_{eR})$. They are, it is customary to say, trapped by the wave in a *potential well.* According to the law of conservation of energy, the following condition must be satisfied for these particles: $e\phi_0=\frac{1}{2}m(\Delta v_{eR})^2$. Since the total range of velocity variation is $2\Delta v_{eR}$, the quantity $\Delta v_{eR}=\left(\frac{e\phi_0}{m}\right)^{1/2}$, the angular frequency of the oscillations of the electrons is $\omega_{eR}=1/\tau_R\simeq K\Delta v_{eR}$, and the *travel time* of a particle trapped in the potential well is $\tau_R\simeq(1/K)(e\phi_0/m)^{-1/2}$. The width of the potential well in velocity space is thus Δv_{eR}, while in coordinate space it is obviously equal to the wavelength λ.

Consequently, the quantity τ_R is the *characteristic time* of the process in question. If τ_R is sufficiently long, which may well be the case if the potential of the field ϕ_0 is low, then the condition $\gamma_e\tau_R\gg 1$ may be satisfied, where γ_e is the coefficient of Landau damping for Langmuir waves. In this case during the travel time of the resonant particle trapped in the potential well the field amplitude has time to become attenuated, and the wave has time to exchange energy with particles. In the opposite case, i.e., when the field potential is quite high and the time τ_R is short, $\gamma_e\tau_R\ll 1$, the wave is not able to exchange energy with particles during the time $\tau_e=1/\gamma_e$, and its amplitude cannot decrease proportionally to $\exp(-\gamma_e t)$. A contradiction! But here it should be recalled that we are using the coefficient of Landau damping γ_e (see (5.8)) derived on the basis of a linearized system of equations. A more comprehensive analysis of the process under review shows that the quantity γ_e given by formula (5.8) is suitable only for a certain time interval $\Delta t\ll\tau_R$. As time passes, the coefficient of Landau damping begins to decrease: the function $\gamma_e(E_0,t)$ becomes nonlinear. Detailed studies indicate that in fact the amplitude of the travelling wave initially decreases, after which for $t\gtrsim\tau_e$ it starts to oscillate with a spatial period $\Delta\simeq(\omega/K^2)(e\phi/m)^{-1/2}$, and that finally the wave acquires a quasi-steady amplitude, becoming a so-called Bernstein–Greene–Kruskal wave (see Al'tshul' & Karpman, 1965; Bernstein *et al.*, 1957).

In the examples considered we have neglected the effective collision frequency ν between particles in comparison with the Landau-damping coefficient when $\gamma<0$ or with the growth rate when $\gamma>0$, i.e., we assumed that $|\gamma|\gg\nu$. In a *collisional plasma* the effect of a field manifests itself as follows. If the mean free time τ_ν of an electron is sufficiently long, then during this time a strong field will impart to it a considerable amount of energy. This increases sizably the energy released by the electron when it collides with an ion or neutral particle, in comparison with the situation in the absence of a field. Thus the plasma temperature ultimately changes as well, and so, accordingly, does the frequency of the collisions between particles. The ratio between the collision frequency and the velocity of the particles is important with regard to this. As we know, in a plasma consisting of electrons and ions, the collision frequency $\nu_{ei} \sim (m/kT_e)^{3/2} \sim 1/v_e^3$ (see formula (4.4)). When the directed velocity $\dot{\mathbf{r}}_e \simeq e\mathbf{E}/m\omega$ imparted to an electron by, for example, an alternating field is larger than, or of the order of, v_e, which may well be the case in the field of a strong wave, the collision frequency $\nu_{ei} \sim 1/\dot{\mathbf{r}}_e^3$. Thus the field causes a reduction in the collision frequency, i.e., it diminishes the attenuation of the amplitude of a wave propagated in or excited in the plasma. This makes the relationship between the field and the temperature of the medium somewhat complex. Consquently, the physical state of the plasma is altered and, in particular, various instability phenomena appear.

The variation in plasma temperature caused by the field in turn leads to a change in the permittivity of the plasma and, of course, it also disturbs the spatial distribution of the electron density of the plasma. One important particular consequence of this, as of other nonlinear effects, is a *violation of the principle of superposition* of the oscillations, i.e., of the additivity of the waves propagated in the plasma. Ultimately this even distorts the structure of the field that is acting on the plasma.

Therefore, the effect of a field on a collisional plasma manifests itself in various kinds of nonlinear phenomena. The nature of these will depend on the time $\tau_\nu \simeq \nu^{-1}$ between particle collisions. If the time during which the plasma properties become altered is short in comparison with τ_ν, then the nonlinear effect on the plasma will be dominant. In the opposite case, the nonlinear effects are weaker, being dependent on the size of the variation in ν caused by the field $\mathbf{E}_0$.

Below we present some results of studies of the phenomena appearing in a plasma under the influence of external sources of field, as well as some which are peculiar to the plasma itself. Interest in investigations of all these subjects has been particularly reinforced during recent years, in connection with the accumulation of numerous experimental data that are not susceptible to interpretation within the framework of the linear theory. A large portion of these data requiring a nonlinear theoretical explanation were obtained during studies of wave processes in the near-Earth plasma using artificial Earth satellites and space probes. On the other hand, new technological possibilities enabling the creation of intense fields in the ionosphere have led to the posing of new problems with regard to the diagnostics of plasma structure and the study of plasma processes caused by the artificial action of *strong pump fields* of powerful radio transmitters.

The first nonlinear effects to be detected in a plasma were apparently found experimentally in the ionosphere more than forty years ago, during signal reception by radio-communication stations, when the phenomenon of radio-wave cross-modulation was discovered. The audio modulation of the transmission from one radio station, which was picked up by reflection of the radiation from the ionosphere, was recorded during reception of another radio station operating on a completely different frequency, but whose emission was reflected from the same region of the ionosphere. This is the so-called Luxemburg–Gorki effect (L'vovich, 1937). Some more intensive theoretical studies of various kinds of nonlinear phenomena in a plasma were apparently also begun about 30 or 40 years ago. The earliest of these were also associated with investigations of the ionosphere (Bailey, 1934, 1938, 1959; Huxley *et al.*, 1948; Fejer, 1955). Theoretical results which were first obtained during studies of nonlinear phenomena in a plasma, or which served as an impetus to the development of new directions of such studies, have been presented or cited in many works. See, for instance Akhiezer & Lyubarskii, 1951; Al'tshul' & Karpman, 1965; Berezin & Karpman, 1964, 1966; Bernstein *et al.*, 1957; Tsytovich, 1967; DuBois & Goldman, 1965, 1967; Gurevich, 1955, 1956a,b; O'Neil, 1965; Silin, 1964, 1965, 1966; Zabusky & Kruskal, 1965; Zakharov & Karpman, 1962; and Gurevich, 1978; Karpman, 1973; Silin, 1973.

The nonlinear effects that may be observed in the outer ionosphere and in the higher regions of the near-Earth plasma can be divided into the following groups. One of these is connected with the variation in plasma temperature caused by a strong radio-frequency (RF) field acting on a *collisional plasma*. These can be called nonlinear phenomena of the *heating type*. Another group of effects is caused by a variation in the plasma properties under the influence of an external source of field; these effects, which are not necessarily associated with heating of the plasma, are also observed in a *collisionless plasma*. They are called nonlinear phenomena of the *parametric type*. One very important class of the nonlinear phenomena observed in the plasma regions of interest to us includes the modulation of wave packets, associated, in particular, with nonlinear Landau damping, the separation of waves into *isolated packets*, and in some cases with the appearance of so-called solitons. The additional pressure on electrons (and sometimes on ions as well) produced in a spatially nonuniform electric field $\mathbf{E}(\mathbf{r})$ also plays an important role in all nonlinear processes. In fact, a number of nonlinear effects are exclusively due to the force exerted by the field pressure. These are usually called nonlinear phenomena of the *striction type*. Of course, these additional pressure forces $\Delta(NkT_e)$ also appear in nonlinear phenomena of the heating and parametric types, in particular due to variations in the temperature and density of the plasma.

7.2. Nonlinear phenomena of the 'heating' type in a collisional plasma

The phenomena in a collisional plasma to be considered in this section pertain to the frequency range in which the effect of an external alternating electric field on the motion of electrons is dominant. In this case a variation of the plasma properties often already manifests itself for relatively low values of the field intensity of the source $\mathbf{E} \simeq \mathbf{E}_0 \exp(i\omega t)$. The ions and the neutral particles present in the plasma in this case participate only in collisions with the electrons. This process is accompained by the transfer of energy from the electrons to the heavy particles, which can be considered as stationary. Here we are interested in conditions when the characteristic time for the energy 'transfer' by electrons (the relaxation time of the temperature T_e of the electrons $\tau_T \simeq (\delta\nu_e)^{-1}$) is much longer than the time of variation of the field $\tau_E \simeq \omega^{-1}$, i.e.,

$$\tau_T \gg \tau_E, \quad \omega \gg \delta\nu_e, \tag{7.1}$$

where δ is the fraction of the energy lost by the electrons during a collision and ν_e is the collision frequency of the electrons. However, the assumption of immobility of the ions means that the ion flux is small compared to the electron flux. This is possible if the following condition is satisfied:

$$\omega^2 \gg (\nu_{en}/\nu_{in})(\omega_H\Omega_H) \simeq (M/m)^{1/2}\omega_L^2, \tag{7.2}$$

where ω_L is the lower-hybrid frequency. The limitations (7.1) and (7.2) are fulfilled, with a great deal to spare, in all the cases of interest to us. Of course, for there to be effects on the ions, considerably stronger electric fields are necessary. Thus it is very difficult to realize such fields experimentally. Nevertheless, it may well be that in certain experiments nonlinear effects were observed in which the motion of the ions was indeed affected, for instance, in cases where instruments aboard Earth satellites recorded plasma oscillations at the lower-hybrid frequency $\omega_L \simeq (\omega_H\Omega_H)^{1/2}$ (see Volume 2), under the influence of high-frequency radio transmitters. In the vicinity of a body the electric fields can be quite strong.

As mentioned above, here it is assumed that the frequency ω of the wave travelling through the plasma $E_0 \exp(i\omega t)$ is higher than all the characteristic frequencies of the plasma $(\Omega_H, \Omega_0, \omega_L)$, associated with the motion of ions. Under these conditions a good measure of whether a field is acting strongly or weakly on a medium is the so-called *characteristic plasma field*, given by the formula

$$E_p = [(3kT_{e0}m\delta_0/e^2)(\omega^2 + \nu_{e0}^2)]^{1/2}. \tag{7.3}$$

The field is taken to be *strong* or *weak*, respectively, if

$$p = E_0/E_p \gtrsim 1 \quad \text{or} \quad p = E_0/E_p \ll 1, \tag{7.4}$$

where E_0 is the amplitude of the wave (Gurevich, 1955, 1956a,b). For the definitions of δ_0 and ν_{e0}, see below.

The value of E_p can be found by equating the mean kinetic energy lost by an electron during a collision to the mean energy imparted to it under steady-state conditions as a result of the action of the alternating electric field. If the interaction of the field with the particle has been established, this relation has the form

$$\tfrac{3}{2}k[T_e(E_0) - T_{e0}]\nu_{e0}\delta(T_e) = e\overline{(\dot{\mathbf{r}}_e \cdot \mathbf{E})} = \frac{e^2E_0^2\nu_e(E_0)}{m[\omega^2 + \nu_e^2(E_0)]}. \tag{7.5}$$

Here the stabilization time for the temperature is

$$\tau_{T_e} = 1/(\nu_{en}\delta_{en} + \nu_{ei}\delta_{ei}), \tag{7.5a}$$

where δ_{en} and δ_{ei} are, respectively, the mean values of the portion of the energy lost by the electron during an inelastic collision with a neutral particle or during an elastic collision with an (Gurevich, 1978). For the ionosphere at heights $z > 100$ to 200 km it can be assumed that $\delta_{en} \simeq 2 \times 10^{-3}$. Note that in a nonisothermal plasma the $(\nu\delta_e)$ terms in (7.5*a*) must have multiplying factors φ_T, the so-called nonisothermality factors. In the plasma regions under review, however, it is usually assumed that $\varphi_T \simeq 1$.

In equation (7.5) and the other expressions above, the following notation is employed. The quantities δ_0 and $\delta(T_e) = \delta[T_e(E_0)]$ correspond to the absence (subscript 0) or presence of a field. The collision frequencies ν_{e0} and $\nu_e(E_0) = \nu_e[T_e(E_0)]$ are defined similarly. The directed velocity $\dot{\mathbf{r}}_e$ imparted to an electron by the electric field $\mathbf{E}$ (see formula (7.5)) has a relaxation time $\tau_e = \nu_e^{-1} \ll \tau_{T_e}$ which is considerably shorter than the stabilization time of the temperature τ_{T_e}. For elastic collisions, which play the main role in the plasma regions of interest here, $\delta_{ei} \simeq \delta_0 \simeq 2m/M$, so if $\nu_{en}\delta_{en} \ll \nu_{ei}\delta_{ei}$, then the relaxation time for the temperature is $\tau_{T_e} \simeq [(2m/M)\nu_{ei}]^{-1}$.

7.2.1. Electron temperature and collision frequency

One of the main consequences of the influence of an electric field on a collisional plasma is that the electron temperature $T_e(E_e)$ depends on the field strength, and thus that the collision frequencies of the electrons $\nu_{ei}(T_e)$ and $\nu_{en}(T_e)$ depend upon the field as well. If the conditions (7.1) are satisfied, when it can be assumed that $T_e(E_0)$ does not depend on time, it follows from (7.5) that for an isotropic plasma ($H_0 = 0$)

$$T_e(E_0) = T_{e0}\left[1 + \frac{\delta_0}{\delta(T_e)}\left(\frac{E_0}{E_p}\right)^2 \frac{(\omega^2 + \nu_{e0}^2)}{[\omega^2 + \nu_e^2(T_e)]} K_\sigma\left(\frac{\omega}{\nu_e}\right)\right] \tag{7.6}$$

(Gurevich, 1958a, 1959). When calculating the temperature, it can be assumed in (7.6) that $\delta(T_e) \simeq \delta_0$ and the value of K_σ can be found from the curve in Fig. 7.9 (see below).

If the main role is played by collisions with neutral particles (see formula (4.6)), then assuming that the effective collision cross-section does not depend on the velocity of the electrons we get the following formula for the electron temperature:

$$T_e(E_0) = T_{e0}\left\{1 + \frac{\omega^2 + \nu_{en0}^2}{2\nu_{en0}^2}\left[\left(1 + \frac{4\nu_{en0}^2}{(\omega^2 + \nu_{en0}^2)}\frac{E_0^2}{E_p^2} K_\sigma\left(\frac{\omega}{\nu_{en0}}\right)\right)^{1/2} - 1\right]\right\} \tag{7.7}$$

(Fig. 7.1). In the plasma regions of interest to us, at heights $z \simeq 200$ to 300 km, the following condition is always satisfied:

$$\omega^2 \gg \nu_{en}^2(E_0). \tag{7.8}$$

Therefore $T_e(E_0)$ depends slightly on the collision frequency and

$$T_e(E_0) = T_{e0}\left\{1 + \frac{e^2E_0^2}{3kTm\delta_0\omega^2} K_\sigma\left(\frac{\omega}{\nu_e}\right)\right\} = T_{e0}\left\{1 + p^2 K_\sigma\left(\frac{\omega}{\nu_e}\right)\right\}. \tag{7.9}$$

The factor $K_\sigma(\omega/\nu_e)$ in formulas (7.6), (7.7), and (7.9) is just a correction factor, obtained from a rigorous kinetic-theoretic study of the corresponding problem. The factor $K_\sigma \simeq 1$ when $\omega^2 \gg \nu_e^2$, and it is a maximum, namely $K_\sigma \simeq 2$, when $\omega^2 \ll \nu_e^2$ and the main role is played by collisions between electrons and ions; $\nu_{ei} \gg \nu_{en}$ (see Fig. 7.9). In practice, since ν_e is always known only approximately and since in addition the condition (7.8) is satisfied, the factor K_σ could be left out of these formulas, except for the gyroresonance region $\omega \simeq \omega_H$ where its role is greater.

The effect of the wave field $E_0 \exp(i\omega t)$ on the collision frequencies ν_{ei} and ν_{en} naturally manifests itself as a change in the electron temperature:

$$\nu_{ei}(E_0) \simeq \nu_{ei0}\left[\frac{T_e(E_0)}{T_{e0}}\right]^{-3/2}, \quad \nu_{en}(E_0) = \nu_{en0}\left[\frac{T_e(E_0)}{T_{e0}}\right]^{1/2}. \tag{7.10}$$

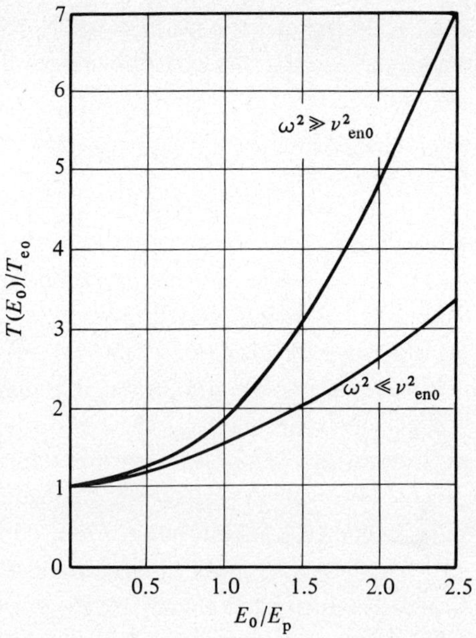

Fig. 7.1. Curves of the ratio of the electron temperature $T_e(E_0)$ in the field of a strong wave (see formulas (7.3) and (7.5)) to the undisturbed temperature T_{e0} in an isotropic plasma. These theoretical curves were calculated taking into account only the collisions ν_{en} of electrons with neutral particles (Gurevich, 1978).

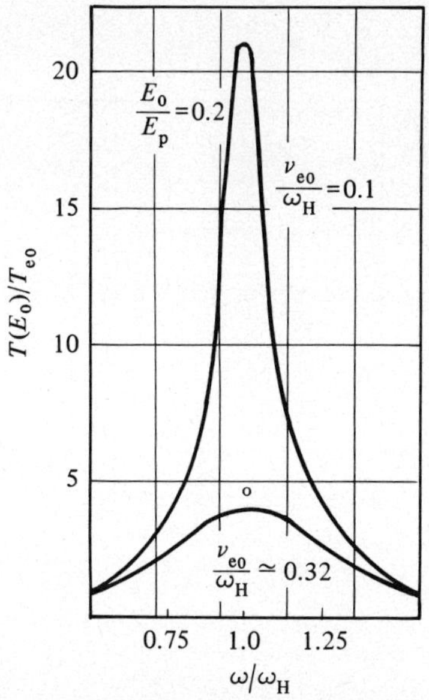

Fig. 7.2. Curves of the electron-temperature ratio $T(E_0)/T_{e0}$ in a magnetoplasma in the vicinity of the electron gyroresonance (Gurevich, 1978).

In a magnetoplasma the temperature variation is especially large near the electron gyroresonance $\omega \simeq \omega_H$. In this frequency range, due to the action of the extraordinary wave,

$$T_e(E_0) = T_{e0}\left\{1 + \frac{e^2 E_0^2 K_\sigma(|\omega - \omega_H|/\nu_e)}{3m\delta_0[(\omega - \omega_H)^2 + \nu_e^2(E_0)]}\right\} \tag{7.11}$$

and the $T_e(E_0)$ relationship has a resonant character. The amount of the increase of $T_e(E_0)$ in the resonance region $\omega \simeq \omega_H$ will be greater, the lower the ratio ν_e/ω_H (Fig. 7.2). As mentioned above, the role of the factor $K_\sigma(\xi = |\omega - \omega_H|/\nu_e)$ is greater in this case.

When collisions between electrons and ions $\nu_{ei}(E_0)$ play the main role, in the frequency range $(\omega - \omega_H)^2 \ll \nu_{ei}^2(E_0)$ the temperature relationship $T_e(E_0)$ exhibits an interesting, important feature (Gurevich, 1959). A similar phenomenon is observed in an isotropic plasma, when $\omega^2 \ll \nu_{ei}^2$ (see (7.7)). This effect becomes possible because the collision frequency $\nu_{ei}(E_0) \sim [T_e(E_0)/T_{e0}]^{-3/2}$ decreases more rapidly than $[T_e(E_0) - T_{e0}]^{-1/2}$. The nature of this effect is illustrated clearly by the curve for $T_e(E_0)/T_{e0}$ as a function of p, which is shown in Fig. 7.3 for $|\omega - \omega_H| \simeq 10^{-2}\nu_{ei}$. In the interval of certain critical values of the field

$$E_{k1} > E_0 > E_{k2} \tag{7.12}$$

the temperature relation $T_e(E_0)$ has three branches. However, only two of these are stable: the lower one and the upper one. As E_0 varies in the

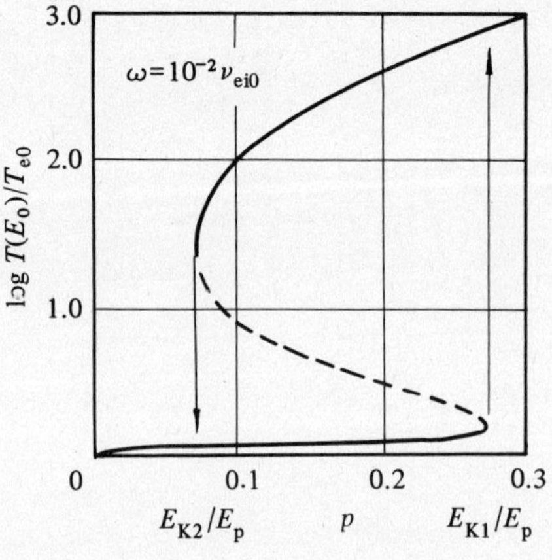

Fig. 7.3. Curve of the electron-temperature ratio in a strong field, in the case where the main role is played by collisions between electrons and ions ν_{ei}. The region of 'hysteresis' of T_e is indicated by the 'critical' values of the field E_{K1} and E_{K2} (see (Gurevich, 1978,) and formulas (7.13) and (7.14)).

interval (7.12), the lower (low-temperature) branch of $T_e(E_0)$ is effective first. For $\omega^2 \ll \nu_{ei0}^2$ the higher critical value of the field E_0 is not frequency dependent:

$$E_{K1} = \frac{2}{3\sqrt{3}} E_p \simeq 0.38 E_p. \tag{7.13}$$

When E_0 reaches the value E_{K1}, the $T_e(E_0)$ relation goes over to the high-temperature branch (the arrow pointing upwards in Fig. 7.3), after which, if E_0 is reduced to

$$E_{K2} = \frac{3^{1/2}}{2^{1/3}} \left(\frac{\omega}{\nu_{ei0}}\right)^{2/3} E_p \simeq 1.37 \left(\frac{\omega}{\nu_{ei0}}\right)^{2/3} E_p \tag{7.14}$$

it returns to the lower branch (the downward arrow). This behavior of the temperature is explained by the fact that for $E_0 > E_{K1}$ the electrons stop imparting to the ions all the energy acquired by them from the field. This is associated with the rapid rise with temperature of the amount of energy $\overline{\dot{\mathbf{r}} \cdot \mathbf{E}} \sim (1/\nu_{ei})(T_e/T_i)^{3/2}$ imparted to the electrons by the field, and the decrease with temperature of the portion of the energy $\delta_0 \nu_{ei}(T_e - T_i) \sim T_e^{1/2}$ imparted to the ions by the electrons. To sum up, with an increase in E_0 there will be a continual rise in the temperature of the electrons, and the state of the plasma becomes unstable. This phenomenon is known as *heating instability* in a collisional plasma. Note here that in an inhomogeneous plasma, in the region where the heating instability takes place, filamentation of the plasma may occur, with linear scales L_N and L_T for the electron density and temperature respectively (see (7.24), Al'pert *et al.*, 1965, and Gurevich, 1978). In a constant field, the heating instability appears only on the low-temperature branch, when

$$E_0 > E_K \simeq E_{K1}/2^{1/2}. \tag{7.15}$$

With an increase in the field intensity, for $E > E_p/\delta^{1/2}$, the mean directed velocity $\dot{\mathbf{r}}_e$ of the electrons gradually becomes greater than their thermal velocity v_e. Ultimately the steady state of the plasma becomes disturbed, since the collision frequency $\nu_{ei}(E_0) \sim (\dot{\mathbf{r}}_e)^{3/2}$ increases rapidly and in a very strong field the role of the collisions decreases markedly. This leads to a continual acceleration of the electrons and to the appearance of the phenomenon of 'run-away' electrons. In an alternating field this phenomenon is absent (Dreicer, 1959; Gurevich, 1960).

In a magnetoplasma ($\mathbf{H}_0 \neq 0$) at frequencies higher than the cutoff frequency ω_- (see (4.25)), where there are three branches of slightly damped high-frequency waves (slow and fast extraordinary waves and a fast ordinary wave, see Figs. 4.2, 4.4, and 4.5), if the condition (7.1) is satisfied the temperature of the electrons in a strong field is

$$T_e(E_0) = T_{e0}\left\{1 + p^2\left[\cos^2\beta + \sin^2\beta(\omega^2 + \nu_e^2)\left(\frac{K_\sigma(|\omega - \omega_H|/\nu_e)}{s(|\omega - \omega_H|)} + \frac{K_\sigma(|\omega + \omega_H|/\nu_e)}{s(|\omega + \omega_H|)}\right)\right]\right\} \quad (7.16)$$

where β is the angle between $\mathbf{E}$ and $\mathbf{H}_0$, while

$$s(|\omega \mp \omega_H|) = 2[(\omega \mp \omega_H)^2 + \nu_e^2(E_0)]. \quad (7.17)$$

When $\omega_H \ll \omega$, the temperature T_e naturally does not depend on ω_H and formula (7.16) becomes (7.6).

In the regions of the near-Earth plasma of interest to us, where $\omega^2 \gg \nu_e^2$, we have $\delta_0 \simeq 7 \times 10^{-5}$ to 10^{-3}, $kT_e \simeq 0.1$ to 10 eV, and the characteristic plasma field

$$E_p \simeq 4.14 \times 10^{-6}[T_{e0}(eV)\delta_0]^{1/2}\omega \simeq (10^{-8} \text{ to } 4 \times 10^{-7})\omega \text{ V m}^{-1}$$

For frequencies $\omega \simeq 3 \times 10^7$ to 10^8 s^{-1}, at which the ionosphere is usually transparent, we obtain

$$E_p \simeq 0.3 \text{ to } 40 \text{ V m}^{-1}.$$

For $E_0 \simeq E_p$ or, for example, when $E_0 \simeq 2E_p$, it follows from (7.6) and (7.10) that

$$T_e(E_0) \simeq 3T_{e0}, \quad \nu_{ei}(E_0) \simeq 0.19\nu_{ei0},$$
$$T_e(E_0) \simeq 5T_{e0}, \quad \nu_{ei}(E_0) \simeq 0.09\nu_{ei0}.$$

Evidently, both the temperatures and the collision frequencies differ substantially from the corresponding values in a weak field, when $E_0 \ll E_p$. For this to be the case, in the indicated frequency range relatively low field intensities are necessary.

Theoretical studies of the electron temperature $T_e(E_0)$ in an anisotropic, spatially inhomogeneous plasma, and of the stabilization of this temperature after turning on or turning off a source of powerful waves, are fraught with difficulty. The appropriate calculations were carried out for a model of an inhomogeneous plasma which closely approximated the lower part of the outer ionosphere in the height interval $z \simeq 200$ to 800 km (Meltz & LeLevier, 1970; Meltz *et al.*, 1974). This study was part of a general program comprising a comprehensive series of experiments to investigate the effects of modifying the ionosphere with the aid of very powerful radio waves (Utlaut, 1970, 1975; Utlaut & Cohen, 1971; Radio Science, 1974). Some results of these experiments and theoretical studies will be presented later in this section, during the description of various types of nonlinear phenomena. Here only data on the temperature will be given.

In the cited works the nonlinear differential equations describing the law of conservation of energy and the ionization balance in the plasma were solved numerically using a perturbation method. Since in the outer ionosphere the magnetic pressure is considerably higher than the gas-kinetic pressure of the plasma (see Table 2.1), in the theoretical model the case is considered where the transfer of particles and energy takes place along the lines of force of the Earth's magnetic field $\mathbf{H}_0$.

Electron heat conduction is neglected. Thus the initial equations become one-dimensional; they are symmetrical with respect to the direction $\mathbf{z}\cos\theta \parallel \mathbf{H}_0$, where θ is the angle between the vertical z, coinciding with the direction of the wave vector $\mathbf{K}$ of a wave incident normally to the ionosphere, and the vector $\mathbf{H}_0$. The inhomogeneity of the ionosphere with respect to the height z is described in these calculations analytically with the aid of a model of the so-called Chapman layer (see, for instance, Al'pert, 1973). Its maximum is located in different cases at heights $z_M F2 \simeq 270$ to 310 km, while the critical frequency $f^{(o)}$F2 varies from 6.3 to 7.8 MHz. In the above-indicated height interval, the undisturbed values of the electron and ion temperatures vary, for the model of the ionosphere adopted in the calculations, within the following limits: $T_e \simeq 800$ to 2800 K and $T_i \simeq 700$ to 1800 K. The energy flux of an incident 'heating' wave is taken to be $W_s = 50\,\mu\mathrm{W\,m^{-2}}$, which corresponds to the beamed power of the radio transmitter used in the given experiments: $W_h = 10^8$ W. Calculations were made for both an ordinary wave and extraordinary wave, i.e., for, respectively, clockwise or counterclockwise rotation of the electric-field vector E_0 in the heating region, if this is in the northern hemisphere.

Fig. 7.4 shows curves of the relative variation in the electron temperature $\delta T_e(z)/T_{e0}$ as a function of height along the direction of maximum heating. The curves were obtained for the ordinary wave, for the time intervals $\Delta t = 10, 30$, and 110 s after turning on a powerful wave (Meltz & LeLevier, 1970). For $\Delta t = 110$ s under the given conditions a steady state of the profile $\delta T_e(z)$ is established. Fig. 7.5 shows the curves of equal values of $\delta T_e(x, z)/T_{e0}$ for this steady state, in the vertical plane of the magnetic meridian (the xz plane). The regions of maximum heating of the ionosphere are seen to be separated approximately by a horizontal distance of 60 km. Also the $\delta T_e(x, z)/T_{e0}$ curves are elongated along the vector $\mathbf{H}_0$. This is because the trajectory of an ordinary wave incident normally on the ionosphere deviates in it somewhat northward from the vertical in the plane of the magnetic meridian, while the extraordinary wave deviates

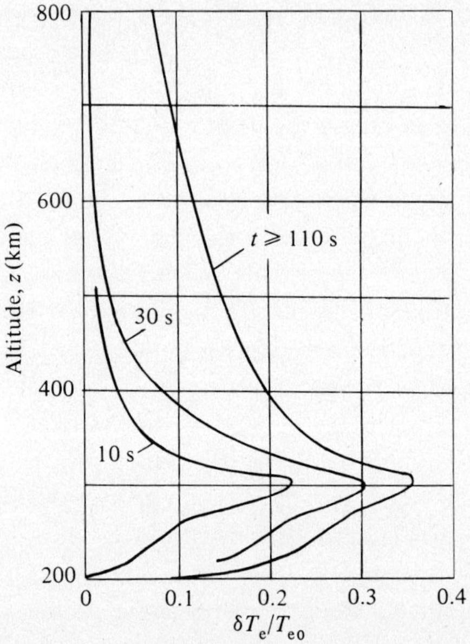

Fig. 7.4. Height dependence of the disturbance of electron temperature $\delta T_e/T_{e0}$ in the field of a strong ordinary wave. The three curves pertain to different intervals of time after turning on a powerful radio transmitter with frequency $f_{0h} = 7.02$ MHz. The critical frequency of the ionosphere is $f^{(o)}F2 = 7$ MHz; the height of the maximum is $z_M F2 = 310$ km (Meltz & LeLevier, 1970).

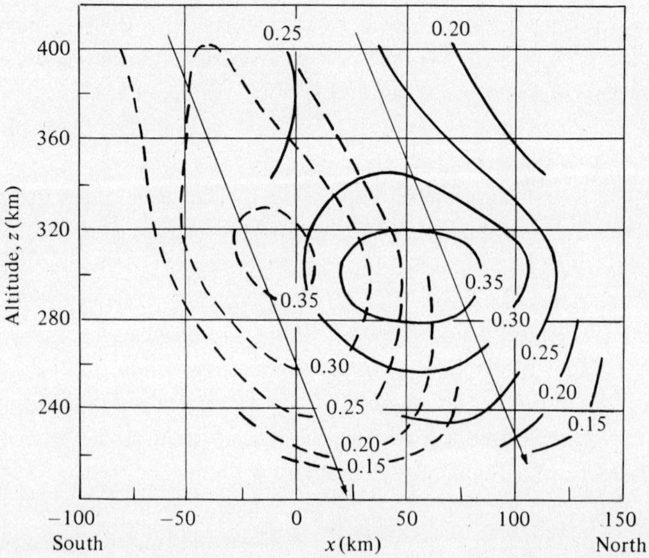

Fig. 7.5. Curves of equal values of the disturbance of the electron temperature in the plane of the magnetic meridian (xz plane) after $\Delta t = 110$ s. The solid and dashed curves indicate, respectively, the values of $\delta T_e(x, z)/T_{e0}$ resulting from the action on the ionosphere of powerful ordinary or extraordinary waves (Meltz & LeLevier, 1970).

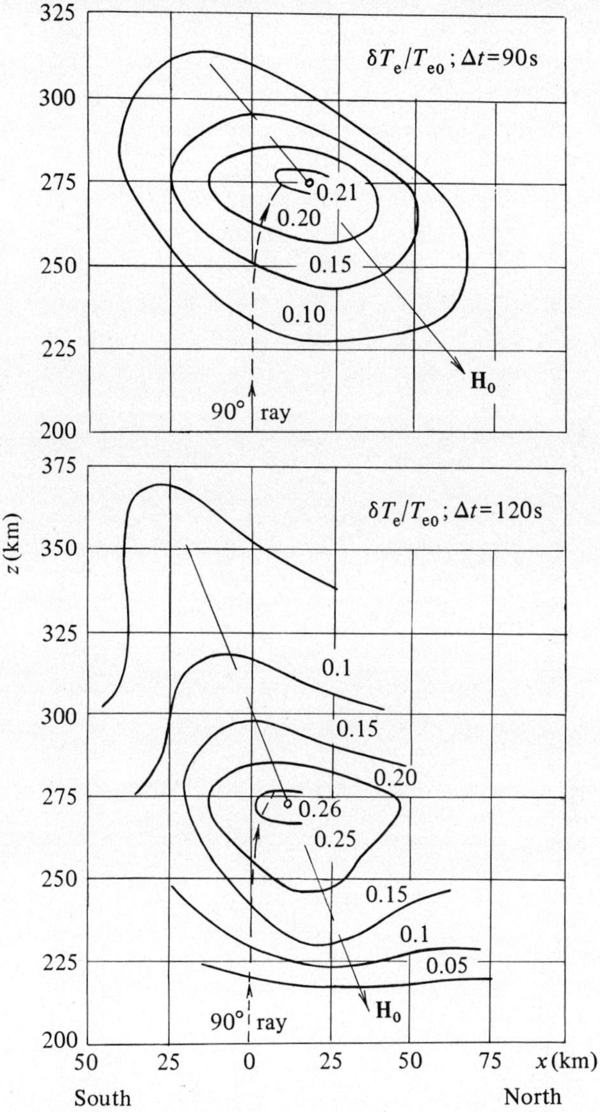

Fig. 7.6. Curves of equal values of the perturbation of the electron temperature $\delta T_e(x, z)/T_{e0}$ in the plane of the magnetic meridian (xz plane) at times $\Delta t = 90$ and 120 s after the start of the action of a strong ordinary wave, with frequency $f_{0h} = 5.94$ MHz ($f^{(o)}$F2 = 6.24 MHz, z_MF2 = 280 km (Meltz *et al.*, 1974).

somewhat southward, the trajectory of the ordinary wave being bent considerably more than that of the extraordinary wave. The data in Figs. 7.4 and 7.5 were obtained for a height of the maximum of the F2 region of the ionosphere $z_M = 310$ km and for ratios of the frequency of the heating radiation f_h to the critical frequencies of the ordinary wave $f^{(o)}F2$ and extraordinary wave $f^{(x)}F2$, respectively, of $f_h/f^{(o)}F2 \simeq 1.003$ and $f_h/f^{(x)}F2 \simeq 1.014$.

Figure 7.6 illustrates the nature of the temporal variation of the contours of $\delta T_e(x, z)/T_{e0}$ in the plane of the magnetic meridian. These curves were calculated for $z_M = 280$ km and $f_h/f^{(o)}F2 \simeq 0.95$. Under the given conditions, after about 120 s a steady state of the temperature is already established. For $\Delta t = 90$ s on the other hand, the spatial distribution of $\delta T_e(x, z)$ is still considerably different. This pattern, namely a relatively

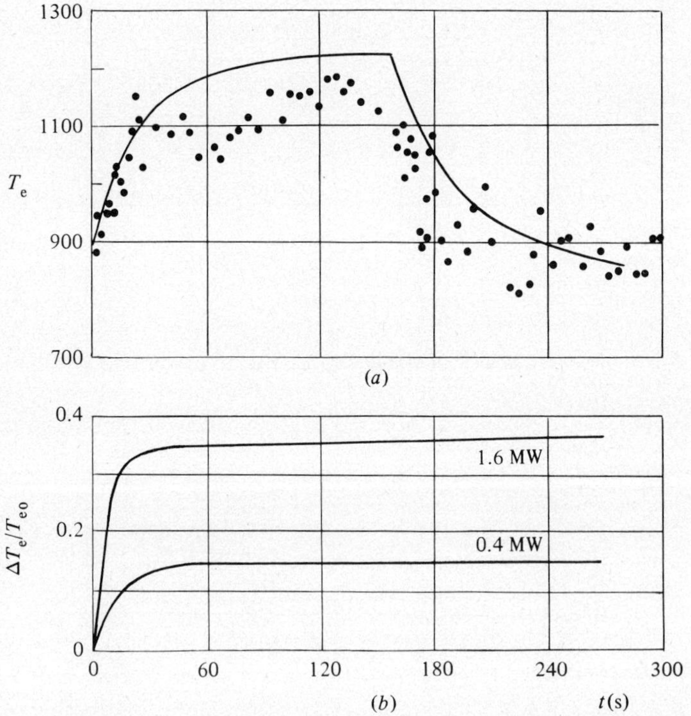

Fig. 7.7. (a) Theoretical (solid curve; Meltz *et al.*, 1974) and experimental (points; Gordon & Carlson, 1974) dependences of the temperature T_e on time in the interval $t = 0$ *to* 150s as a result of the action of a strong radio wave, and also for $t > 150$ s after turning off the transmitter. (b) Theoretical dependences of the relative perturbation $\delta T_e(t)/T_{e0}$ on time for two values of the transmitter power W_h. The model of the ionosphere is the same as in Fig. 7.6 (Meltz *et al.*, 1974).

rapid saturation of the electron temperature of the ionosphere $T_e(t)$ under the influence of an RF field, is verified by experimental data. This follows from Fig. 7.7*a*, which shows the theoretical curve of $T_e(t)$ calculated for the time interval (t_0 to t_h), from the time $t_0 = 0$ of the turning on of the source (the beginning of the *heating* of the plasma) to the time $t_h = 150$ s when it is turned off, as well as for $t > t_h$, the time of *cooling* of the plasma (Meltz *et al.*, 1974). Fig. 7.7*a* also shows the results of measurements of $T_e(t)$ at Arecibo using the incoherent-scattering method (Gordon & Carlson, 1974). These data, like the theoretical relations for $\delta T_e(t)/T_{e0}$, calculated for different source intensities W, when the heating wave is the ordinary wave and $f_h/f^{(o)}F2 = 0.95$ (Fig. 7.7*b*; Rao & Thome, 1974), indicate that the dependence of the temperature on time is quite nonlinear. As time goes by, T_e first increases rapidly and then reaches saturation, becoming stationary at the level $T_{e,stat}$. Thus the function $T_e(t)$ in the time intervals $t_0 \leqslant t \leqslant t_h$ and $t > t_h$, respectively, during which the plasma is first heated and then cools down (after the powerful radio-wave transmitter is turned off), can be described adequately by the following formulas:

$$\frac{T_{e,stat} - T_e(t)}{T_{e,stat}} \simeq 1 - \exp\left(-\frac{t}{\tau_T}\right), \quad t = (t_0 - t_h),$$
$$\frac{T_{e,stat} - T_e(t)}{T_{e,stat}} \simeq \exp\left(-\frac{t}{\tau_T}\right), \quad t > t_h, \tag{7.18}$$

where τ_T is the relaxation time of the temperature (see 7.1) and (7.5*a*), (Gurevich, 1978). The corresponding curves calculated by Gurevich show a good fit with the experimental points in Fig. 7.7.

As the intensity of the wave that is heating the plasma is increased, the plasma temperature is seen to vary quite smoothly (Fig. 7.8*a*). The experimental height dependences $T_e(z)$ for various values of W_h (Fig. 7.8(*b*); Medved, 1969) also demonstrate that T_e rises uniformly as the intensity W_h of the heating wave increases. It should be mentioned here that the above-indicated calculations of the large-scale temperature variations $T_e(E, t)$ were carried out without taking into account the self-focusing effect, which appears in the heated region of the plasma (see below). The oscillations of $T_e(t)$ which were observed experimentally are, in the opinion of the authors of (Meltz *et al.*, 1974), a consequence of this effect.

(*a*) *Tensor elements.* In a number of cases a strong electric field can affect considerably the *refractive index n* and *attenuation coefficient* κ of a plasma. Thus the results of kinetic calculations should be used to determine

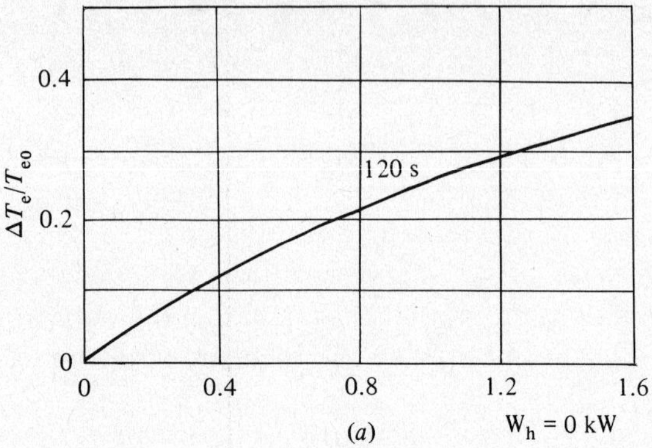

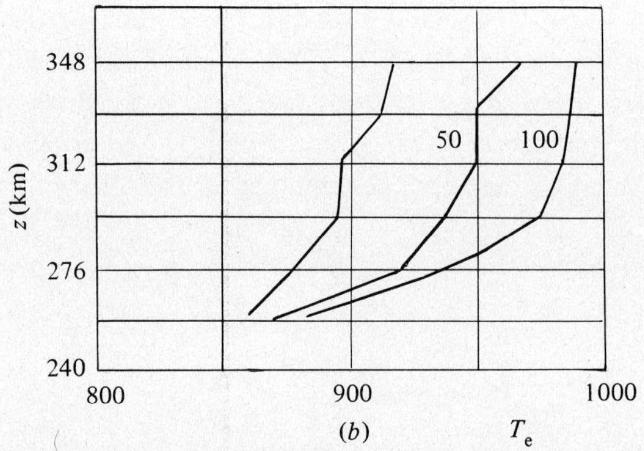

Fig. 7.8. (*a*) Theoretical dependence of $\delta T_e/T_{e0}$ on the power W_h (in megawatts) of a strong wave 120 s after the transmitter starts operating (Meltz *et al.*, 1974). (*b*) Experimental height dependence of the electron temperature, obtained for various powers W_h of a wave disturbing the ionosphere at the frequency $f_h^{(o)} = 5.1$ MHz under conditions where $f^{(o)}$F2 = 6 MHz (Gordon & Carlson, 1974).

these. The quantity $(n - i\kappa)$ is then expressed more conveniently in terms of the complex tensor elements: $\varepsilon_s - i\sigma/\varepsilon_0\omega = \varepsilon_s^*$. In an isotropic plasma the relation between ε^* and $(n - i\kappa)$ has the simple form

$$\left.\begin{aligned} (n - i\kappa)^2 &= \varepsilon - i\sigma_s/\varepsilon_0\omega, \\ (n^2 - \kappa^2) &= \varepsilon, \quad 2n\kappa = \sigma/\varepsilon_0\omega, \end{aligned}\right\} \tag{7.19a}$$

where, taking the effect of the field E_0 into account, kinetic calculations give

$$\varepsilon = 1 - \frac{\omega_0^2}{\omega^2 + \nu_e^2(E_0)} K_\varepsilon\left(\frac{\omega}{\nu_e}\right), \quad \frac{\sigma}{\varepsilon_0\omega} = \frac{\omega_0^2 \nu_e(E_0)}{[\omega^2 + \nu_e^2(E_0)]} K_\sigma\left(\frac{\omega}{\nu_e}\right). \tag{7.19b}$$

For a magnetoplasma, n_{12} and κ_{12} are determined with the aid of the general formulas (4.2) and (4.3), in which the complex tensor elements have the following form:

$$\left.\begin{aligned}
\varepsilon_1^* &= \varepsilon_1 - \mathrm{i}\frac{4\pi\sigma_1}{\omega} \\
&= 1 - \frac{\omega_0^2}{\omega}\left[\frac{(\omega-\omega_H)K_\varepsilon\left(\frac{|\omega-\omega_H|}{\nu_e}\right)}{s(|\omega-\omega_H|)} + \frac{(\omega+\omega_H)K_\varepsilon\left(\frac{|\omega+\omega_H|}{\nu_e}\right)}{s(|\omega+\omega_H|)}\right] \\
&\quad - \mathrm{i}\frac{\omega_0^2\nu_e}{\omega}\left[\frac{K_\sigma\left(\frac{|\omega-\omega_H|}{\nu_e}\right)}{s(|\omega-\omega_H|)} + \frac{K_\sigma\left(\frac{|\omega+\omega_H|}{\nu_e}\right)}{s(|\omega+\omega_H|)}\right],
\end{aligned}\right\} \qquad (7.20)$$

$$\left.\begin{aligned}
\varepsilon_2^* &= \varepsilon_1 - \mathrm{i}\frac{4\pi\sigma_2}{\omega} \\
&= \frac{\omega_0^2}{\omega}\left[\frac{(\omega-\omega_H)K_\varepsilon(|\omega-\omega_H|/\nu_e)}{s(|\omega-\omega_H|)} - \frac{(\omega+\omega_H)K_\varepsilon(|\omega+\omega_H|/\nu_e)}{s(|\omega+\omega_H|)}\right] \\
&\quad - \mathrm{i}\frac{\omega_0^2\nu_e}{\omega}\left[\frac{K_\sigma(|\omega-\omega_H|/\nu_e)}{s(|\omega-\omega_H|)} - \frac{K_\sigma(|\omega+\omega_H|/\nu_e)}{s(|\omega+\omega_H|)}\right],
\end{aligned}\right\} \qquad (7.21)$$

$$\varepsilon_3^* = \varepsilon_3 - \mathrm{i}\frac{4\pi\sigma_3}{\omega} = \left[1 - \frac{\omega_0^2 K_\varepsilon(\omega/\nu_e)}{\omega^2 + \nu_e^2(E_0)}\right] - i\left[\frac{\omega_0^2\nu_e(E_0)K_\sigma(\omega/\nu_e)}{\omega[\omega^2 + \nu_e^2(E_0)]}\right]. \qquad (7.22)$$

In formulas (7.20)–(7.22) the quantities $s(\ldots)$ are defined by formulas (7.17), while the dependences of K_ε and K_σ on the dimensionless parameters $\xi = \omega/\nu_{ei}$, $|\omega - \omega_H|/\nu_{ei}$, $|\omega + \omega_H|/\nu_{ei}$ are given in Fig. 7.9 (see Fried, 1959).

At frequencies quite far away from gyroresonance, where $K_\varepsilon \simeq K_\sigma \simeq 1$ and $(\omega - \omega_H)^2 \gg \nu^2$, the complex tensor elements are

$$\left.\begin{aligned}
\varepsilon_1^* &= 1 - \frac{\omega_0}{\omega^2 - \omega_H^2}\left[1 + \mathrm{i}\frac{\nu_e(E_0)}{\omega}\frac{(\omega^2 + \omega_H^2)}{(\omega^2 - \omega_H^2)}\right], \\
\varepsilon_2^* &= \frac{\omega_H\omega_0^2}{\omega(\omega^2 - \omega_H^2)}\left[1 + \mathrm{i}\frac{2\nu_e(E_0)\omega}{(\omega^2 - \omega_H^2)}\right], \\
\varepsilon_3^* &= \left[1 - \frac{\omega_0^2}{\omega^2}\right] - \mathrm{i}\left[\frac{\omega_0^2}{\omega^2}\frac{\nu_e(E_0)}{\omega}\right].
\end{aligned}\right\} \qquad (7.23)$$

For $E_0/E_p \ll 1$, the formulas (7.23) naturally become (4.8) and (4.9), if in the latter the effect of the ions is neglected; in a collisionless plasma, the formulas (7.20)–(7.22) become (4.1).

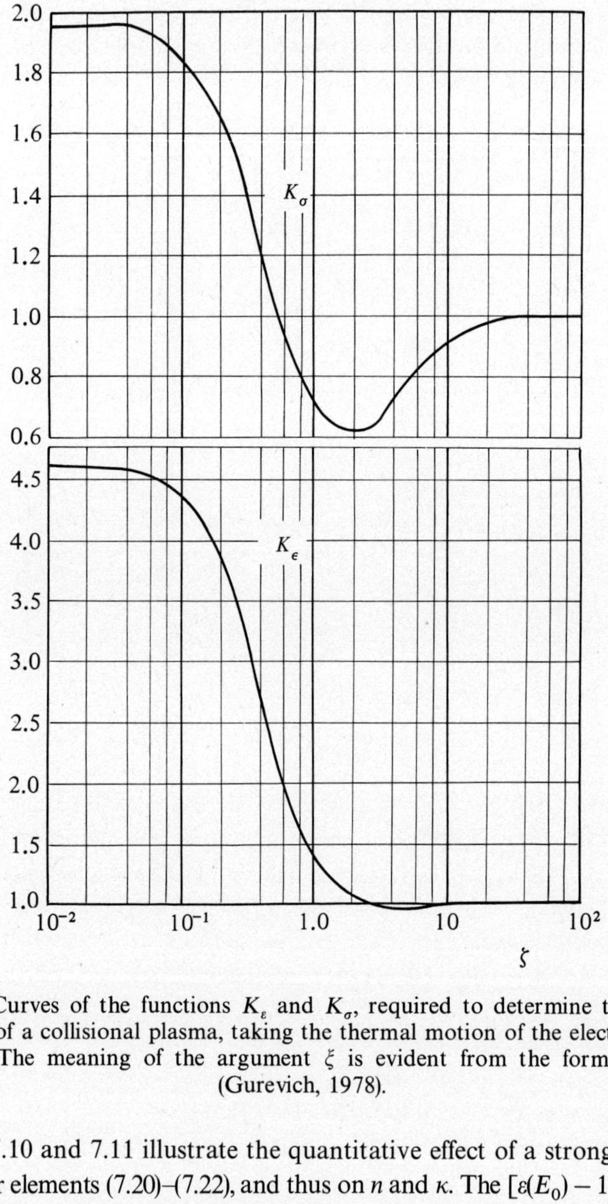

Fig. 7.9. Curves of the functions K_ε and K_σ, required to determine the tensor elements of a collisional plasma, taking the thermal motion of the electrons into account. The meaning of the argument ξ is evident from the formula (7.20) (Gurevich, 1978).

Figs. 7.10 and 7.11 illustrate the quantitative effect of a strong field on the tensor elements (7.20)–(7.22), and thus on n and κ. The $[\varepsilon(E_0)-1]/(\varepsilon_0-1)$ and $\sigma(E_0)/\sigma_0$ relations were plotted taking into account only the collision frequency ν_{en} for electrons with neutral particles (Gurevich, 1955, 1956a,b). The curves in Fig. 7.10 pertain to an isotropic plasma ($H_0=0$) (see (7.19)), while the $\sigma_1(E_0)/\sigma_{10}$ relations in Fig. 7.11 ((7.20)–(7.22)) were calculated for the values $p=10^2$ and 10^6 in an anisotropic plasma ($\mathbf{H}_0 \neq 0$), for the

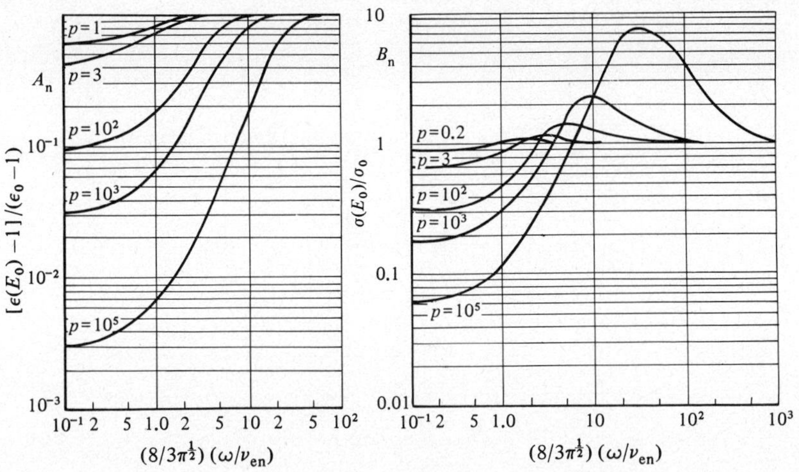

Fig. 7.10. Curves of the permittivity $\varepsilon(E_0)$ and of the conductivity $\sigma(E_0)$ of an isotropic plasma, taking into account the effect of an electric field ($p = E_0/E_p$, see formulas (7.3). and (7.5)).

two limiting cases $\beta = 0$ and $\beta = \pi/2$ (see (7.16)). It is seen that in a strong field the tensor elements can vary by as much as two or three orders of magnitude, depending on the value of ω/ν_{en}. For $\beta = \pi/2$, when the extraordinary wave becomes circularly polarized, the tensor element σ_1 exhibits sharply resonant behavior (see above, (7.11) and Fig. 7.2).

(*b*) *Some conclusions.* The heating of the plasma, together with the consequent variation in the frequency of the collisions between electrons and other particles, naturally alters the absorption of electromagnetic waves propagating in the plasma. In the plasma regions of interest to us here, where the main role is played by electron–ion collisions and $\omega^2 \gg \nu_{ei}$, the effect of a strong wave tends to reduce the absorption coefficient $\kappa_e \simeq (\nu_{ei}/\omega)(\omega_0^2/\omega^2)$, since the temperature $T_e(E_0)$ rises (see formulas (7.10) and (7.6)). Ultimately the heating wave itself is less attenuated – it appears to *pave a way* for itself through the plasma. In turn, the nature of its amplitude and phase modulation is changed, i.e., the information carried by the wave becomes distorted, and its propagation velocity (i.e. the phase velocity of the wave) is altered as well, since there is a corresponding change in the refractive index. Consequently, some new effects are observed: *self-modulation*, *self-action*, and *cross-modulation* of electromagnetic waves propagated in the plasma, partly also due to a violation of the additivity of the wave processes taking place in the plasma.

The heating of the plasma also changes the nature of the *ionization*

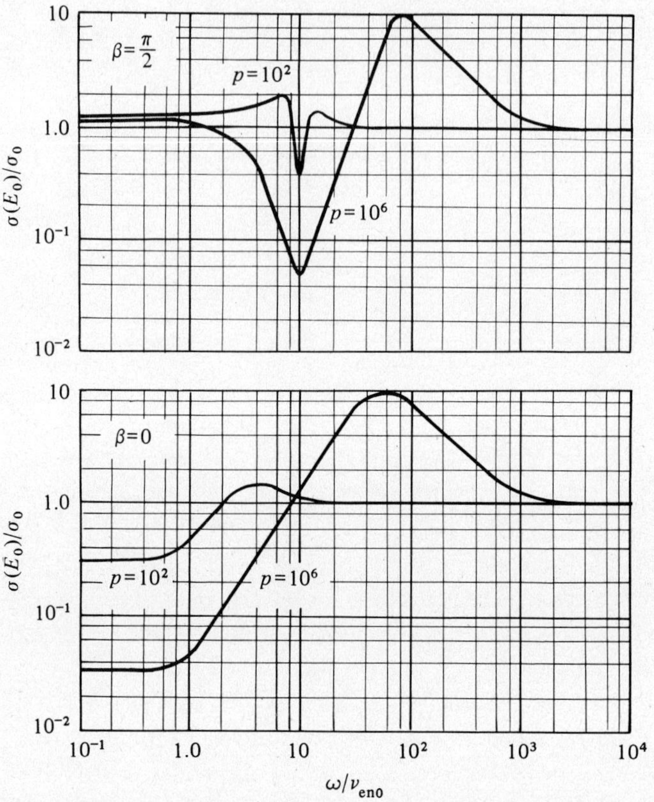

Fig. 7.11. Curves of the conductivity $\sigma(E_0)$ of a magnetoplasma, taking into account the effect of an electric field ($p = 10^2$ and 10^6) for two values of the angle $\beta = 0$ and $\pi/2$ between the electric field $\mathbf{E}_0$ of a strong wave and the constant magnetic field $\mathbf{H}_0$ ($\omega_H/\nu_{en} = 10$, $p = E_0/E_p$ see (7.3) and (7.5)).

balance (recombination of electrons) and of the *transport of particles* (diffusion and heat conduction). This in turn alters the spatial distribution of the electron density $N(x, y, z)$. The region of intense heating of the plasma and of maximum release of the energy of the heating wave becomes inhomogeneous. Thus a large-scale inhomogeneity appears in the plasma, the linear dimensions of which may, as will be seen below, reach many tens of kilometers. As nonlinear processes develop in this region, elongated irregularities of various scales are created. The spectrum of their linear sizes varies from fractions of a meter or several meters (Debye length D_e, wavelength λ) to hundreds of meters or several kilometers. The maximum size of an irregularity is determined by the rate of diffusion of electrons along $\mathbf{H}_0$. In turn, there will also be an inverse effect of the structural changes of the plasma on the properties of waves propagated through it. *Focusing of the radiation* appears, and *beams of electromagnetic waves* are produced. The distribution of the field becomes very nonuniform.

These processes are accompanied by an increased *instability* of the plasma, which in turn leads to the formation of different types of inhomogeneities and waves in the plasma. Inhomogeneities tend to develop particularly intensely in the vicinity of a plasma resonance. There is also an enhancement of the parametric instability inherent to a collisionless plasma, which may lead, for instance, to *decay* of the wave into a plasma wave and an ion-acoustic wave. Some oscillations peculiar to the heating nonlinearity are excited as well. It is, however, quite difficult to describe theoretically this entire complex of nonlinear phenomena in a collisional plasma. Such a description requires the successive consideration of a number of particular cases. An idea of the nature of the observed phenomena can often be arrived at only via a numerical analysis of the corresponding equations, which is in general the case for most nonlinear problems in a plasma. The following subsections will describe briefly some results of investigations of the indicated effects of the *heating type.*

7.2.2. Large-scale changes in electron density

The disturbance of the electron density $\delta N(E)/N_0$ due to the effect of the field $\mathbf{E}$, taking into account the equation of ionization balance and including transport phenomena, can be described in the case under review, where $\omega^2 \gg \nu_{ei}^2$ and $\nu_{ei}^2 \gg \nu_{en}^2$, using formulas containing the characteristic lengths

$$L_N = (D_A \tau_N)^{1/2}, \quad L_T = \left(\frac{\kappa_e \tau_T}{N_{e0}}\right)^{1/2} = (D_T \tau_T)^{1/2} \tag{7.24}$$

and, more generally than p^2 (see (7.4)), the quantity

$$C_p = p^2 \frac{\phi_p}{\phi_T} = \left(\frac{E_0}{E_p}\right)^2 \frac{\phi_p}{\phi_T}, \tag{7.25}$$

which is a measure of the strength of the applied field (see (7.5), Gurevich, 1967, 1972).

The meanings of the quantities defined in formulas (7.24) and (7.25) are as follows. The length L_N, which is the characteristic linear scale of the irregularities of the electron density, depends on the diffusion of the particles and on the ionization of the plasma (i.e. on the establishment of its ionization balance). Here D_A is the coefficient of ambipolar diffusion, while

$$\tau_N = \frac{\partial}{\partial N}(\alpha N^2)^{-1} = \frac{1}{\alpha}\left(\frac{1}{N} - \frac{1}{N_0}\right) \simeq \frac{1.7}{\alpha_{eff} N_0} \tag{7.24a}$$

is the recombination time, which is so defined in (7.24) that after τ_N we have $N = N_0 \exp(-1)$ (in (7.24*a*) α_{eff} is the effective recombination coefficient). The length L_T is a characteristic linear scale of the irregularity which depends on the process of establishment of the plasma temperature; τ_T is the relaxation time of the electron temperature (see (7.1)), and $\kappa_e = D_T N$ is the electron heat conductivity.

In general, the quantities D_A and D_T are tensors, and their formulas are quite complicated. However, in order to interpret the experimental data to be considered below, it is often sufficient to know just their longitudinal (along $\mathbf{H}_0$) or transverse components. These are, respectively,

$$\left.\begin{aligned} &D_{A\parallel} \simeq v_i^2/\nu_{in}, \quad D_{T\parallel} \simeq v_e^2/\nu_{ei}, \\ &D_{A\perp} \simeq v_i^2 \nu_{en}^2/\Omega_H \omega_H \nu_{ei}, \quad D_{T\perp} \simeq \tfrac{3}{2}\nu_{ei} v_e^2/\omega_H^2. \end{aligned}\right\} \tag{7.24b}$$

When writing the formulas (7.24*b*), it was assumed that $\nu_{ei}^2 \gg \nu_{en}^2$.

It is easy to see that the coefficient of longitudinal ambipolar diffusion is meaningless if ion–neutral collisions are absent; thus $D_{A\parallel} \to \infty$ if $\nu_{in} \to 0$. In this case the problem of particle transport requires special study.

The factor ϕ_p in (7.25) depends on the type of polarization of the wave propagating in the plasma. For simplicity, let us present a formula for the case in which the angle β between the field $\mathbf{E}$ and the magnetic-field vector $\mathbf{H}_0$ is $\pi/2$. In this case

$$\phi_p = \frac{\omega^2}{2}\left[\frac{1}{(\omega - \omega_H)^2 + \nu_{ei}^2(E)} + \frac{1}{(\omega + \omega_H)^2 + \nu_{ei}^2(E)}\right] \tag{7.26}$$

and, if $\omega \gg \omega_H, \nu_{ei}$, then the factor $\phi_p \simeq 1$. The factor ϕ_T in (7.25) characterizes the nonisothermal nature of the plasma. With sufficient accuracy, it can be assumed for the F2 region of the ionosphere of interest to us here that $\phi_T \simeq 1$.

In the works cited above (Gurevich, 1967), calculations were made of the perturbation of the electron density in the ionosphere along the axis of a symmetrical beam of waves incident upon it, the field of which possesses at the base of the ionosphere a uniform Gaussian distribution:

$$E_0^2(x, y, z) = E^2(y, z)\exp(-x^2/d^2), \tag{7.27}$$

where $d = \theta_0 z/\sin\theta$ characterizes the width of the beam at a height z in the x direction; θ_0 is a small angle defining the degree of directivity of the transmitter (a measure of the gain of the transmitting antenna), and θ is the angle between the z-axis and the vector $\mathbf{H}_0$. Such a distribution of the field describes closely the results of experiments in which beams of waves directed toward the ionosphere were used to intensify the field in it. For $\theta \simeq \pi/2$, i.e., when the axis of the beam is normal to the magnetic field, we can use the following formulas for the perturbation of the electron density $\Delta N(z)$ as a function of height (see Gurevich, 1978).

Respectively, for

$$\left.\begin{aligned}
&(1)\quad L_T<d,\quad L_N<d,\quad \Delta N(z) = -N_0\left[2\kappa_T C_p\left(\frac{L_N}{d}\right)^2\right],\\
&(2)\quad L_T>d,\quad L_N<d,\quad \Delta N(z) = -N_0 C_p \kappa_T\left(\frac{L_N}{L_T}\right)^2\\
&\qquad\qquad \times\left[1-\frac{\pi^{1/2}}{2}\left(\frac{d}{L_T}\right)-2\left(\frac{L_N}{d}\right)^2\right],\\
&(3)\quad L_T<d,\quad L_N>d,\quad \Delta N(z) = -N_0 C_p \kappa_T\left[1-2\left(\frac{L_T}{d}\right)^2-\frac{\pi^{1/2}}{2}\left(\frac{d}{L_N}\right)\right],\\
&(4)\quad L_T>d,\quad L_N>d,\quad \Delta N(z) = -N_0\left[\frac{\pi^{1/2}L_N}{(L_N+L_T)}-\kappa_T\frac{d}{L_T}\right]\frac{C_p d}{2L_T},
\end{aligned}\right\} \tag{7.28}$$

where κ_T, the so-called thermal-diffusion ratio, is defined as

$$\kappa_T = T_{e0}/(T_e + T_{i0}). \tag{7.28a}$$

Table 7.1 gives a model of the outer ionosphere and the values of the quantities necessary to determine the relative perturbation of the electron density $\Delta N(z)/N_0$, in the daytime and in the middle latitudes, which are close to those used in the theoretical discussion to be presented below. Note that for $z > 400$ km the characteristic length gradually loses its significance: it becomes considerably larger than the plasma region itself. This is because here, as noted previously, the definition of the coefficient of ambipolar diffusion no longer applies, since ν_{in} decreases rapidly.

The height dependences $(1/C_p)(\Delta N(z)/N_0) = (1/p^2)(\phi_T/\phi_p)(\Delta N(z)/N_0)$ obtained in Gurevich, 1967, for $z \gtrsim 200$ km are shown in Fig. 7.12. It is seen that, in the region of the ionosphere being considered, where $\nu_{ei}^2 \gg \nu_{en}^2$, the effect of a strong wave is to reduce the electron density, and the plasma seems *to be ejected* from this region by the field $\mathbf{E}$. Clearly, this is a case of a large-scale inhomogeneous formation: a rarefied cloud

Table 7.1. *Characteristics of the outer ionosphere*

z(km)	200	300	400	600	800	1000
$N(\text{cm}^{-3})$	5×10^5	2×10^6	1×10^6	3×10^5	1×10^5	3×10^4
$N_n(\text{cm}^{-3})$	1.2×10^{10}	2×10^9	3×10^8	2×10^7	2×10^6	4×10^5
T_e°	1000	1500	1600	1700	1900	2300
$M_0 = \bar{M}_i/M(H_1)$	22	16	10	3	2	1.5
$v_e(\text{cm s}^{-1})$	1.7×10^7	2×10^7	2.2×10^7	2.3×10^7	2.4×10^7	2.6×10^7
$v_i(\text{cm s}^{-1})$	8×10^4	1.2×10^5	1.6×10^5	3×10^5	4×10^5	5×10^5
$\nu_{ei}(\text{s}^{-1})$	700	1500	700	200	60	14
$\nu_{en}(\text{s}^{-1})$	140	30	4	0.3	3×10^{-2}	7×10^{-3}
$\nu_{in}(\text{s}^{-1})$	4	0.9	0.2	2×10^{-2}	3×10^{-3}	8×10^{-4}
$\alpha_{eff}(\text{cm}^3\,\text{s}^{-1})$	10^{-8}	2×10^{-10}	2×10^{-11}	10^{-12}	10^{-12}	10^{-12}
$\tau_N(\text{s})$	340	4×10^3	8×10^4	5×10^6	1.7×10^7	5×10^7
$\tau_T(\text{s})$	3	6	12	14	30	88
$D_{A\parallel}(\text{cm}^2\,\text{s}^{-1})$	1.6×10^9	1.5×10^{10}	1.2×10^{11}	4.5×10^{12}	5×10^{13}	3×10^{14}
$D_{T\parallel}(\text{cm}^2\,\text{s}^{-1})$	4×10^{11}	2.7×10^{11}	7×10^{11}	2.6×10^{12}	9.6×10^{12}	4.8×10^{13}
L_N(km)	7.4	77.5	970	4.7×10^4	$\simeq 3 \times 10^5$	$\simeq 10^6$
L_T(km)	11	13	29	60	170	650

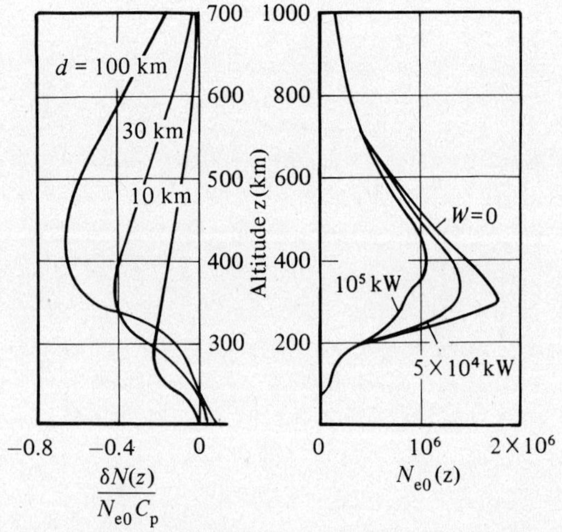

Fig. 7.12. Theoretical height profiles $N(z)$ of electron density in the ionosphere for three values of the transmitter power W and curves of the height dependence of the perturbation of electron density $\delta N(z)/N_{e0}C_p$ for three widths of a beam of strong waves (parameter d) Gurevich, 1967).

of plasma of complex form. At lower altitudes, on the other hand, where electron–neutral collisions play a role, the electron density increases. The corresponding data are given in Gurevich, 1967. Figure 7.12 also shows calculated height dependences of the electron density in the ionosphere taking into account the weakening of the field of the wave as it moves through the plasma. The transmitter, located at the Earth's surface, was assumed to have an equivalent dipole power which for an antenna directivity $\theta_0 = 0.1$ radians was equal to W_E. This means that the mean transmitter power $W_0 \simeq (D^2/\pi)W_E \simeq 3.2 \times 10^{-3} W_E$.

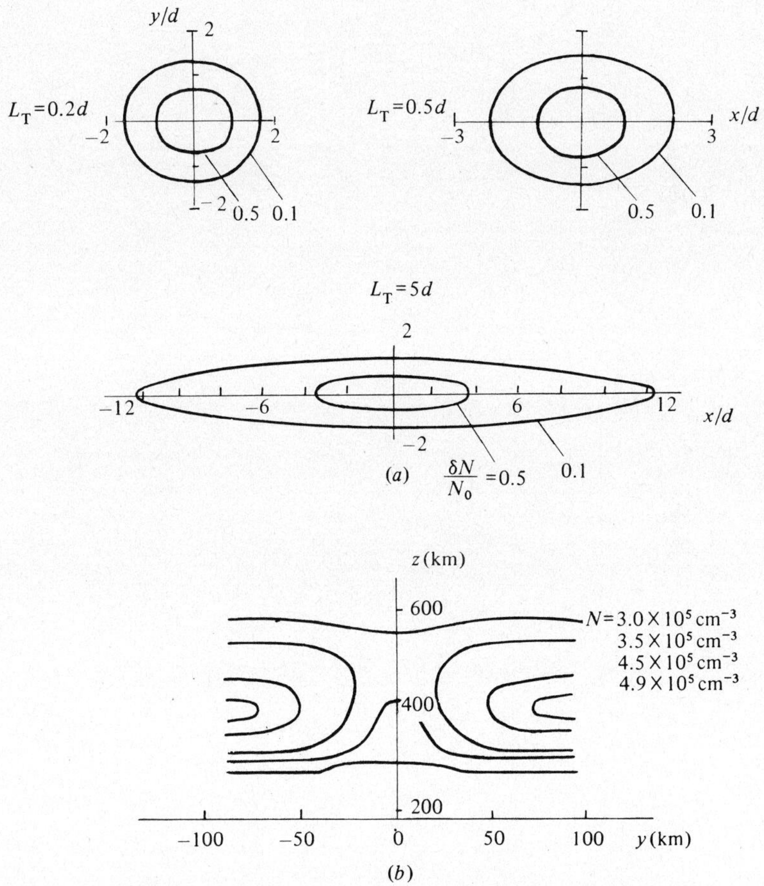

Fig. 7.13. (*a*) Theoretical curves of equal values of the perturbation of electron density $\delta N(xy)/N_0$ in the horizontal (xy) plane (see text). (*b*) Theoretical curves of equal values of electron density $N(y, z)$ in the vertical (yz) plane. Transmitter power $W_h = 5$ MW; the angular width of the wave beam $\theta_0 = 0.1$ radians (Gurevich, 1967).

Fig. 7.13 shows the variation of the distribution of the electron density in the vertical (yz) and horizontal (xy) planes under the influence of a strong wave. In the xy plane the disturbance $\delta N(x, y)$ extends symmetrically along the vector $\mathbf{H}_0$, which is directed along the x-axis and makes an angle $\pi/2$ with the upward vertical. The lines of equal values of $\delta N(x, y)$ show that the large-scale inhomogeneity has a quite complicated shape.

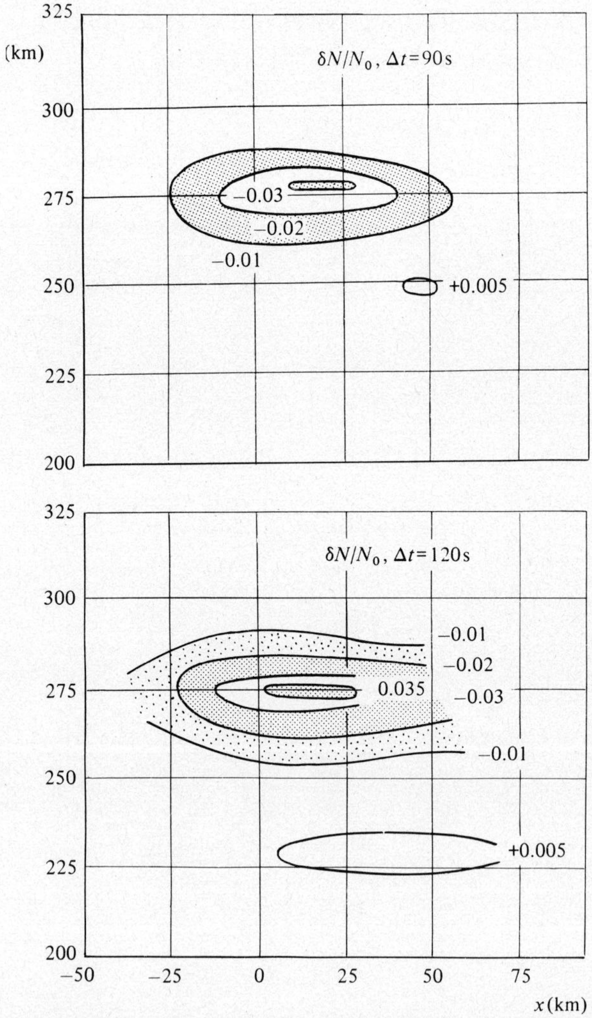

Fig. 7.14. Theoretical curves of equal values of the perturbation of electron density $\delta N(x, z)/N_0$ in the plane of the magnetic meridian, $\Delta t = 90$ and 120 s after the beginning of the action of a strong ordinary wave, with frequency $f_h^{(o)} = 5.94$ MHz; $f^{(o)}F2 = 6.24$ MHz, $z_M F2 = 280$ km (Meltz *et al.*, 1974).

The curves of equal values of the relative perturbation of the electron density $\delta N(x, z)/N_0$ in the plane of the magnetic meridian (xz plane) for an angle $\theta = 67^\circ$ between the Earth's magnetic field $\mathbf{H}_0$ and the vertical z were calculated in detail in Meltz & LeLevier, 1970; Meltz *et al.*, 1974. The models used in these calculations are described briefly in the previous subsection 7.2.1 presenting the relations for the electron temperature T_e when modified by a strong wave. For the effect of a strong extraordinary circularly-polarized wave at a frequency f_h, the ratio of which to the critical frequency $f^{(o)}F2$ is 0.95, Fig. 7.14 shows the relations $\delta N(x, z)/N_0$ at times $\Delta t = 90$ s and 120 s after turning on the transmitter. After 120 s, the perturbation of the electron density has apparently not yet become completely stable. This is also implied, in particular, by the theoretical

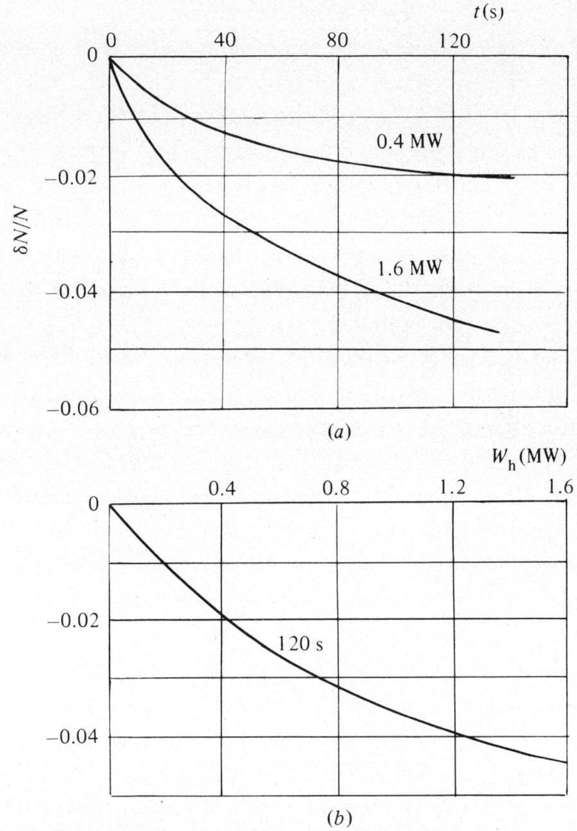

Fig. 7.15. Theoretical curves of the perturbation of electron density: (*a*) as a function of time for different transmitter powers W_h; (*b*) as a function of W_h at 120 s after turning the transmitter on (Meltz *et al.*, 1974).

relations for $\delta N(t)/N_0$ (Fig. 7.15*a*) for different intensities. Note that the perturbation of the electron density proceeds more slowly than the stabilization of the electron temperature T_e (see Fig. 7.7*b*). This is also confirmed by experimental data. For instance, variations $\delta N/N_0 \simeq 0.14$ were observed 10 minutes after the heating up of the ionosphere (Gordon & Carlson, 1974) as well as values $\delta N/N_0 \simeq 0.07$ (Utlaut & Violette, 1974). It is true that in both cases the experimental values of $\delta N/N_0$ were considerably higher than the theoretical values in Figs. 7.14 and 7.15. The curve for $\delta N(W_E)/N_0$ as a function of transmitter power, calculated for $\Delta t = 120$ s (Fig. 7.15*b*), shows that $\delta N(W_E)$ varies quite uniformly, as does $\delta T_e(W_E)$ (see Fig. 7.7*b*).

7.2.3. Cloud-like inhomogeneous structure of the perturbed plasma region. Field-aligned structure

An important consequence, and one of the most interesting as well, of the effect of a strong wave on an inhomogeneous plasma such as the ionosphere is the creation in it of a large-scale inhomogeneous structure and the formation of irregularities of different sizes. The generation of a wide spectrum of irregularities in the ionosphere is naturally associated with an enhancement, in the field of the strong wave, of the plasma instabilities related to the resonance branches of the oscillations known from the linear theory, as well as with the appearance of new kinds of instabilities of nonlinear character. The complicated nature of the gradient of the electron density in the plasma may well, for instance, increase the amount of drift instability. On the other hand, focusing of the field and the concentration of particles in the large-scale inhomogeneous structure leads to the appearance of instabilities of the self-focusing type, these being a consequence of the nonlinearity of the processes taking place in the plasma. In the field of a strong wave, the plasma resonance instability also acquires some completely different properties. The parametric decay instability in a collisional plasma takes place differently. Some theoretical conclusions pertaining to these subjects will be presented below.

With regard to the near-Earth and interplanetary plasma, however, it is mainly experiments that have taken precedence in the discovery of new nonlinear phenomena. This is especially true for the nonlinear phenomena observed during experiments on the modification of the plasma region closest to the Earth, the ionosphere, where nonlinear effects of the heating type play a major role. See, for instance, Utlaut, 1970; Utlaut & Cohen, 1971; Utlaut & Violette, 1972; Gordon & Carlson, 1974; Radio Science, 1974; Utlaut, 1975; and also Getmantsev *et al.*, 1973, 1974; Belikovich *et al.*, 1974; Mityakov *et al.*, 1976; Belikovich *et al.*, 1976a,b; and Shlyuger, 1974a,b. Many of these phenomena were not predicted by theory, and they have not as yet been given any complete, or even satisfactory, explanation. At the same time, a number of them agree quite well with the results of theoretical calculations, as was seen in a foregoing sections of this chapter.

For the reasons indicated, it will be useful to present here a brief account of the main results of the corresponding experiments, together with only some of the attempts at their theoretical explanation. We will examine phenomena observed in a plasma when the main role is played by Coulomb interactions between charged particles ($\nu_{ei} \gg \nu_{en}$), since throughout this monograph it is mainly the properties of such a plasma that are of interest to us. With regard to the ionosphere, this means that we confine ourselves here to the F region and to the outer ionosphere, or plasmasphere, i.e., to heights $z > 200$ km.

The data presented indicate the vital, heuristic role which is played by, or may be played by, nonlinear effects in the near-Earth and interplanetary plasma. Naturally, however, a detailed examination of all these topics would go far beyond the scope not only of this chapter, but of the present work as a whole.

(*a*) *Artificial sporadic layer.* In the first experiments on ionospheric effects using intense radio waves it was found that at heights $z \simeq 200$ to 300 km in the F2 layer a large-scale formation with a cloud-like structure was produced, which scattered effectively a wave incident upon it (Utlaut *et al.*, 1970). Close to a powerful radio transmitter sending circularly-polarized waves upward at frequencies f_h less than or of the order of the critical frequency $f^{(o)}$F2, simultaneous recordings were made with ionosondes of the altitude-frequency characteristics of the ionosphere (ionograms). These ionograms showed the frequency dependences of the group-delay times Δt_v of radio waves reflected from the ionosphere. The time Δt_v is expressed in terms of the so-called virtual height of the ionosphere $z_v = c\Delta t_v/2$. For reflection from the F layer of the ionosphere ($z > 200$ km), two traces of the reflected waves are clearly evident on the ionograms: an ordinary wave (subscript o) and an extraordinary wave (subscript x), which travel along different trajectories at different velocities. If the height dependence of the electron density has two peaks (N_MF1 and N_MF2), i.e., if there are two F layers in the ionosphere, as is often observed in the daytime, then the ionogram has the form shown in Fig. 7.16*a*. At the peaks of the F1 and F2 layers, penetration of the waves takes place,

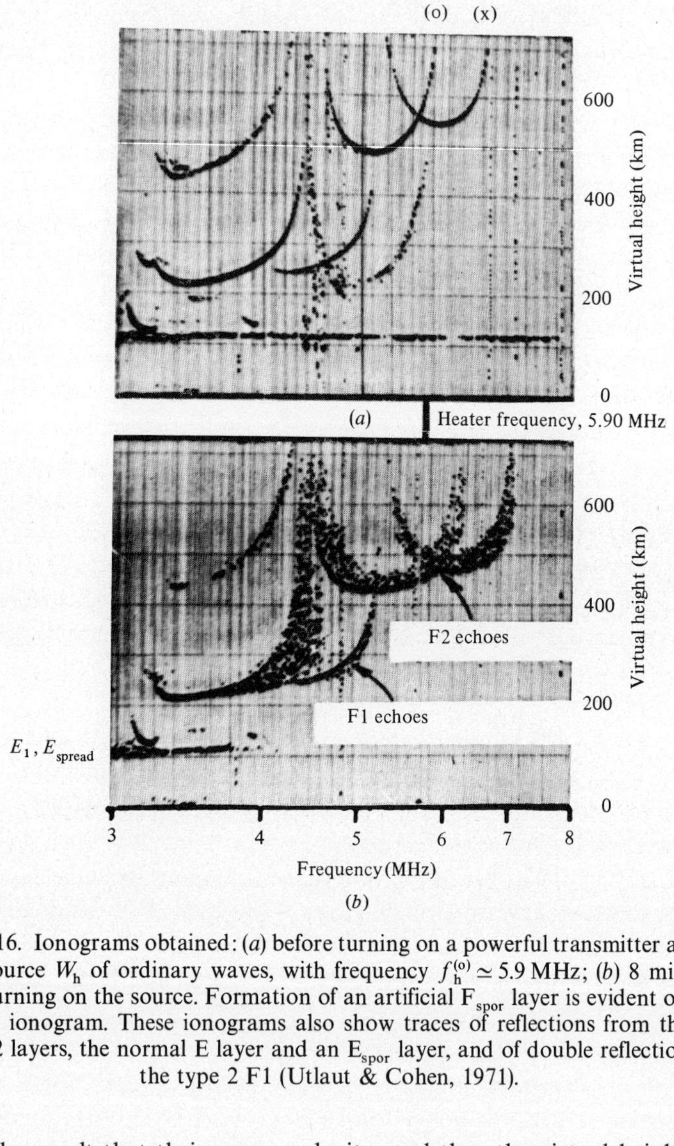

Fig. 7.16. Ionograms obtained: (*a*) before turning on a powerful transmitter acting as a source W_h of ordinary waves, with frequency $f_h^{(o)} \simeq 5.9$ MHz; (*b*) 8 minutes after turning on the source. Formation of an artificial F_{spor} layer is evident on the second ionogram. These ionograms also show traces of reflections from the F1 and F2 layers, the normal E layer and an E_{spor} layer, and of double reflections of the type 2 F1 (Utlaut & Cohen, 1971).

with the result that their group velocity, and thus the virtual heights z_v of both waves, increase rapidly. In such cases four reflection traces from the F region are observed. In the presence of doubly-reflected waves from F1 and F2 and reflections from the lower ionosphere (E layer), the number of traces increases. As on the ionogram shown here, there will be an additional trace of reflections from the E layer ($z \simeq 100$ km), plus a branch of double reflections of the ordinary wave from the F1 layer ($z \simeq 150$ to

200 km). Obviously, if the height dependence of the electron density is smooth, and if there are no ionized clouds in the ionosphere, the traces on the ionogram will become quite thin and under the given conditions will assume the U-shape shown in Fig. 7.16. However, in a number of cases the F region is full of ionized clouds. Then, in addition to the regular reflection of radio waves, their partial reflection from these clouds is also recorded, i.e., wave scattering occurs, and the traces on the ionogram become blurred. In such cases a sporadic layer F_{spor} (or F_{spread}) is said to exist in the ionosphere, and the ionogram has the form in Fig. 7.16*b*. However, this ionogram *records* the appearance of the *artificial* F_{spor} *layer* after the powerful radio transmitter with a frequency $f_h \gtrsim f^{(o)}F2$ has been turned on. The F_{spor} layer was generated in the heated large-scale inhomogeneity which appeared in the F layer.

Naturally, this phenomenon was of interest to investigators. Since it soon became evident that the artificially created cloud-like inhomogeneity in the ionosphere acts as a mirror, which under certain conditions also scatters effectively waves with frequencies $f = 80$ to 100 MHz, that is, with wavelengths λ of 3 or 4 m, the practical significance of this effect also became clear. It can be utilized as a new means of transmitting ultrashort radio waves over great distances in some required direction and sector. However, a more profound study of this effect became of particular interest when the details of the processes taking place in the perturbed region of the ionosphere became known. Some of these turned out to be quite unexpected. Here, however, it will only be possible to outline briefly and schematically the results of these studies.

The chief technique employed to study experimentally the modified region of the ionosphere is to measure the properties of the radio waves scattered by it over a wide frequency range, both at a single point (*backscattering*) and at separated observation points (*side scattering of radio waves*). Measurements were also made of the luminosity of the modified region (the emission of the 6300 Å line of atomic oxygen), which begins to increase markedly two or three minutes after turning on the strong wave. These data will not be examined here, however (see Biondi *et al.*, 1970, 1972; Haslett & Megill, 1974).

The results of numerous experiments showed that the artificial F_{spor} layer is already formed only a few seconds after the transmitter is turned on, for effects on the ionosphere of both a strong ordinary wave and a strong extraordinary wave, at frequencies $f_h \simeq 0.5 f^{(o)}F2$ to $1.05 f^{(o)}F2$, where $f^{(o)}F2$ is the critical frequency of the ordinary wave in the F2 region. However, the perturbed region approaches a steady state only after two or three minutes. This is evident from various data to be given below, which also imply that certain effects in this region are observed mainly via the action of the ordinary wave on the ionosphere. This wave corresponds to the HF branch of longitudinal electron plasma waves (see Figs. 4.3–4.7).

The vertical thickness of the F_{spor} layer covers the range $\Delta z \simeq 10$ to 100 km. The horizontal extent of this region is close to, or somewhat larger than, the area irradiated by the beam of upward-moving radio waves. After the transmitter has been turned off, the *lifetime* τ *of the* F_{spor} *layer* depends on the state of the ionosphere. In the daytime, when the recombination of electrons is accompanied by the production of new electrons, we have $\tau(F1) \simeq 1$ min and $\tau(F2) \simeq 5$ to 10 min. After sunset, when the ionization source has just disappeared, we have $\tau(F2) \simeq 10$ to 20 min, while at night $\tau(F2) \simeq 1$ hr.

In the first experiments it was also established that the generation of inhomogeneities and their form are essentially regulated by the Earth's magnetic field $\mathbf{H}_0$. The disturbed region F_{spor} apparently constitutes a unique kind of channel directed parallel to the vector $\mathbf{H}_0$. On the other hand, individual inhomogeneous formations (ionized clouds) have a shape elongated (along $\mathbf{H}_0$) with a linear size ρ_0 that can vary within quite large limits (see below). The longitudinal dimensions of the clouds $\rho_{0\parallel}$ may be tens or even hundreds of times greater than their transverse dimensions $\rho_{0\perp}$. A somewhat strange, and completely unexpected, discovery in the disturbed region was the presence of irregularities with a transverse size $\rho_{0\perp} \simeq 1$ m, or even less. All these properties of the modified region of the ionosphere, like its very formation, were investigated during the course of a large series of experiments, some of which were very ingenious.

Using a radar device provided with a circular-scanning antenna system, forming a very narrow beam of waves (pencil beam) $\theta_0 \simeq 0.035$ radians $\simeq 2°$ wide at a frequency $f_s \simeq 10$ MHz, it was possible to obtain 'sky maps' of the scattering region F_{spor} and to trace out the dynamics of its structure. Some of the results of these mapping experiments are given in Figs. 7.17 and 7.18. On the sky maps in the horizontal plane (black squares in Figs. 7.17 and 7.18), the centres of which are located at the zenith of the observation point, the bright circles mark the region in which the wave beam is reflected. On the first map in Fig. 7.17 (top left) the single bright circle portrays specular reflection of the beam from the undisturbed smooth ionosphere. This circle is somewhat off-centre because the reflecting surface is tilted, due to the effect of slow internal gravity waves which were passing through the studied region of the ionosphere at that time. The velocity

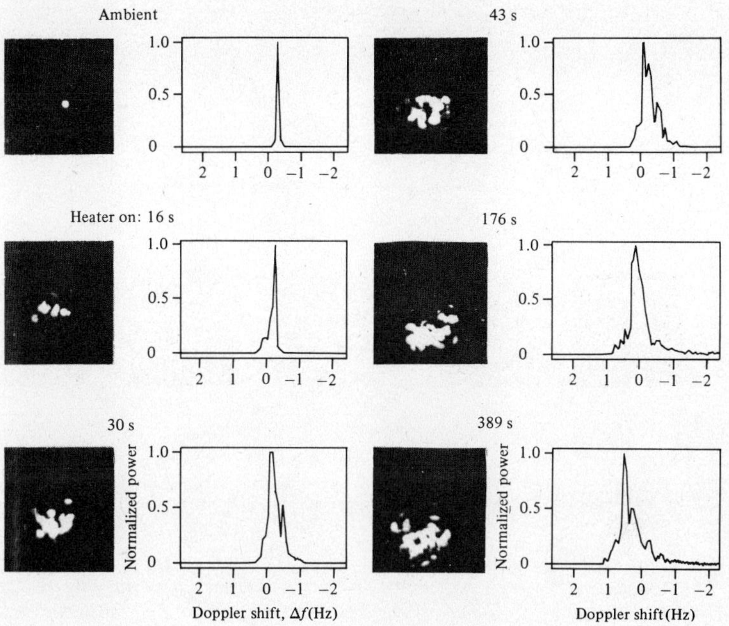

Fig. 7.17. Sequence of radio pictures of the reflecting region of the ionosphere: 'sky maps' and spectra of the Doppler frequency shifts Δf of the reflected waves, illustrating the formation of an artificial F_{spor} layer. The sky maps were obtained with a circular-scanning antenna; the height of reflection of the radio-wave beam was kept constant (Meltz *et al.*, 1974).

of these waves also accounts for the slight Doppler shift in the frequency Δf_s of the reflected wave relative to that of the transmitted wave; the difference between these frequencies was also determined during these experiments. The corresponding curves for the reflected-wave intensity as a function of Δf_s are shown next to the sky maps. Sixteen seconds after the action of the heating wave, the picture on the map has already become complicated, and the angular and Doppler spectra of the reflected waves have grown considerably broader. This rapid development of the process is maintained for about a minute, after which it slows down and after about two or three minutes (after 176 s in Fig. 7.17) the quantitative characteristics of the scattered waves become nearly steady. The form of the picture and the value of Δf_s in Fig. 7.17 continue to vary gradually, mainly because the individual irregularities in the scattering region are in motion, and also because of the above-mentioned gravity waves. When examining Fig. 7.16, it should be noted that F_{spor} is shifted at the end of its formation process more to the south of the point of specular reflection

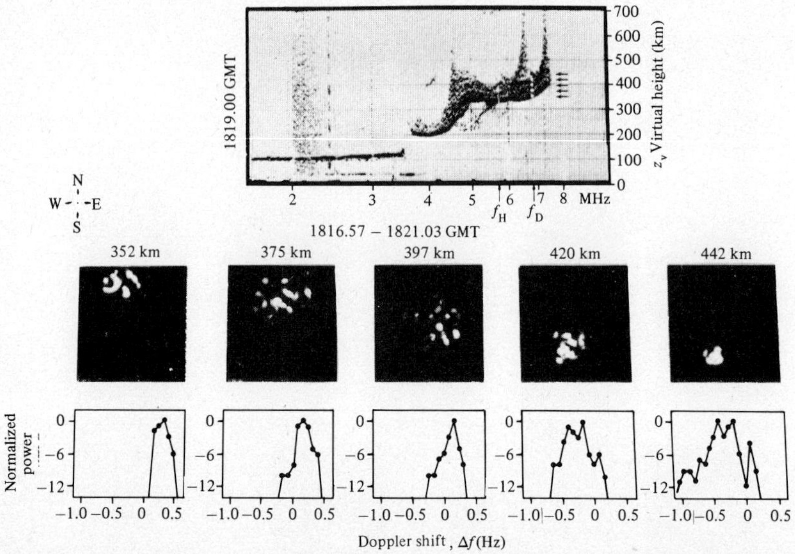

Fig. 7.18. Sequence of 'sky maps' (radio pictures of the F_{spor} layer) and spectra of the Doppler frequency shifts Δf of the scattered waves for F_{spor} sounding with a circular-scanning antenna at a frequency $f_D^{(x)} = 6.85$ MHz of the extraordinary wave. The frequency of the strong ordinary wave $f_h^{(o)} = 5.75$ MHz; the critical frequencies of the ionosphere were $f^{(o)}F2 = 6.75$ MHz and $f^{(x)}F2 = 7.4$ MHz (see ionogram). The maps were made for different heights at equal intervals of the virtual height $\Delta z_v = 22.5$ km (arrows on the ionogram). Under the conditions of the experiments the extraordinary wave $f_D^{(x)}$ was reflected from the same height as the heating wave $f_h^{(o)}$ (Meltz *et al.*, 1974).

of the wave beam. Direct recordings of the delay times of reflected pulses (echo signals) in these experiments indicated that no more than 10 s later the pulses started to widen considerably, attesting to an earlier start of the disturbance of the ionosphere.

All the sky maps in Fig. 7.17 were obtained for a constant delay time of the echo signals, i.e., for approximately the same height of reflection. If in the same manner the ionosphere is 'scanned' at a fixed frequency after about 20 minutes, when the creation of F_{spor} has been completed, and if the reflection height of the wave beam is increased, then we obtain the evolution of the sky-map pictures and the frequency spectra depicted in Fig. 7.18. This same figure shows an ionogram of the ionospheric region being considered; the arrows indicate the virtual heights z_v to which each of the sky maps corresponds. With a change in z_v, the F_{spor} pictures are seen to shift from north to south. The Doppler shift of the reflected-wave frequencies changes sign accordingly. An analysis of the results of these experiments indicated that the height variations of the angular and frequency spectra agree with the model of a channel formed parallel to the vector $\mathbf{H}_0$, within which the F_{spor} layer is generated (Meltz *et al.*, 1974).

It is important to note here that the artificial F_{spor} layer observed on the ionograms is a consequence of backscattering of the radio waves sounding the ionosphere, mostly from the large-scale part of the spectrum of field-aligned irregularities, generated in the disturbed region of the plasma. Their linear size $\rho_{0\perp} \simeq 10^2$ m to 1 km and $\rho_{0\parallel} \simeq 1$ km to 10 km, while the fluctuations in their electron density $\delta N/N_0 \lesssim 10^{-2}$ (see for instance, Bowhill, 1974). Apparently, it is precisely these irregularities that appear in the F_{spor} layer during the initial stage of its development. In the literature a possible mechanism for their creation has been considered: the effect of the self-action of the strong wave, leading to a so-called *self-focusing* instability of the heating type (Georges, 1970; Perkins *et al.*, 1974; Vas'kov & Gurevich, 1975b; see also Cragin & Fejer, 1974). However, the theory describing the generation of irregularities of this type has not yet been developed sufficiently. Consequently, it is difficult at present to ascertain the true mechanism of excitation of the large-scale part of the irregularity spectrum. Irregularities smaller than these, as well as very small-scale ones ($\rho_{0\perp} \lesssim 1$ m), which as mentioned above are observed in the perturbed part of the ionosphere (see, for instance, Fig. 7.23, below), can be assumed to appear due to the development of the drift instability, in turn a consequence of the large-scale inhomogeneities (Borisov *et al.*, 1976, 1977). Of course, there may also be some weakly interacting mechanisms at work as well, whose effect is to excite different parts of the wide spectrum of irregularities in the artificial F_{spor} layer.

(*b*) *Anomalous absorption of radio waves.* During the experiments described here, in a number of cases a marked decrease in the reflectivity of the disturbed region of the ionosphere was observed (Cohen & Whitehead, 1970). Later it was found that this phenomenon appears in the frequency range $f_h \lesssim f^{(o)}$F2, soon after the strong ordinary wave is turned on, and that it is apparently caused by the same processes that explain several properties of the spectra of the scattered radio waves, a brief description of which will be given below.

The attenuation – or, rather, the complete disappearance – of the reflected waves is illustrated by the ionograms in Fig. 7.19 (Utlaut & Cohen, 1971; Utlaut & Violette, 1972). The lower ionogram was taken $\Delta t = 16$ s after the strong ordinary wave at a frequency $f_h^{(o)} = 6.25$ MHz

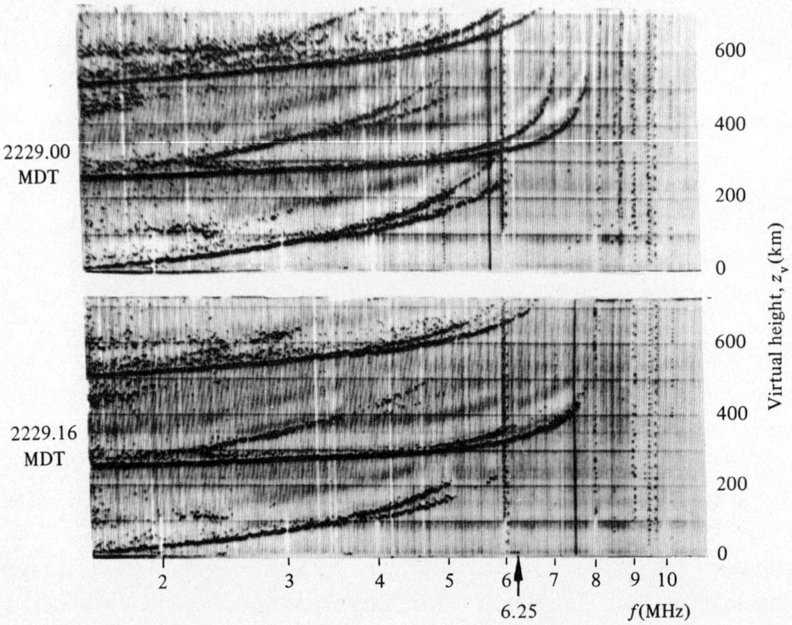

Fig. 7.19. Ionograms illustrating the anomalous absorption of an ordinary wave in the frequency range $f \simeq 6$ to 7 MHz. The lower ionogram was taken at the time $22^h29^m16^s$, MDT i.e., 16 s after turning on a strong ordinary wave at the frequency $f_h^{(o)} = 6.25$ MHz (Utlaut, 1975).

was turned on, and the upper one immediately after it was turned on. On the lower ionogram, the branch describing the ordinary wave is seen to be completely absent over the entire frequency range $f \simeq 6$ to 7 MHz. This anomalous drop in the reflectivity of the ordinary wave from the F_{spor} layer sets in very rapidly, being observed only $\Delta t \simeq 0.1$ s after the switching on of a powerful radio transmitter. Fig. 7.20 illustrates the variation in the intensity of an ordinary wave reflected from the ionosphere, as a function of the power W_h of a source generating an F_{spor} layer (Belikovich *et al.*, 1976a). With an increase in the power W_h, the reflected waves at first increase in intensity, and then, beginning at some power value, they rapidly become weaker again. With a further rise in the source power W_h, the intensity of the reflected signals reaches an almost constant level. Naturally, this anomalous attenuation of the reflected ordinary wave is not associated with any increase in the absorption coefficient κ_e of the ionosphere. In the given case, on the contrary, κ_e actually decreases, since the reflection is from the region of the ionosphere where the strong wave reduces the number of collisions, because there the electron–ion collisions

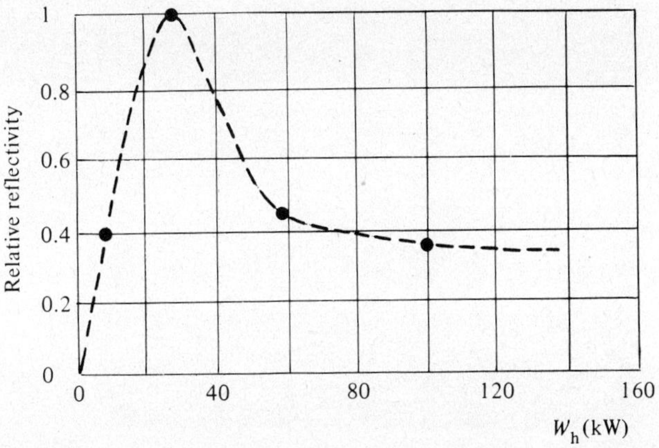

Fig. 7.20. Intensity of an ordinary wave reflected from the ionosphere (wave frequency $f_h^{(o)} = 5.75$ MHz), as a function of the power W_h of the transmitter (Belikovich *et al.*, 1976a).

predominate, with $\nu_{ei} \sim T_e^{-3/2}$ (see above). Consequently, the attenuation of the reflected ordinary wave is collisionless in nature. The main cause of this effect is apparently an *anomalous resonance absorption* of the plasma, arising in the field of a strong wave. Due to the anomalous absorption, very-small-scale irregularities are created in the F_{spor} layer (see Vas'kov & Gurevich, 1977). The following subsection, 7.2.3(*c*), will describe some of the properties of these irregularities, and it includes formulas which, as shown in the reference cited above, characterize quite well the results of the corresponding observations.

(*c*) *Very-small-scale irregularities.* The structure of the individual irregularities elongated along $\mathbf{H}_0$, these being generated in the F_{spor} layer, was observed in the experiments, which included remote sensing of the disturbed region of the ionosphere by irradiating it obliquely with a narrow beam of radio waves (a pencil beam) having a frequency much higher than the critical frequency $f^{(o)}$F2 of the ionosphere (Thome & Blood, 1974; Fialer, 1974; Minkoff *et al.*, 1974a). In these experiments it was established that the intensity of the reflected waves depends markedly on the angle of incidence of the wave beam at the F_{spor} layer. The cross-section σ_s in the opposite direction ($\theta = \pi$) is a maximum when the centre of the beam of waves is incident in a direction perpendicular to the magnetic field $\mathbf{H}_0$ (for $\theta = \pi/2$). If the point of reception of the scattered waves is far away from the transmitter, then the maximum value of σ_s corresponds to the condition of 'specular reflection' from the direction $\mathbf{H}_0$ (see Fialer, 1974;

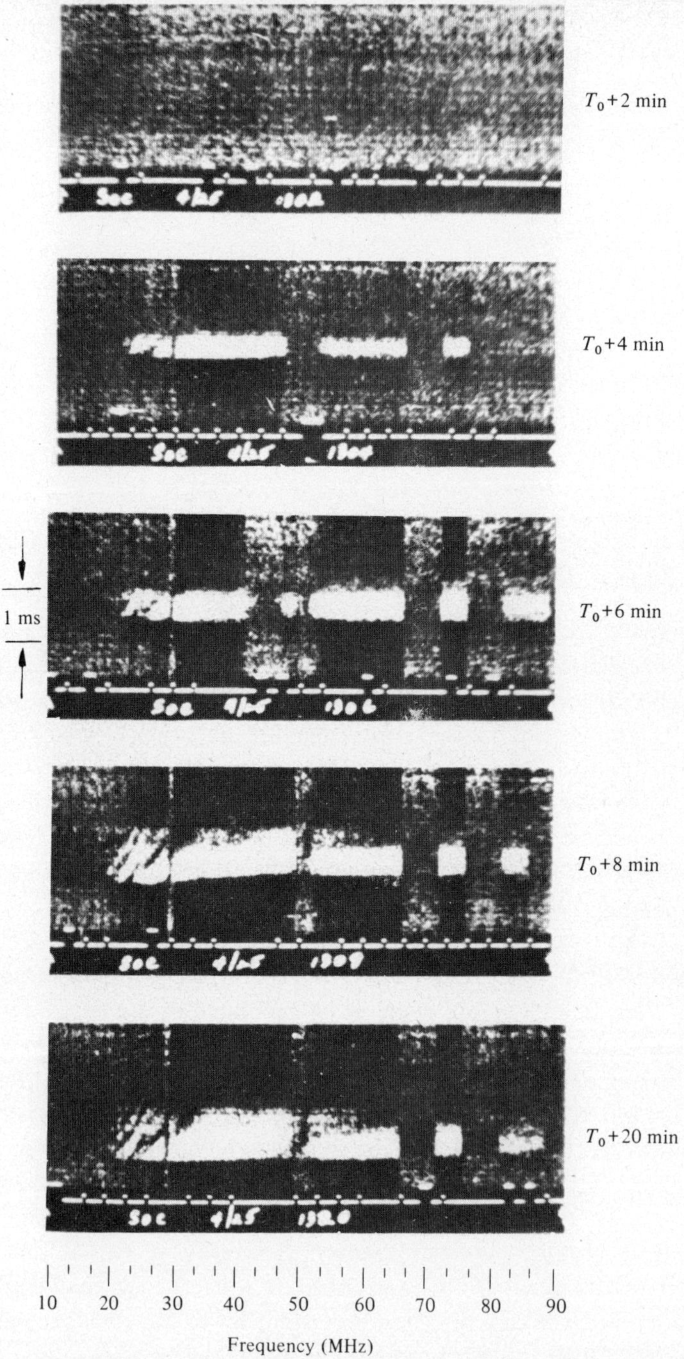

Fig. 7.21. Series of spectograms for waves scattered specularly from a heated F_{spor} region in the ionosphere. The spectograms were taken 2, 4, . . . , 20 minutes after turning on a strong wave (time T_0), using a radar set with a swept frequency. The measurements were carried out at a distances of several hundred kilometers from F_{spor}, with oblique irradiation of this layer (Fialer, 1974).

Minkoff *et al.*, 1974b; and Carpenter, 1974). Such a dependence of the cross section σ_s on an inhomogeneous formation of cloud-like structure signifies that, as was demonstrated earlier (Booker, 1956), the ionized clouds have an elongated shape. Their longitudinal $\rho_{0\parallel}$ and transverse $\rho_{0\perp}$ dimensions must satisfy the condition $\rho_{0\parallel} \gg \rho_{0\perp}$. It is precisely this case that was realized in the experiments described. Moreover, many new phenomena taking place in the modified region of the ionosphere were discovered in these experiments, and a number of their more subtle properties were investigated.

The following data, for example, illustrate the development of a disturbed region created artificially in the ionosphere, as well as its relaxation in time when a source of strong waves is repeatedly turned on and off, after a preliminary heating of it with the aid of UHF waves ($f_v \gg f_h$). In experiments using a radar system with a swept frequency at various distances from the artificial F_{spor} layer, it was found that over a wide frequency range radio waves begin to be scattered only two or three minutes after turning on the heating radio transmitter. This process becomes established after six or eight minutes. In particular, this result is implied by the series of frequency–time dependences of the scattered-wave intensities given in Fig. 7.21.

A very gradual increase in the intensity of the scattered waves and in the size of the scattering region still takes place about 20 minutes after the source is first switched on. If the source is then turned off and on again repeatedly, generating an F_{spor} layer at an interval of, for instance, 5 s, total appearance and disappearance of the scattered waves are observed with a relaxation time $\tau_s \simeq 4$ s at a frequency $f_s \simeq 40$ MHz and with $\tau \simeq 2$ s at $f_s \simeq 80$ MHz, during the course of a few minutes after the preliminary heating of the ionosphere. Fig. 7.22 illustrates the rapid relaxation of the change in the intensity of the scattered UHF radio

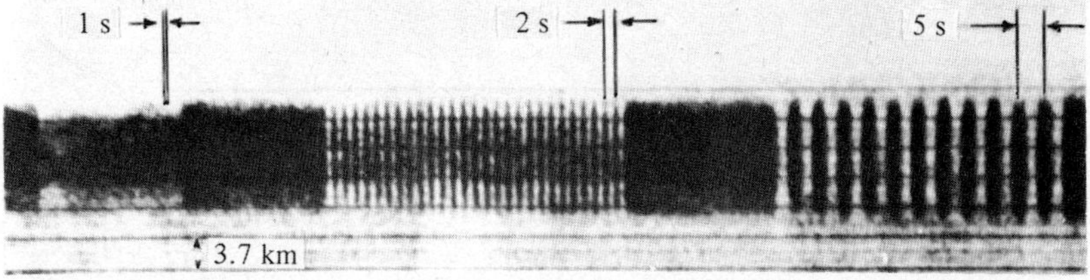

Fig. 7.22. Spectrogram of scattered waves at the frequency $f_s = 157.5$ MHz for oblique irradiation of an F_{spor} layer at a distance $r = 925$ km. The powerful radio transmitter generating the F_{spor} layer was modulated with square pulses of width $\Delta t \simeq 1$, 2, and 5 s (Minkoff *et al.*, 1974a).

waves. This figure gives the corresponding recording at a frequency $f_s = 157.5\,\text{MHz}$ ($\lambda_s = 1.9\,\text{m}$) for an oblique distance $r \simeq 925\,\text{km}$ from the F_{spor} layer, when the heating radio transmitter is turned on and off periodically, every 1, 2, or 5 s. The intensity of the scattered waves, as the recording in Fig. 7.22 shows, drops to almost nothing and then recovers during some fractions of a second. The relaxation time for this process, as a function of the frequency, has also been studied by Thome and Blood (1974). The following data were obtained during the above experiments:

$$\tau_s \simeq 1\,\text{s}, \quad f_s \simeq 54\,\text{MHz},$$
$$\tau_s \simeq 10\,\text{s}, \quad f_s \simeq 30\,\text{MHz},$$
$$\tau_s \simeq 60\,\text{s}, \quad f_s \simeq 15\,\text{MHz}.$$

The observed variation in τ_s with the frequency apparently indicates that small irregularities are generated and disappear in the F_{spor} layer more rapidly than large irregularities. This is because the backscattering from an irregularity is quite intense when the size of the latter $\rho_0 \gtrsim \lambda_s$. Under the conditions being considered, this occurs if $\rho_{0\perp} \gtrsim \lambda_s$ and if, after a time τ_s, diffusion of the perturbation of the electrons along $\mathbf{H}_0$ causes the longitudinal dimension of the irregularity $\rho_{0\parallel}$ to become much greater than $\rho_{0\perp}$.

On the basis of the measured dependence of the cross-section σ_s on the frequency f_s of the scattered waves and on the power W_h of the heating radio transmitter, a model of the irregular cloud-like scattering region of the ionosphere was constructed (Rao & Thome, 1974). The method for calculating σ_s given in Booker, 1956, was used. This model was applied, in particular, to determine the transverse sizes $\rho_{0\perp}$ of elongated irregularities. The scattering cross-sections obtained in Booker, 1956, and Minkoff *et al.*, 1974a, are shown in Fig. 7.23 (points); the upper diagram shows the experimental dependence of σ_s on the frequency f_s (and λ_s) of the radio waves used for observations. The lower diagram (taken from Rao & Thome, 1974) shows the dependence of the experimental data on the *transverse component* of the wave vector of the scattered-wave spectrum, which determines the transverse wavelengths of the spectrum $\lambda_\perp \simeq \rho_{0\perp}$. This length scale was obtained theoretically in Rao & Thome, 1974. The method used by these authors to analyse the results of various experiments made it possible to establish that in the artificial F_{spor} layer small-scale irregularities are created, the transverse dimensions of which vary within the following limits:

$$(0.5-1)\,\text{m} \lesssim \rho_{0\perp} \simeq (8-10)\,\text{m}$$

The dependence of the cross-section σ_s on the power W_h of the heating

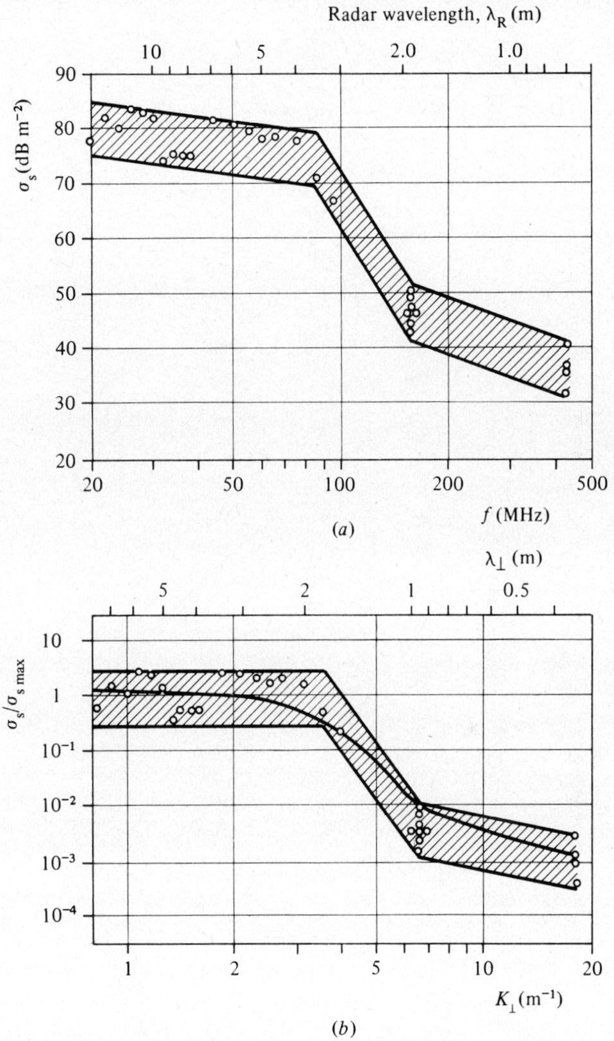

Fig. 7.23. (*a*) Experimental dependence (points) of the cross-section σ_s on frequency, obtained for oblique irradiation of an F_{spor} layer (Fialer, 1974; Minkoff *et al.*, 1974a). (*b*) Dependence of $\sigma_s/\sigma_{s\max}$ (points) on the transverse component $K_\perp$ of the wave vector of the scattered-wave spectrum (results of processing experimental data). The scale $\lambda_\perp$ on this diagram determines the transverse linear extent $\rho_\perp$ of the irregularities (Rao & Thome, 1974).

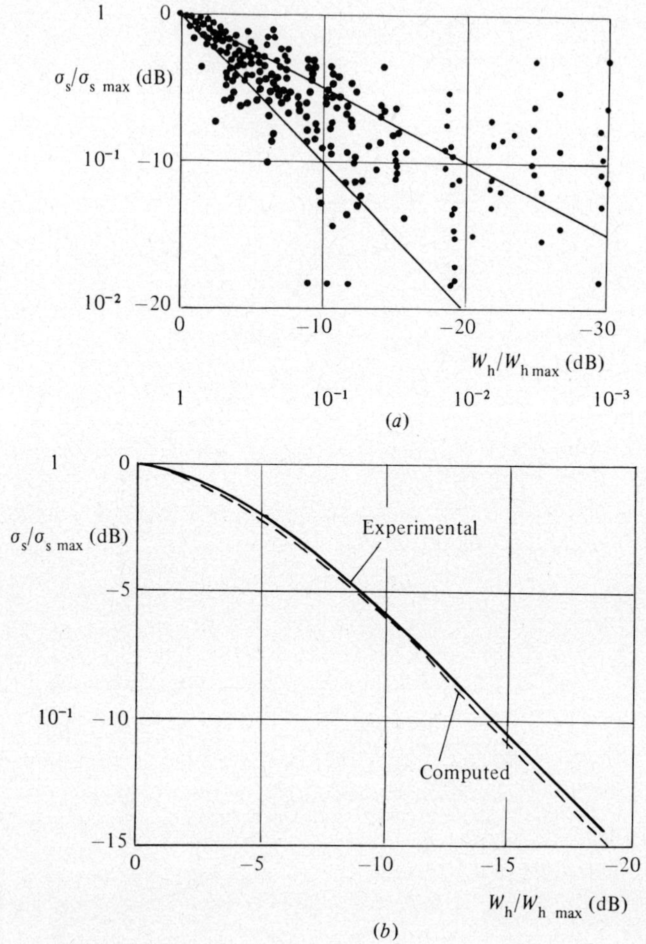

Fig. 7.24. Relative cross-section $\sigma_s/\sigma_{s\max}$ as a function of the relative power $W_h/W_{h\max}$ of a transmitter generating an artificial F_{spor} layer (Fialer, 1974; Rao & Thome, 1974).

transmitter is an important characteristic of the generation process. Fig. 7.24 shows the ratio $\sigma_s/\sigma_{s\max}$ as a function of $W_h/W_{h\max}$ (data taken from Booker, 1956). From case to case, the cross-section for the scattering from F_{spor} in the frequency range $f_s \gg f^{(o)}F(2)$, f_h is seen to vary considerably (Fig. 7.24*a*). The results of model calculations of $\sigma_s/\sigma_{s\max}$ as a function of $W_h/W_{h\max}$ obtained by Rao & Thome, 1974, show good agreement with the experimental data (Fig. 7.24*b*).

Various experiments indicate that the scattering cross-section for an artificially created scattering region of the ionosphere varies within the following limits, depending on the conditions:

$$\sigma_s \simeq (10^4\text{–}10^9)\,\mathrm{m}^2.$$

Consequently, F_{spor} is actually a reasonably good 'mirror', and the fact that it can be produced by artificial means is naturally of great practical interest, all the more so since the phenomena in it (to be described later) and the properties of these phenomena all go to show that under certain circumstances this mirror only slightly distorts the waves scattered by it.

(*d*) *Scattered-wave spectra.* A number of interesting phenomena were revealed by studies of radio waves scattered by a modified region of the ionosphere. The spectra of these waves have the following significant characteristic: they differ considerably depending on whether *specular* (perpendicular) scattering relative to the vector $\mathbf{H}_0$ is being recorded, or, alternatively, *scattering of radio waves at some oblique angle with respect to* $\mathbf{H}_0$. An analysis of these data indicated that this is because, in the two cases, the scattering is due to different mechanisms of plasma instability *vis-à-vis* Langmuir waves, which coexist in the same or different parts of the F_{spor} layer. Let us consider these data briefly.

The frequency f_h of the strong radio waves used to produce the F_{spor} layer is generally much lower than the frequency f_r of the beam of waves used to study its structure. Thus it is possible to examine in detail the f_s spectra of the scattered waves, which consist of bands centred on the following frequencies:

$$f_{s1} = f_r, \quad f_{s2} = f_r \pm f_h \simeq f_r \pm f_0$$

where f_0 is the Langmuir frequency of the scattering region of the plasma. Note that the effects to be described below are observed only if the ionosphere is modified by a circularly-polarized ordinary wave.

The band of waves scattered around the frequency $f_{s1} = f_r$ is usually called the *centre scattering line*, while the two bands around the frequencies f_{s2} are called the *plasma lines*. Depending on whether *specular scattering* relative to the direction of vector $\mathbf{H}_0$ or *side scattering*, i.e., scattering at some oblique angle to $\mathbf{H}_0$, is being recorded, the spectra of both types of line will be very narrow (first case) or the f_{s2} lines will be wide with a complicated structure (second case).

Specular scattering. In various experiments it was established that for specular scattering, both when the transmission and reception points coincided (backscattering) and when these points were different, the centre line at $f_{s1} = f_r$ and the plasma line at $|f_{s2} - f_r| = f_0$ were both very narrow.

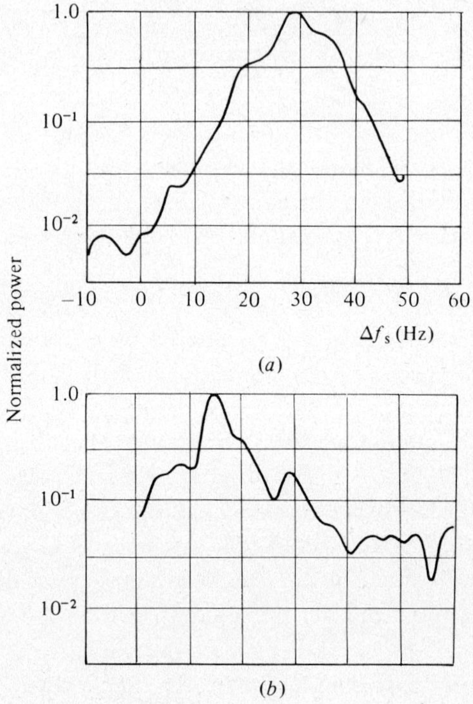

Fig. 7.25. Spectra of the 'centre', (*a*), and 'plasma', (*b*), lines for waves backscattered specularly from an F_{spor} layer at the frequency $f_r = 157.5$ MHz, with the frequency of the heating transmitter $f_h^{(o)} = 6$ MHz; in (*b*), the vertical graduations are at intervals of 10 Hz as in (*a*), but the absolute values of the frequencies are not known to this order of accuracy. The observations were carried out at an oblique distance from F_{spor} $r \simeq 925$ km; the beam of incident waves made an angle $\theta = \pi/2$ with the magnetic-field vector $\mathbf{H}_0$ (Minkoff *et al.*, 1974a).

For example, for observations at an oblique distance $r = 925$ km from F_{spor} and a zenith angle $\theta \simeq 15°$, the spectra shown in Fig. 7.25 were obtained for the centre line at $f_{s1} = f_r = 157.5$ MHz and the plasma line at $|f_{s2} - f_r| = 6$ MHz (Minkoff *et al.*, 1974a). The bandwidths of the lines are respectively, $\Delta f_{s1} = 6.7 \pm 2$ Hz and $\Delta f_{s2} \simeq 8.2 \pm 2$ Hz. Inspection of Fig. 7.25 shows a shift in the spectrum of the centre line relative to the frequency of the incident beam of waves, $f_s - f_r = 0$. This is a Doppler shift in the frequency of the scattered waves, caused by the northward drift of the F_{spor} layer at a velocity $v \simeq 30$ m s^{-1} which was observed during these experiments. On the spectrogram of the plasma line, the Doppler shift in frequency due to the F_{spor} drift is not observed, since the absolute value of $f_h = f_0$ was measured less accurately than Δf_s.

Experiments were conducted at frequencies $f_1 = 49.8$ MHz and $f_2 = 423.3$ MHz, in which the points of irradiation, of heating of the ionosphere, and of reception of the scattered waves formed on the Earth's surface a triangle with sides $r \simeq 1600, 1100$, and 1000 km. Fig. 7.26 shows the spectra corresponding to the recorded values of $\Delta f_{s1} \simeq 10$ to 15 Hz and $\Delta f_{s2} \simeq 80$ to 120 Hz. When interpreting Fig. 7.26, it should be kept in mind that $\Delta f_{s2}/\Delta f_{s1} \simeq f_2/f_1$.

A more detailed analysis of the spectra of specularly scattered waves showed that both lines apparently are accompanied by noise bands with a width $\Delta f_{sh} \simeq 100$ Hz (Minkoff & Kreppel, 1976). However, these properties of the spectra require additional verification.

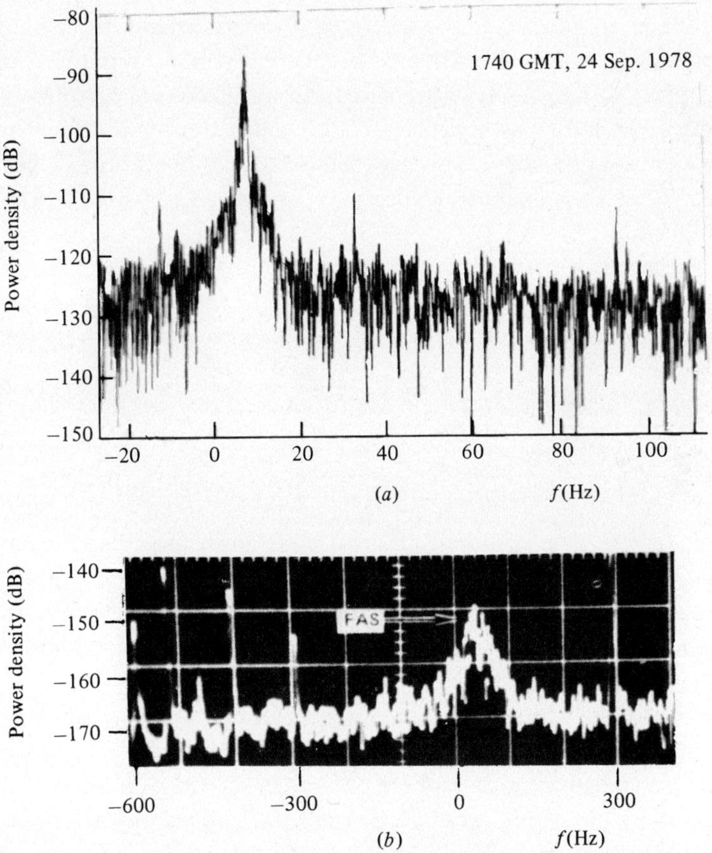

Fig. 7.26. Spectra of the 'centre' line of waves scattered specularly from an F_{spor} layer, for oblique irradiation of the layer and oblique reception, at different points located several hundred kilometers from the wave source. (a) $f_r = 49.8$ MHz; (b) $f_r = 423.3$ MHz (Carpenter, 1974).

The data presented show that both bands of specularly scattered waves have the same spectral characteristics. The characteristic relaxation time τ for the increase and decrease of their intensity when a powerful radio source is turned on and off are also the same, being of the order of some fractions of a second ($\tau = 10$ ms). However, the properties of the two modes also exhibit some very significant differences. For instance, the maximum scattering intensity of the centre line f_{s1} is governed only by the angle θ between the bisectrix of the directions of the beam of incident and scattered waves and the vector $\mathbf{H}_0$. For backscattering, the condition $\theta = \pi/2$ has to be satisfied very precisely. The scattering of the plasma line f_{s2}, on the other hand, is quite strong within an angle interval $\Delta\theta \simeq 5°$, and it takes place from the region of greatest heating of the F_{spor} layer, where longitudinal plasma waves are generated. Consequently, it turns out that the lines at f_{s1} and f_{s2} are scattered in different regions of the F_{spor} layer, i.e., at different heights. It was pointed out in Minkoff & Kreppel, 1976, that their widths Δf_s increase when scattering from the region of maximum heating is being recorded. The thickness of the regions which scatter the central and plasma lines are, respectively, of the order of 40 to 50 km and 10 to 20 km.

The difference in the frequency dependences of the scattering cross-sections for the two lines is important also. The scattering cross-section σ_{s1} of the centre line increases with the frequency, whereas σ_{s2} of the plasma line does just the opposite. Thus in Minkoff *et al.*, 1974a, the following ratios of the σ_s were obtained, respectively, for these lines:

$$\frac{\sigma_{s1}(157\,\text{MHz})}{\sigma_{s1}(435\,\text{MHz})} = \tfrac{1}{5}, \quad \frac{\sigma_{s2}[(435 \pm \text{f}_0)]\,\text{MHz}}{\sigma_{s2}[(157 \pm \text{f}_0)]\,\text{MHz}} = \tfrac{1}{12}\,.$$

Note with respect to this that never in these experiments were bands of scattered waves observed at the harmonics of f_0, i.e., at frequencies $f_{s2} = f_r \pm s f_0$ (where s is an integer). This fact may well prove to be significant for the theoretical interpretation of the phenomena being considered.

It is easy to see why the scattered-wave spectra $f_{s2} = f_r \pm f_h$ are observed when the ionosphere is being perturbed only by means of the ordinary wave $f_h^{(o)}$. The appearance of these lines is associated with the resonance excitation by the strong heating wave at f_h, of plasma waves at $f_0 \simeq f_h^{(o)}$. An extraordinary wave $f_h^{(x)}$ is known to be reflected in the ionosphere at a height where, from the condition of its reflection $n^{(x)} = 0$, it follows that

$$f_h^{(x)} = (f_0^2 + f_h^{(x)} f_H)^{1/2}$$

(see (4.24)). Thus the extraordinary wave does not reach the ionospheric regions where $f^{(x)} = f_0$ and where the excitation of plasma waves is possible.

The narrowness of the plasma line at $f_{s2} \simeq f_r + f_0$ is evidence of a linear mechanism for conversion of the incident transverse waves into longitudinal resonant oscillations of the plasma, which can also scatter beams of waves with frequencies $f_r \gg f_h \simeq f_0$. A similar phenomenon is observed during investigations of the incoherent scattering of radio waves from the ionosphere. However, under the influence of the strong wave, the process of wave conversion is enhanced as result of the heating of electrons, due in turn to the absorption of energy of the incident wave by the longitudinal plasma oscillations. Therefore, the perturbation of electron density in the longitudinal waves gradually increases, and the generation of irregularities is intensified accordingly. Thus, a resonant *plasma instability* with a markedly nonlinear nature ultimately develops in the plasma. Such physical premises as these formed a basis for the theory explaining the generation of very-small-scale irregularities in the F_{spor} layer, the anomalous absorption of ordinary waves, and a number of other phenomena (Vas'kov & Gurevich, 1975a,b, 1977; Gurevich, 1978). The results of these studies can be summed up briefly as follows.

Assume that in the region of radio-wave reflection there are very-small-scale irregularities extended along $\mathbf{H}_0$, and that they involve a perturbation of the electron density which is of the order of or greater than some threshold value $(\delta N/N)_0$. Then the electric polarization of the irregularities under the influence of the strong f_h wave causes an intense excitation of longitudinal oscillations at the upper-hybrid frequency, and maximum development of the nonlinear resonant instability takes place in the region of the ionosphere where the resonance condition $f_h = (f_0^2 + f_H^2)^{1/2}$ is satisfied. Here the energy of the excited oscillations goes to develop this process, and to enhance the generation of irregularities with linear dimensions

$$\left.\begin{aligned} \rho_{0\|} &= (4D_{T\|}\tau_{Td})^{1/2} = \rho_{0\perp}\frac{\omega_H}{\nu_{ei}}, \\ \rho_{0\perp} &= \left(D_e\frac{c}{\omega_H}\right)^{1/2} = \left(\frac{v_e c}{2\omega_0\omega_H}\right)^{1/2}, \quad \tau_{Td} \simeq \frac{\rho_{0\perp}^2}{4D_{T\perp}}, \end{aligned}\right\} \tag{7.29}$$

where D_e is the Debye length for the electrons, and $D_{T\|}$ and $D_{T\perp}$ are, respectively, the longitudinal and transverse values of the diffusion coefficients, associated with electronic heat conduction (see (7.24)), and

τ_{Td} is the characteristic time for dissipation of the irregularities. Note that in Vas'kov & Gurevich, 1977, the plasma instability is shown to be enhanced in the vicinity of the second-harmonic gyroresonance $f = 2f_H$, which should lead to an intensification of the entire portion of the spectrum of irregularities, the linear dimensions of which are less than or equal to $\rho_{0\perp}$.

The threshold and mean-square perturbations of the electron density in the irregularities being considered are given by the following formulas:

$$\left.\begin{aligned} \left(\frac{\delta N}{N}\right) &\simeq \text{Const.}\frac{\mathrm{d}N}{\mathrm{d}z}\frac{\rho_{\|}}{N}\left(\frac{\omega_H}{\omega_h}\right)^{1/2}\left(\frac{v_e}{c}\right)^2\frac{NkT_e}{E_0^2}, \\ \overline{\left(\frac{\delta N}{N}\right)_c^2} &\simeq \frac{\lambda}{2\pi^2}\frac{1}{N}\frac{\mathrm{d}N}{\mathrm{d}z}\left(\frac{\omega_H}{\omega_h}\right)^{1/2}\frac{\omega_H + \omega_h}{|\omega_H - \omega_h|}, \end{aligned}\right\} \tag{7.29a}$$

where $\mathrm{d}N/\mathrm{d}z$ is the height gradient of the electron density in the region of wave reflection and const. $\simeq 90$. By substituting into formulas (7.29) and (7.29*a*) the values of the various quantities characterizing the conditions under which the experiments were conducted, we arrive at the following values for the parameters of the very-small-scale irregularities:

$$\rho_{0\perp} \simeq 0.3 \text{ to } 0.5\,\mathrm{m}, \quad \rho_{0\|} \simeq 1\,\mathrm{km}, \quad \tau_{Td} \simeq 10^{-1}\,\mathrm{s}$$
$$(\delta N/N) \simeq 10^{-3}, \quad (\delta N/N)_c \simeq 10^{-2}.$$

On the whole, these values show a good fit with the various experimental data obtained during studies of the properties of the artificial F_{spor} layer.

The anomalous resonance absorption can, as was shown by Gurevich, 1978, and Scarf *et al.*, 1972, be taken into account using the coefficient of wave reflection $E_r/E_0 = \exp(-K/2)$, where $K = K_v + K_{res}$. Here

$$K_{res} = \left(\frac{\delta N}{N}\right)_r^2 \frac{\pi(\omega\omega_H)^{1/2}}{c[(1/N)(\mathrm{d}N/\mathrm{d}z)]_r}\left(1 - \frac{\omega}{\omega_H}\right)\left(1 + \frac{\omega_H}{\omega}\right)^3, \tag{7.29b}$$

where r denotes the point at which the condition for upper-hybrid resonance $\omega_0^2(z_r) = \omega_h^2 - \omega_H^2$ is satisfied. Formula (7.29*b*) is valid when the transverse dimensions of the irregularities $\rho_\perp < \lambda/2\pi$. The quantities E_r and E_0 are, respectively, the amplitudes of the reflected and incident waves, and $K_v = 2(\omega_h/c)\int k\mathrm{d}z$, where $\kappa(z)$ is the coefficient of absorption of the wave in the ionosphere.

Oblique scattering. The spectra of the scattered waves when they are not reflected specularly from $\mathbf{H}_0$ differ considerably from those described above. In this case as well, of course, we consider irradiation of the ionosphere with a beam of circularly polarized ordinary waves. The

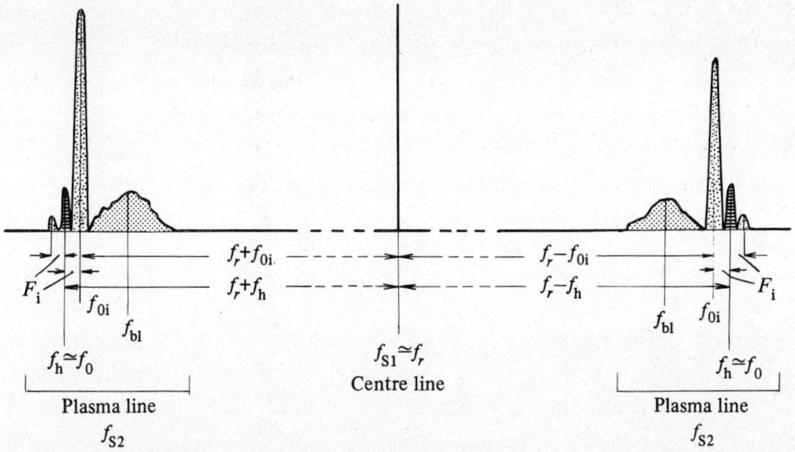

Fig. 7.27. Spectra of the backscattered waves for vertical irradiation of an F_{spor} layer under conditions when the beams of the incident and of the reflected waves are not normal to the direction of the Earth's magnetic field vector $\mathbf{H}_0$ (i.e., $\theta \neq \pi/2$, oblique scattering). Along the horizontal axis, the frequency is increasing from right to left.

corresponding experimental results (see, for instance, Carlson *et al.*, 1972; Gordon & Carlson, 1974; Kantor, 1974) were obtained with the aid of an installation for studying the incoherent scattering of radio waves (Gordon & LaLonde, 1961).

As mentioned above, only the spectrum of the f_{s2} line, i.e., the spectrum of the so-called 'plasma line,' possesses other features. Here the term 'line' is somewhat less appropriate, however, since actually it consists of several quite wide lines. The spectrum of these scattered waves is shown schematically in Fig. 7.27. To the right and to the left of the centre line $f_{s1} = f_r$ which is at the frequency f_r of the beam of incident waves, in a number of cases the scattered-wave spectrum contains two additional lines, at $f_r + f_h$ and at $f_r - f_h$. The latter are shifted quite precisely relative to the centre line by the frequency of the strong wave f_h, which is (approximately) equal to the plasma frequency f_0 of the scattering region of the plasma. However, these lines are not always recorded, and, in any case, they are always considerably less intense than the principal lines of the spectrum. Moreover, it is precisely these components that are intimately associated with the resonance frequency f_0 of the plasma waves. They are thus known as 'growing lines.'

The main component of this set of lines, the amplitude of which is considerably greater than that of the $(f_r \pm f_h)$ lines, is a line that is shifted relative to the latter by the frequency of the ion-acoustic waves. This line can be called the 'decay line'. It appears as a result of parametric decay instability, which apparently plays a significant role in the creation of these wave spectra (Perkins *et al.*, 1974). The frequency shift of these spectral components relative to the centre line f_r can be designated as f_{0i}. In the above-mentioned experiments (Carlson *et al.*, 1972; Kantor, 1974), the following differences were observed:

$$|f_h - f_{0i}| \simeq |f_0 - f_{0i}| = 3.5 \text{ to } 6\,\text{kHz}.$$

However, the properties of the spectral components at frequencies $(f_r + f_{0i}) = f_{s2+}$ and $(f_r - f_{0i}) = f_{s2-}$ where found to be asymmetrical. The amplitude of the spectral line at f_{s2-} is often less than the amplitude at f_{s2+}. Moreover, the frequencies f_{s2+} and f_{s2-} are generally not identical: $f_{s2+} > f_{s2-}$ and on the average $(f_{s2+} - f_{s2-}) \simeq 0.4\,\text{kHz}$ (Hall *et al.*, 1964). The widths of these two lines Δf_{s2} differ little, however, and on the average $\Delta f_{s2+} \simeq \Delta f_{s2-} \simeq 2\,\text{kHz}$. Recall here that for specular scattering $\Delta f_{s2} \simeq 10\,\text{Hz}$!

Just to the side of the two f_{0i} lines the spectrum also has 'broad lines', the frequencies of which will be designated as f_{bl}. The maximum amplitude of their spectra corresponds to a frequency $|f_{0i} - f_{b1}| >$ or $\gg F_r$. In the above experiments the observed values were:

$$|f_{0r} - f_{b1}| = F_{is} \simeq 7 \text{ to } 20\,\text{kHz}.$$

The widths of these broad lines Δf_{b1} were found to vary from 2 to 10 kHz. Although their amplitudes are always less than those of the decay lines f_{0i}, the energy contained in them (i.e. the area of the spectrum) is often twice as great. In some cases weak 'image decay lines' are also recorded. They are shifted relative to the $f_r \pm f_{0i}$ lines (see Fig. 7.27).

As an illustration, let us present here experimental spectra of mode f_{s2}, taken under different conditions during irradiation of the F_{spor} layer at a frequency $f_r = 430\,\text{MHz}$ (Fig. 7.28). One of the f_{s2} spectra has three components: $f_0 \simeq 8.195\,\text{MHz}$, f_{0i}, and f_{b1} (cf. Fig. 7.27). In the other spectrum the $f_0 \simeq f_h$ lines are absent, and the broad-line spectrum f_{bl} is less intense. In both cases the f_{bl} spectra are asymmetrical.

Conclusions. Consequently, the scattered-wave spectra are seen to be quite complicated, with a structure that at first sight would appear to be explainable in terms of the effect of some nonlinear mechanism involving the parametric decay instability (see below, Section 8.3.2). In this case there may well be a nonlinear transformation of the incident transverse wave into longitudinal high-frequency plasma oscillations near ω_0 and ion-acoustic oscillations near ω_s. Nevertheless, such a theoretical expla-

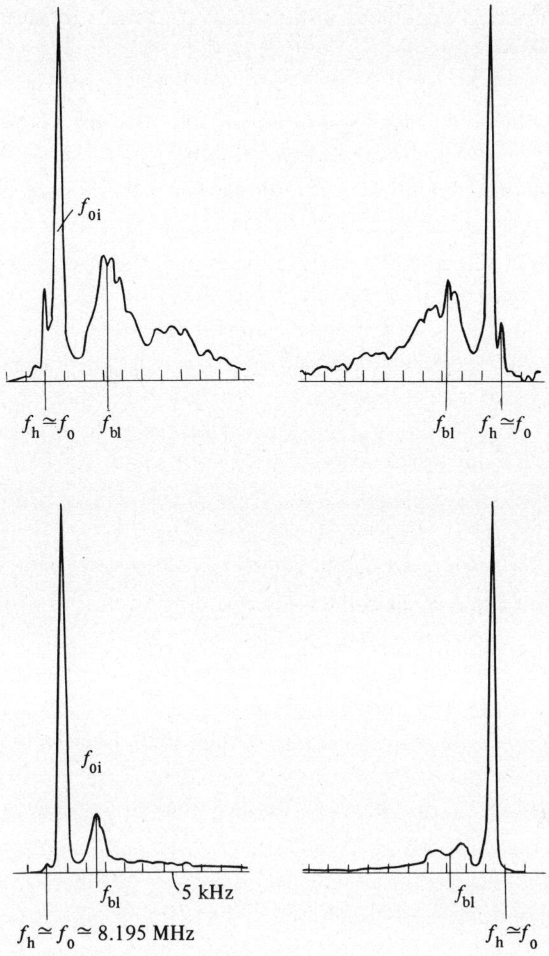

Fig. 7.28. Experimental spectra of the 'plasma line' f_{s2} for scattering, i.e., when the angle θ between the beam of waves incident upon the perturbed region of the ionosphere and the direction of the vector $\mathbf{H}_0$ is not equal to $\pi/2$. The observations were carried out at a frequency $f_r = 430$ MHz, while the frequency of the heating transmitter was $f_h^{(o)} = 8.195$ MHz (Gordon & Carlson, 1974).

nation (see, for example, Perkins *et al.*, 1974) has not as yet been found to be satisfactory *vis-à-vis* the results of a number of experiments. This disagreement between theory and experiment does not, however, mean that parametric instability plays no part at all in the phenomena being considered, even during the initial stage of their development. The problem will be solved satisfactorily only when we have available the findings of detailed comprehensive experimental studies of this intricate complex of phenomena, findings which will enable their complete and thoroughgoing explanation to be found.

7.2.4. Self-action and focusing of waves and other effects of changes in their field structure

If a wave is strong enough, it can alter significantly its own propagation in a plasma. As mentioned above, this may manifest itself as a variation in the coefficient of wave absorption, in the wave phase, or in the nature of the amplitude and phase modulation; it may also cause the appearance of harmonics of the carrier frequency ω of the wave and a number of other effects. Moreover, the field of a strong wave, if it is more intense than some critical value, changes the ionization balance of the plasma. Fast electrons cause additional ionization, and instability may well appear, leading to excitation in the plasma of natural low-frequency oscillations and to amplitude *self-modulation* of the wave which produced these oscillations. In plasma regions where the refractive index is quite low ($n^2 - \kappa^2 \to 0$), for instance in the region of wave reflection by the ionosphere, specific phenomena appear due to the standing waves resulting from the nonlinear interaction between the incident wave and the partially or totally reflected wave. It should also be kept in mind that in itself a nonlinear interaction between ordinary and extraordinary waves propagated in a plasma is bound to affect the structure of their field. Finally, the *self-action* of a spatially-narrow *beam of waves* is also important. Such beams may be created either by artificial means, with the aid of a directional antenna, or else they may appear in the plasma due to the trapping of electromagnetic waves in a cone around the direction of the vector $\mathbf{H}_0$ of the constant magnetic field (see Chapter 8). Depending on the conditions in the plasma, a beam of waves may become compressed (*self-focusing of the beam*) or may expand (*defocusing of the beam*).

Let us describe here some computational results, presented in the form of diagrams, which will illustrate quantitatively some of the more interesting effects observed in a plasma, where the role of electron – neutral collisions is minor (Gurevich, 1958b, 1978).

The decrease in the amplitude of the electric field E_0 of a wave in a plane-stratified (in the z direction) isotropic plasma (ionosphere) is determined in the field of a strong wave by the attenuation coefficient

$$\kappa[z, \omega, E_0(z)] = \kappa_0(z, \omega)\left(\frac{N(T_e)}{N_0}\right)\left(\frac{\nu_e(T_e)}{\nu_{e0}}\right)\frac{\omega^2}{[\omega^2 + \nu_e^2(T_e)]}, \qquad (7.30)$$

where

$$\kappa_0(\omega, z) = \tfrac{1}{2}\left(\frac{\nu_{e0}}{\omega}\right)\frac{\omega_0^2}{\omega^2 n(\omega, z)} \tag{7.31}$$

is the attenuation coefficient in a weak field ($p \ll 1$) when $\omega^2 \gg \nu_{e0}^2$. Assuming, as previously, that everywhere $\omega^2 \gg \nu_e^2(E_0)$ (see formula (7.8)), we obtain

$$\kappa[\omega, z, E_0(z)] = \tfrac{1}{2}\left(\frac{\nu_e(\omega)}{\omega}\right)\left(\frac{N(E_0)}{N_3}\right)\frac{\omega_0^2}{\omega^2 n(\omega, z)}. \tag{7.32}$$

When only electron ion collisions play a role, the quantities $\nu_e[E_0(z)]$ and $N[E_0(z)]$ are determined by formulas (7.9), (7.10), and (7.29).

Solution of the wave equation in the approximation of geometrical optics, when $N(z)$ varies quite slowly and it is possible to neglect the disturbance of the electron density $\Delta N(E_0)$, leads to the equation

$$\frac{\tau-1}{\tau+1}\exp\left[2\left(\tau+\frac{\tau^3}{3}\right)\right] = \frac{\tau_0-1}{\tau_0+1}\exp\left\{2\left[\tau_0+\frac{\tau_0^3}{3}-K(\omega, z)\right]\right\}, \tag{7.33}$$

giving the distribution of the field $E_0(z)$ taking into account the nonlinear properties $\kappa[\omega, z, E_0(z)]$. In implicit form (7.33) defines $E_0(z)$ in terms of the dimensionless quantity τ:

$$\left.\begin{aligned} &\tau(z, E_0) = \left(\frac{T_e(E_0)}{T_{e0}}\right)^{1/2} \simeq \left(1+\frac{E_0^2(z)}{E_p^2}\right)^{1/2}, \quad \tau_0 = \left(\frac{T_e[E_0(0)]}{T_{e0}}\right)^{1/2}, \\ &E_0(0) = E_0(z=0), \quad K(z) = \frac{\omega}{c}\int_0^z \kappa_0(\omega, z)\mathrm{d}z. \end{aligned}\right\} \tag{7.34}$$

The solution of formula (7.33) gives us the field within the plasma:

$$E_0(z) = E_0(0)P(\tau_0)\exp[-K(z)] \tag{7.35}$$

where

$$P(\tau_0) = \frac{2E_p}{E_0(0)}\left(\frac{\tau_0-1}{\tau_0+1}\right)^{1/2}\exp\left(\tau_0-1+\frac{\tau_0^3-1}{3}\right), \tag{7.36}$$

can be called the *self-action factor*.

Figure 7.29 shows curves of $E_0(z)/E_0(0)$ as a function of $K(z)$ (for fixed values of $p = E_0/E_p$), and of p (for fixed values of $K = K_s$), obtained with the aid of formulas (7.33)–(7.36). These curves illustrate the marked drop in wave attenuation (increase of $E_0(z)$, at sufficiently high values of p.

The *phase perturbation* $\Delta\phi$ in the field of a strong wave is described by

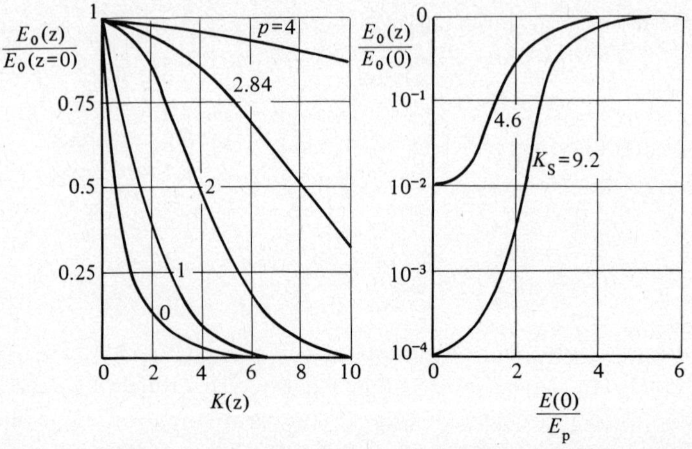

Fig. 7.29. Curves of $E_0/E_0(0)$ illustrating the effect of the self-action of a wave in the ionosphere (rise in intensity of the field when the wave is strong). (Gurevich, 1978.)

the formula

$$\Delta\varphi \simeq -\frac{\pi}{2}\left(\frac{\pi}{2}\right)^{1/3}\left(\frac{\omega}{\nu_{\mathrm{ei}0}}\right)^{1/3}\left(\frac{\nu_{\mathrm{ei}}}{\omega}\right)\left(\frac{1}{\tau^2}-\frac{1}{\tau_0^2}\right). \tag{7.37}$$

In the quasi-steady state, when $\Omega_0 \ll \delta_0 \nu_{\mathrm{e}0}$, the *amplitude-modulation index* M_0 of a wave incident upon the plasma boundary ($z = 0$), for which

$$E_{z=0} = E_0(1 + M_0 \cos \Omega_0 t) \tag{7.38}$$

decreases considerably. Thus, for instance, when $M_0 \ll 1$ and, as previously, $\omega^2 \gg \nu_{\mathrm{ei}}^2(E_0)$, we have

$$M[E_0(z)] = M_0 \frac{\tau[E_0(z)]}{\tau_0[E_0(0)]} \tag{7.39}$$

Since for large distances z the quantity $\tau \to 1$, a wave quite deep inside the plasma will be *markedly demodulated* and $M = M_0/\tau_0$. In general, for arbitrary values of M_0, the modulation index

$$M(z) = \frac{E_{\max}(z) - E_{\min}(z)}{E_{\max}(z) + E_{\min}(z)} = M(E_0) \tag{7.40}$$

can be studied only with the aid of numerical calculations (see Fig. 7.30*a*). The results presented here, computed with the aid of formulas (7.39) and (7.40), pertain to a slightly ionized plasma, such as the D and E regions of the ionosphere, where it is already necessary to take electron – neutral collisions into account when considering such phenomena (see Gurevich

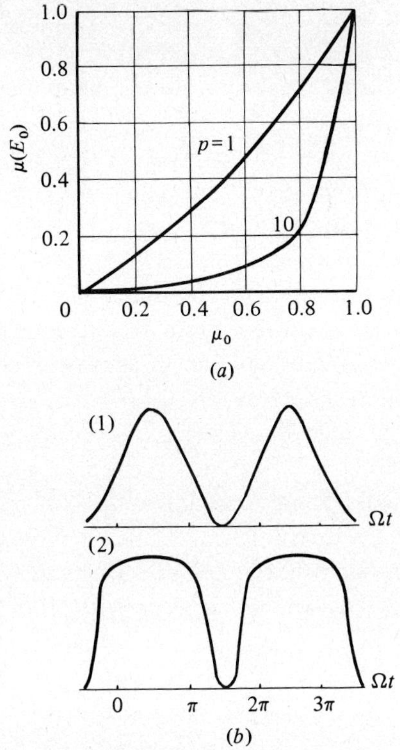

Fig. 7.30. Curves characterizing the variation of the phase refractive index and the nature of the wave modulation in the ionosphere under the influence of a strong radio frequency electric field (Gurevich, 1978).

& Shlyuger, 1975). Inspection of the figure shows that the demodulation of the incident wave diminishes as $M_0 \to 1$.

In the wave field, of course, there will be a distortion of the form of the modulation function $\cos \Omega t$. The modulating signal will contain high harmonics $\cos 2\Omega t, \cos 3\Omega t \ldots$ Figure 7.30*b* illustrates the change, within the plasma interior, of the form of the modulation function (curve 1), when $E_0 E_p = 10$ (curve 2). Due to the nonlinear variation of the wave phase (see (7.37)), *phase modulation* appears in the field of the amplitude-modulated (see (7.38)) wave. If $M \ll 1$, then

$$\Delta\phi(t, E_0) \simeq \Delta\phi_0(M_0 = 0) + M_0 \beta \cos \Omega t + \ldots, \tag{7.41}$$

where the phase-modulation index

$$\beta = E_0(0) \frac{\partial(\Delta\phi)}{\partial E_0(0)}. \tag{7.42}$$

The index β is not large, and in the general case, if $\tau_0 \gg 1$, its maximum value is $\beta_{max} \simeq M_0/2\sqrt{2}$ (Gurevich, 1958b).

The above-considered effects of wave self-action were calculated for the case of a plane wave propagating through the plasma. For propagation of a *narrow beam of waves*, on the other hand, which can be represented as an aggregate of plane waves with different wave vectors $\mathbf{K}_s = (\omega/v_\phi)\mathbf{s}$, where $\mathbf{s}$ is the unit vector in the direction of propagation of the sth plane wave, the picture is more complicated. The nonlinearity of the plasma violates the principle of superposition of the initial wave beam. Thus, the waves ultimately combine into a beam of different form.

In plasma regions where the electron density is reduced as a result of the strong wave, the beam of waves is *compressed*, or *focused*. It is precisely this case which is of interest to us here; it corresponds to conditions for which $v_{ei}^2 \gg v_{en}^2$. Moreover, a beam structure of a certain type created in a plasma gradually changes: additional maxima appear and a single beam may decay into several beams. For such an evolution of the wave beams, certain conditions have to be satisfied, which depend both on the initial form of the beams and on the properties of the plasma. Obviously,

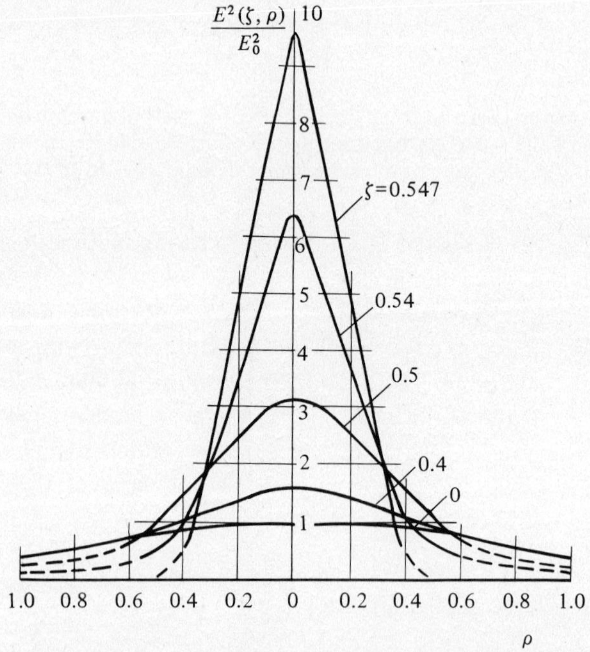

Fig. 7.31. Calculated effect of wave-beam focusing in the ionosphere on the field of a strong wave (Gurevich, 1978).

studies of the nonlinear development of a wave beam in a plasma involve the solution of some very complicated nonlinear equations. The computational results (see Gurevich, 1978) presented here give a general qualitative and quantitative picture of the *self-action* of a beam of waves in a plasma.

Fig. 7.31 portrays the computational results for a parallel beam of waves propagated in an isotropic plasma along the z-axis, which is perpendicular to the boundary between the plasma and free space. The relative intensity of the incident beam of waves has a Gaussian distribution at the plasma boundary:

$$E^2(\rho_0, z=0) = E_0^2 \exp(-\rho^2), \quad \rho^2 = \frac{\rho_0^2}{d^2} = (x^2+y^2)/d^2. \qquad (7.43)$$

The numbers next to the curves in the figure denote the dimensionless distances to which the beam has penetrated into the plasma, reckoned from $z=0$, namely

$$\zeta = (z/d)(\beta E_0^2)^{1/2}, \qquad (7.44)$$

where d is the characteristic width of the incident wave beam in the xy plane. The quantity β in (7.44) is the nonlinearity parameter; it governs the perturbation of the plasma dielectric permittivity as a result of the field, i.e.,

$$\Delta\varepsilon(z, E^2) = -\left(\frac{\omega_0}{\omega}\right)^2 \frac{\varepsilon_0(z)}{N} \frac{\partial N}{\partial E^2} E^2(z) = 2\beta\varepsilon_0(z)E^2(z). \qquad (7.45)$$

The computational results presented here pertain to the case when $\Delta\varepsilon \ll \varepsilon_0(z)$, where $\varepsilon_0(z)$ is the permittivity in the absence of a field. As Fig. 7.31 shows, with an increase in ζ, i.e., as the beam penetrates into the plasma, it gradually becomes focused. At some critical value ζ_k, the beam intensity along the z-axis reaches a maximum. For the case in Fig. 7.31, $\zeta_k = 0.547$, while $E^2(\zeta, K)/E_0^2 = 9.6$. With a further rise in ζ, the beam at first becomes slightly deformed, but then, at quite high values of $\zeta > \zeta_k$, scattering of the beam may appear.

Fig. 7.32 portrays a more complicated case of the variation of the structure of a beam in a plasma. For an initial beam shape described by the formula

$$E^2(\rho, z=0) = E_0^2(1-\rho^2)(\rho^4 - \rho^2 - 1) \qquad (7.46)$$

a beam penetrating the plasma is first compressed and then, at some height corresponding to transition through the critical value ζ_k, where $\zeta > \zeta_k$ (for (7.46), $\zeta_k = 0.38$), the beam starts to exhibit secondary maxima, i.e., it

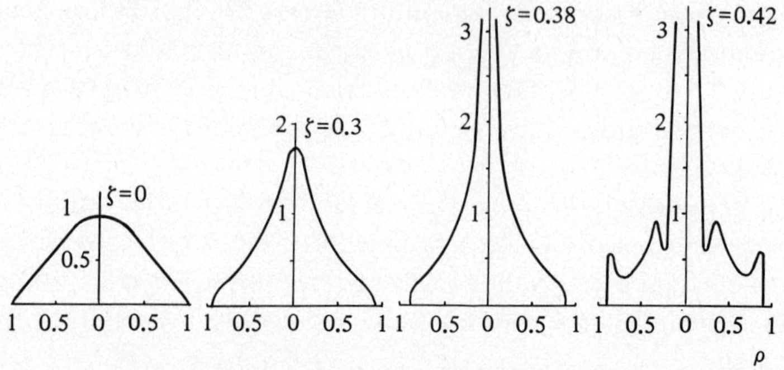

Fig. 7.32. Calculated results illustrating the effect of focusing and the variation of the form of a wave beam (decay of the beam) in the field of a strong wave; the ordinate is $E^2(\zeta, \rho)/E_0^2$, as in Fig. 7.31 (Gurevich, 1978).

starts to decay. In the transition region $\zeta \simeq \zeta_k$, the beam intensity at the z-axis, as in the case considered in Fig. 7.31, is about an order of magnitude greater than its initial intensity. It should be noted that the distance between maxima for the beam described by formula (7.46) is of the order of $0.2d$.

The examples cited indicate that the deformation of a beam in plasma, and beam decay in particular, depend considerably on the initial form of the beam. With regard to this, it is interesting that a beam of parabolic shape $E^2(\rho, z=0) = E_0^2(1-\rho^2)$, all the rays of which emanate from a single point $z = -z_0$ (as $z_0 \to \infty$, the beam becomes parallel) does not become focused if z_0 is less than a certain critical value z_k. However, when $z_0 > z_k$, the beam begings to be markedly compressed and, strictly mathematically, focused to a point; at this point its intensity goes to infinity.

When a beam of waves penetrates a magnetoplasma ($\mathbf{H}_0 \neq 0$), the structure of a beam symmetrical about the z-axis will also vary in form. Within the plasma the horizontal sections of the beam (circular at $z=0$) become elongated along the magnetic-field vector $\mathbf{H}_0$.

The *self-focusing* of a strong wave in the ionosphere leads in some cases to the formation of a 'bulge' in the region of wave reflection, due to the considerable variation in the trajectory of the wave. Here, if the frequency of the strong wave f_h is close to the critical frequency $f^{(o)}$ F2, then the centre of the bulge has a 'hole' in it, through which the strong wave can leak and continue beyond the ionosphere. (Note that the possibility of such wave leakage through a region in which a strong wave acts to reduce the electron density $N(y, z)$ is also implied by the calculated results in Fig. 7.13, above; however, these results were obtained without

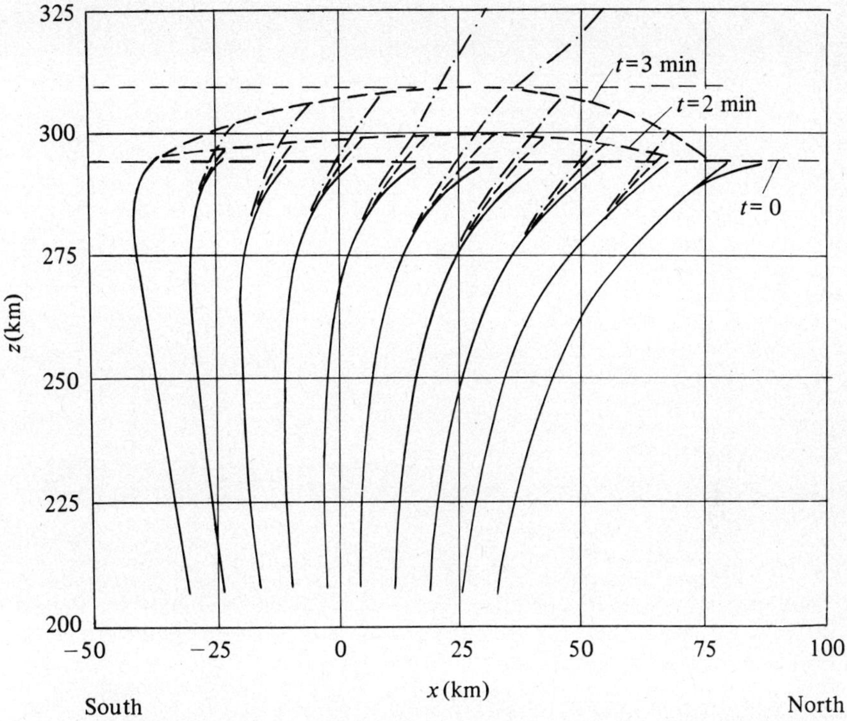

Fig. 7.33. Calculated results illustrating the formation of a 'bulge' and a 'hole' for a family of trajectories of a strong wave ($f_h = 5.9$ MHz) near the level of its reflection from the ionosphere. The calculations were carried out for $f^{(o)}$F2 = 6.00 MHz and z_MF2 = 310 km (Meltz *et al.*, 1974).

taking self-focusing of the wave into account). Figure 7.33 illustrates such propagation of a strong wave, with a frequency $f_h = 5.94$ MHz. These results were obtained for a model of the ionosphere in which the critical frequency $f^{(o)}$F2 = 6.00 MHz, with z_MF2 = 310 km (Meltz *et al.*, 1974). Fig. 7.33 shows the trajectories of the strong wave two minutes and three minutes after it is turned on. The reflecting surface (horizontal at $t = 0$) gradually gets a bulge in it, and then a 'burn-through' is formed, via which the waves 'leak' through the maximum of the ionosphere. A similar effect was apparently observed experimentally, during studies of the backscattering of radio waves from the ionosphere (Minkoff *et al.*, 1974a).

The effects of focusing in the field of a strong wave naturally cause a complex redistribution of its energy flux in the reflection region (Fig. 7.34). Figure 7.34 shows the calculated distribution of the wave energy for a model of the ionosphere in which $f^{(o)}$F2 = 6.34 MHz and z_MF2 = 250 km,

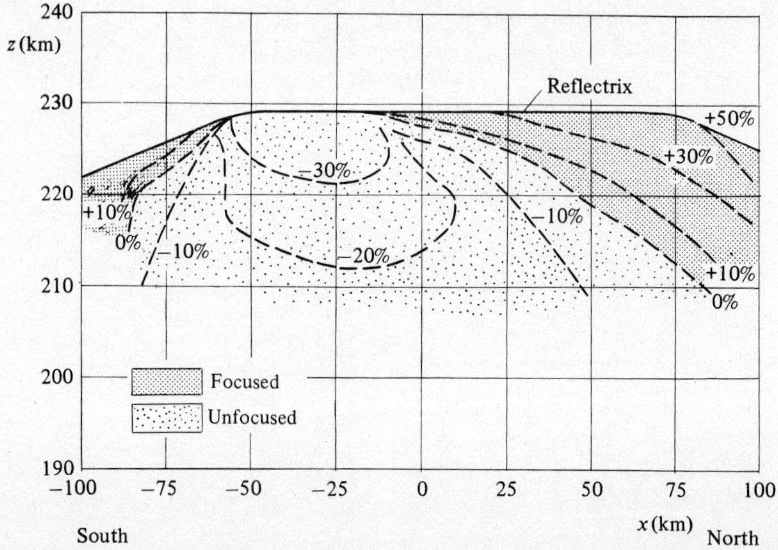

Fig. 7.34. Curves illustrating the redistribution of the energy flux (Poynting vector) near the level of reflection of a strong wave ($f_h = 5.77$ MHz) from the ionosphere $f^{(o)}F2 \simeq 6.34$ MHz, $z_M F2 \simeq 250$ km. (Issledovaniya Kosmicheskogo Prostranstva, 1974.)

for a frequency of the strong wave $f_h = 5.77$ MHz. In a certain region close to the level of total wave reflection (reflectrix) *energy focusing* of up to $+50\%$ takes place, while in its central part, on the other hand, the defocusing may be as much as -30%.

It should be mentioned here that the *nonlinear interaction* of several waves, one of which is a strong wave, may lead to *cross-modulation* of radio waves, a *change in the shape* of a packet of weak waves, the appearance of *combination waves* with frequencies $(\omega_h \pm \omega_0)$, the focusing of a weak wave ω, etc. We will not consider these effects here, however (see Gurevich, 1978), since they usually pertain to the lower (D and E) regions of the ionosphere.

7.3. Parametric nonlinear phenomena

In this section we will consider the effects of certain mechanisms of plasma instability relative to the Langmuir oscillations in the strong field of a wave. This is often called *parametric decay instability* in the field of a *strong 'pump' wave* (DuBois & Goldman, 1965, 1967; Silin, 1964, 1965; Nishikawa, 1968). Essentially this phenomenon takes place as follows. In a strong field under certain conditions a transverse wave is transformed into a longitudinal wave or some other type of transverse wave. A high-frequency wave can thereby excite simultaneously both high-frequency ($\omega \simeq \omega_0$) and low-frequency ($\omega \simeq \Omega_0$) Langmuir waves (Aliev & Silin, 1965) or some other type of plasma oscillation. Thus the wave *decays* into several waves (see below). Such resonance phenomena, observed in a plasma, may occur without the participation of collisions between the particles. They find an analogy in the theory of oscillations, when in a concentrated or distributed system under certain conditions a growth in the oscillations of the system is observed. This so-called parametric resonance occurs if a parameter of the oscillating system, for instance the length of a pendulum or the capacitance of a condenser, varies periodically in time (see Mandel'shtam, 1955); hence the following terms used in plasma theory: parametric resonance, parametric instability, etc.

As a typical parameter determining whether a field is strong or weak, it is convenient in the given case to take

$$\rho_{\mathrm{p}} = 2\pi T_{\mathrm{e}}/\lambda = e\mathbf{k}\cdot\mathbf{E}_0/m\omega_{\mathrm{p}}^2, \tag{7.47}$$

where $\mathbf{r}_{\mathrm{e}} = e\mathbf{E}_0/m\omega_{\mathrm{p}}^2$ is the displacement of an electron in the wave field $E_0 \exp(\mathrm{i}\omega_{\mathrm{p}} t)$ and $\mathbf{K}$ is the wave vector. A field is considered to be strong or weak, respectively, if

$$\rho_{\mathrm{p}} \gg 1 \text{ or } \rho_{\mathrm{p}} \ll 1. \tag{7.48}$$

Note that (7.47) and (7.48) are equivalent to the conditions

$$\left.\begin{aligned} &\lambda \ll \dot{r}_{\mathrm{e}} T_{\mathrm{p}}, \quad \lambda \gg \dot{r}_{\mathrm{e}} T_{\mathrm{p}} \\ &\dot{r}_{\mathrm{e}} \gg v_{\phi}, \quad \dot{r}_{\mathrm{e}} \ll v_{\phi}. \end{aligned}\right\} \tag{7.49}$$

In the formulas (7.49), $\dot{r}_{\mathrm{e}}$ is the directed velocity of an electron under the influence of the wave field. The oscillation period of the wave is $T_{\mathrm{p}} = 2\pi/\omega_{\mathrm{p}}$ and its phase velocity is $v_{\phi} = \omega_{\mathrm{p}}/K$.

The main features of parametric nonlinearity in a collisionless plasma are:

(1) A variation in the resonance branches of the oscillations;

(2) A sensitive instability of the plasma, leading to the excitation of both longitudinal (potential) and transverse (nonpotential) oscillations in relatively weak fields;

(3) A 'decay' character of the instability, making possible the creation of several oscillation spectra with different frequencies in the field of a single harmonic pump wave.

Below we will consider briefly the results of some theoretical studies of these effects.

7.3.1. Nonlinear spectra of the resonance branches of longitudinal waves

Under the influence of an electric field $E = E_0 \exp(i\omega_p t)$, a magnetoplasma has four resonance branches instead of the three, $\omega_1(\theta), \omega_2(\theta)$ and $\omega_3(\theta)$, described by the dispersion equation (4.48) and the formulas (4.49), (4.52), and (4.53) (see Aliev & Silin, 1965, Domrin, 1967, Silin, 1973). If the frequency of the electric field is considerably higher than the characteristic frequencies of the plasma, i.e., if $\omega_p \gg \omega_0, \omega_H$, then assuming satisfaction of the conditions (4.3*a*) making it possible to neglect the thermal motion of the particles, the dispersion equation describing the resonance branches will have the form

$$\left.\begin{aligned} &1 - \frac{\omega_0^2}{\omega^2}\cos^2\theta - \frac{\omega^2\sin^2\theta}{\omega^2 - \omega_H^2} - \frac{\Omega_0^2}{\omega^2}\cos^2\theta - \frac{\Omega_0^2\sin^2\theta}{\omega^2 - \Omega_H^2} \\ &\quad + \Omega_0^2\omega^2[1 - J_0^2(\rho_p)]\left(\frac{\sin^2\theta}{\omega_H^2 - \omega^2} - \frac{\cos^2\theta}{\omega^2}\right)\left(\frac{\sin^2\theta}{\Omega_H^2 - \omega^2} - \frac{\cos^2\theta}{\omega^2}\right) = 0, \end{aligned}\right\} \tag{7.50}$$

where J_0 is a Bessel function. It is easy to show that when $\rho_p \to 0$ equation (7.50) becomes (4.48), since $J_0(\rho_p) \to 1$. For weak fields, moreover, $\rho_p \ll 1$, and using the expansion of the square of the Bessel function $J_0^2(\rho_p)$:

$$J_0^2(\rho_p) \simeq 1 - \tfrac{1}{2}\rho_p^2 + \ldots \simeq 1 - \tfrac{1}{2}\left[\frac{e\mathbf{K}\cdot\mathbf{E}_0}{m\omega_p^2}\right]^2 + \ldots$$

we find that the last term on the left-hand side of (7.50) is small.

Equation (7.50) defines the four resonance branches $\omega_{1,2,3,4}(\theta)$ depicted schematically in Fig. 7.35 for the case when $\omega_0 > \omega_H$. For the additional branch at $\theta = 0$, the frequency is

$$\omega_4(0) = \omega_0^2[1 - J_0^2(\rho_p)] \tag{7.51}$$

The lower-hybrid frequency at $\theta = \pi/2$ differs from that for $E_0 = 0$: thus

$$\omega_{Lp}^2 = \frac{\omega_H\Omega_H}{1 + \omega_H^2/\omega_0^2} + \frac{\Omega_0^2[1 - J_0^2(\rho_p)]}{1 + \omega_H^2/\omega_0^2} = \omega_L^2 + \frac{\Omega_0^2[1 - J_0^2(\rho_p)]}{1 + \omega_H^2/\omega_0^2}. \tag{7.52}$$

Obviously, when $\rho_p \to 0$ and $\Omega_0^2 \gg \Omega_H^2$, a condition which has been assumed to be satisfied everywhere here (see (4.12)), formula (7.52) naturally becomes formula (4.54) (see also (4.22)), which defines the lower-hybrid frequency ω_L without taking into account the effect of the electric field. The upper-hybrid frequency, taking the effect of the electric field into account, is

$$\omega_{Up}^2 = \omega_H^2 + \omega_0^2 + \omega_{Lp}^2. \tag{7.53}$$

Generally ω_{Lp}^2 is much lower than ω_0^2 and ω_H^2, so the upper-hybrid

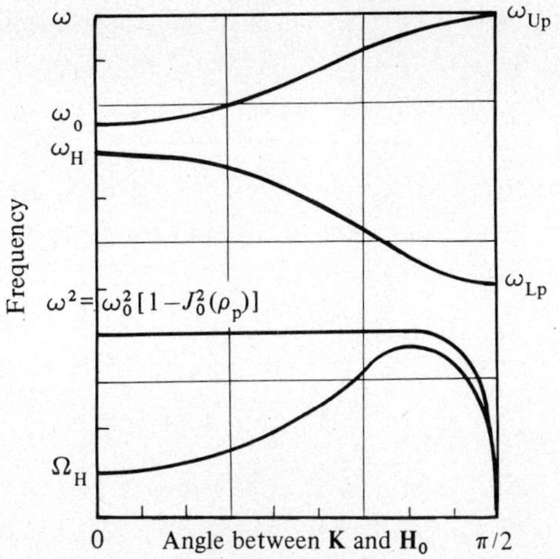

Fig. 7.35. The four resonance branches of a magnetoplasma in the field of a strong wave (Domrin, 1967).

frequency depends only slightly on the effect of the field: $\omega_{\mathrm{Up}} \simeq \omega_{\mathrm{U}}$ (see (4.22) and (4.50)). It should be noted that at frequencies $\omega \gg \omega_{\mathrm{L}}$, the branches of the resonance frequencies $\omega_1(\theta)$ and $\omega_2(\theta)$ generally depend little on the field E_0, since the motion of the ions relative to the electrons does not play a role here. In this frequency range these resonance branches are described by formula (4.49) with a $\pm$ sign ahead of the kernel (see also Fig. 4.6 and (4.54)).

Now let us present some limit formulas for the resonance frequencies, where in some of the formulas (if necessary) the thermal motion of the particles is taken into account (Aliev & Silin, 1965; Domrin, 1967).

In the frequency range $\omega \ll \Omega_{\mathrm{H}}$, if the condition

$$\sin^2\theta(\Omega_0^2/\Omega_{\mathrm{H}}^2)[1 - J_0^2(\rho_{\mathrm{p}})] \gg 1 \tag{7.54}$$

is satisfied, i.e., for angles quite far away from $\theta = 0$, it follows from (7.50) that

$$\left.\begin{aligned} \omega_3^2(\theta) &\simeq \Omega_{\mathrm{H}}^2 \cot^2\theta, \\ \omega_4^2(\theta) &\simeq \omega_0^2[1 - J_0^2(\rho_{\mathrm{p}})]\cos^2\theta. \end{aligned}\right\} \tag{7.55}$$

In the opposite case, (7.54),

$$\left.\begin{aligned} \omega_3^2(\theta) &= \Omega_0^2[1 - J_0^2(\rho_{\mathrm{p}})]\cos^2\theta, \\ \omega_4^2(\theta) &= \frac{\omega_0^2}{1 + (\Omega_0/\Omega_{\mathrm{H}})^2\sin^2\theta}\left[1 + \frac{m}{M}J_0^2(\rho_{\mathrm{p}})\right]. \end{aligned}\right\} \tag{7.56}$$

If thermal motion of the particles is taken into account, when

$$\left(\frac{K_\perp v_e}{\omega_H}\right)^2, \left(\frac{K_\perp v_i}{\Omega_H}\right)^2 \ll 1, \quad \left(2\pi\frac{\rho_{ei}}{\lambda}\right)^2 \ll 1, \quad v_i \ll \frac{\omega}{K_\parallel} \ll v_E, \qquad (7.57)$$

then the $\omega_3(\theta)$ branch in the frequency range $\omega \ll \Omega_H$ describes the spectrum of *modified ion-acoustic oscillations* of the plasma:

$$\omega_3^2(\theta) = \cos^2\theta \left\{ \frac{\Omega_0^2\left[1 - \dfrac{J_0^2(\rho_p)}{1+(KD_e)^2}\right]}{1 + \left(\dfrac{\Omega_0}{\Omega_H}\right)^2 \sin^2\theta \left[1 - \dfrac{J_0^2(\rho_p)}{1+(KD_e)^2}\right]} + 3(Kv_i)^2 \right\} \qquad (7.58)$$

(Domrin, 1967; Silin, 1973). In a cold plasma ($T=0$) formula (7.58) naturally becomes (7.55) and (7.56). In a strong field ($\rho_p \gg 1$)

$$\omega_3^2(\theta) = \cos^2\theta \left[\frac{\Omega_0^2}{1 + (\Omega_0/\Omega_H)^2 \sin^2\theta} + 3(Kv_i)^2\right]. \qquad (7.59)$$

The temporal decay rate of these waves due to Landau damping is

$$\gamma_3 = -\frac{\pi^{1/2}}{2(2)^{1/2}} \frac{\omega^4}{(K_\parallel v_i)^3} \left\{ \exp\left[-\left(\frac{\omega}{2^{1/2} K_\parallel v_i}\right)^2\right] + \left(\frac{v_i}{v_e}\right)^3 \frac{\Omega_0^2 \omega_0^2 J_0^2(\rho_p) \cos^2\theta}{\omega^4 (1 + K^2 D_e^2)} \right\}. \qquad (7.60)$$

In the range $\Omega_H < \omega \ll \omega_H$, assuming that the conditions

$$v_i/v_\phi \gg 1, \quad 2\pi\rho_e/\lambda \simeq 1 \qquad (7.61)$$

are satisfied, i.e., when the limitations (4.3*a*) are not valid, we have

$$\omega_3^2 = \Omega_0^2\left[1 - \frac{J_0^2(\rho_p)}{1+(KD_e)^2}\right],$$

$$\gamma_3 = -\frac{\pi^{1/2}}{2} \frac{\Omega_0 (KD_e)}{1 + (KD_e)^2} \frac{I_n(z_e) J_0(\rho_p) \exp(-z_e)}{(K_\parallel v_e)} \qquad (7.62)$$

where I_n is a modified Bessel function of the first kind, $z_e \simeq (K_\perp \rho_e)^2$, while D_e is the electronic Debye length, and $K_\perp = 2\pi/\lambda_\perp$ and $K_\parallel = 2\pi/\lambda_\parallel$ are, respectively, the longitudinal and transverse components of the wave vector relative to the direction of the external constant magnetic field $\mathbf{H}_0$.

In the field of a high-frequency wave, an isotropic plasma ($\mathbf{H}_0 = 0$) has two resonance branches: a high-frequency branch $\omega_1(\rho_p)$ and a low-frequency branch $\omega_4(\rho_p)$. Taking into account the thermal motion of the particles, we have

$$\left.\begin{aligned} \omega_1^2 &= \omega_0^2\left[1 + \frac{m}{M} J_0^2(\rho_p)\right] + 3(v_e K)^2, \\ \gamma_1 &= -e^{-3/2} \frac{\pi^{1/2}}{2(2)^{1/2}} \frac{\omega_0^4}{(Kv_e)^3} \exp\left[-\left(\frac{\omega}{Kv_e}\right)^2\right], \end{aligned}\right\} \qquad (7.63)$$

and

$$\left.\begin{aligned} \omega_4 &= \Omega_0^2[1 - J_0^2(\rho_p)] + 3K^2[v_i^2 + v_s^2 J_0^2(\rho_p)], \\ \gamma_4 &= -\frac{\pi^{1/2}}{2(2)^{1/2}} \frac{\omega_0^4}{(Kv_e)^3} \left\{ \frac{m}{M} J_0^2(\rho_p) \exp\left[-\left(\frac{\omega}{Kv_e}\right)^2\right] \right. \\ &\quad \left. + \left(\frac{v_e}{v_i}\right)^3 \exp\left[-\left(\frac{\omega}{Kv_i}\right)^2\right]\right\} \end{aligned}\right\} \tag{7.64}$$

where γ_1 and γ_4 characterize the Landau damping by electrons, and $v_s = (kT_e/M)^{1/2}$ is the velocity of nonisothermal sound. Both resonance branches become attenuated slightly when their phase velocities are much higher than the thermal velocities of, respectively, the electrons and the ions. The first branch describes the usual HF electron Langmuir oscillations, the frequency of which has a small correction. A new quality of the nonlinear properties of a plasma is revealed in this case by an analysis of the properties of the low-frequency oscillations. This is seen most clearly if we neglect the thermal motion of the particles, which in this instance does not alter the essence of the phenomenon. Actually, when $\rho_p \ll 1$ and $v_s, v_i \ll V_{ps}$, we have

$$\omega_4^2 = (\mathbf{K}\cdot\mathbf{V}_{ps})^2 = \Omega_0^2\left(\frac{e\mathbf{K}\cdot\mathbf{E}}{2^{1/2} m\omega_p^2}\right)^2. \tag{7.65}$$

The oscillations of frequency ω_4 describe sound waves arising in the plasma as a result of the pressure caused by the oscillatory motion of the electrons. The phase velocity of these waves is

$$\mathbf{V}_{ps} = \frac{e\Omega_0}{2^{1/2} m\omega_0^2}\mathbf{E}_0. \tag{7.66}$$

It is evident from (7.66) that, when $\mathbf{E}_0 \to 0$, the oscillations described by (7.65) are absent; such waves are known as anisotropic sound (Aliev & Silin, 1965).

In the nonisothermal plasma $T_e \gg T_i$, the low-frequency oscillations describe a wider class of waves, namely ionic sound:

$$\left.\begin{aligned} \omega_4^2 &= \Omega_0^2\left[1 - \frac{J_0^2(\rho_p)}{1 + (KD_e)^2}\right] + 3(Kv_i)^2, \\ \gamma_4 &= -\frac{\pi^{1/2}}{2(2)^{1/2}} \frac{\omega^4}{(Kv_i)^3}\left\{\exp\left[-\left(\frac{\omega}{Kv_i}\right)^2\right] + \left(\frac{v_i}{v_e}\right)^3 \frac{\Omega_0^2\omega_0^2 J_0^2(\rho_p)}{\omega^4[1 + (KD_e)^{-2}]^2}\right\} \end{aligned}\right\} \tag{7.64a}$$

On condition that $\rho \ll 1$, we have

$$\omega_4^2 = (\mathbf{K}\cdot\mathbf{V}_{sp})^2 + (Kv_s)^2. \tag{7.65a}$$

7.3.2. Parametric excitation of longitudinal waves

With a reduction in the frequency of the external electric-field source $E_0 \exp(i\omega_p t)$, i.e., for values of ω_p of the order of the *HF resonance frequencies* $\omega_1(\theta)$ and $\omega_2(\theta)$ (see (4.49)), parametric instability of the plasma makes its appearance. Then, maximum values of the growth rates of the corresponding longitudinal oscillations $\omega_1(\theta)$ and $\omega_2(\theta)$ are observed at all the harmonics of the external source, i.e., when

$$\omega_1(\theta) = s\omega_p \text{ or } \omega_2(\theta) = s\omega_p \tag{7.67}$$

($s = 1, 2, 3, \ldots$). The maximum frequency at which instability appears will be

$$\omega_p = \omega_U \left\{ 1 + \left(\frac{m}{M}\right)^{2/3} \left(\frac{\omega_0}{1 + (\omega_H/\omega_0)^2}\right)^{2/3} \right\} \tag{7.68}$$

where $\omega_U = (\omega_H^2 + \omega_0^2)^{1/2}$ is the upper-hybrid frequency (see (4.22)). The growth rates for these oscillations are, respectively,

$$\left.\begin{aligned} \gamma_1 &= -\frac{4^{1/3}}{2(2)^{1/2}}\left(\frac{m}{M}\right)^{1/3} (s^2\omega_p\omega_0^2)^{1/3} \left[J_0^2(\rho_{pH}) \frac{(s\omega_p)^2 - \omega_H^2}{2(s\omega_p)^2 - (\omega_H^2 + \omega_0^2)} \right]^{1/3}, \\ \gamma_2 &= -\frac{(s\omega_p)^{4/3}}{2\omega_0^{1/3}} (4s)^{1/3} \left(\frac{m}{M}\right)^{1/3} \left[J_s^2(\rho_{pH}) \frac{(s\omega_p)^2 - \omega_H^2}{2(s\omega_p)^2 - (\omega_H^2 + \omega_0^2)} \right]^{1/3}. \end{aligned}\right\} \tag{7.69}$$

In an anisotropic plasma ($\mathbf{H}_0 \neq 0$), instead of the quantity ρ_p, which is a measure of the strength or weakness of the field (see (7.47)), the following quantity is used:

$$\begin{aligned} \rho_{pH} = \rho_p f(\mathbf{E}, \mathbf{K}, \mathbf{H}_0) = \Bigg\{ & \left[\frac{e}{m} \left(\frac{K_\parallel E_\parallel}{(s\omega_p)^2} + \frac{K_x E_x + K_y E_y}{(s\omega_p)^2 - \omega_H^2} \right) \right]^2 \\ & + \left[\left(\frac{e}{m}\right)^2 \frac{\omega_H^2}{(s\omega_p)^2} \frac{(E_x K_y - K_x E_y)^2}{[(s\omega_p)^2 - \omega_H^2]^2} \right] \Bigg\}^{1/2}. \end{aligned} \tag{7.70}$$

The function f can be written as

$$\left.\begin{aligned} f^2(\mathbf{E}_0, \mathbf{K}, \mathbf{H}_0) = & \left[\cos\theta \cos\chi + \sin\theta \sin\chi \cos\phi \frac{(s\omega_p)^2}{(s\omega_p)^2 - \omega_H^2} \right]^2 \\ & + \left[\sin\theta \sin\chi \sin\phi \frac{s\omega_p \omega_H}{(s\omega_p)^2 - \omega_H^2} \right]^2, \end{aligned}\right\} \tag{7.70a}$$

where θ is the angle between $\mathbf{H}_0$ and $\mathbf{K}$; where χ is the angle between $\mathbf{H}$ and $\mathbf{E}_0$, and where ϕ is the angle between the vectors $\mathbf{K} \times \mathbf{H}_0$ and $\mathbf{H} \times \mathbf{E}_0$.

Since the γ_{12} are real quantities, it follows from (7.69) that oscillations can be excited if the following limitations are placed on the source frequency. When $\omega_0 > \omega_H$ or $\omega_0 < \omega_H$, the following conditions, respectively, must be satisfied:

$$(s\omega_p)^2 < \omega_H^2, \quad 2(s\omega_p)^2 < \omega_0^2 + \omega_H^2$$
$$(s\omega_p)^2 > \omega_H^2, \quad 2(s\omega_p)^2 > \omega_0^2 + \omega_H^2. \tag{7.71}$$

In an unstable region the resonance values θ_r of the angle are found using the formula

$$\cos^2\theta_r = \left(\frac{s\omega_p}{\omega_H}\right)^2\left(1 + \frac{\omega_H^2 - s^2\omega_p^2}{\omega_0^2}\right). \tag{7.72}$$

The excitation of HF oscillations at $\omega_1(\theta)$ or at $\omega_2(\theta)$ is also possible on the condition that

$$s\omega_p = \omega_1(\theta) + \omega_2(\theta) \tag{7.73}$$

if $\omega_p = \omega_1(\theta)$ or $\omega_p = \omega_2(\theta)$ (Aliev & Zyunder, 1969; Silin, 1973). More precisely, the frequency range in which this kind of parametric resonance may be observed is specified by the inequality

$$\omega_L < s\omega_p < (\omega_H + \omega_0). \tag{7.74}$$

In this case the formulas for the growth rates and the correction to the resonance frequencies are extremely complicated, and they will not be given here (see Aliev & Zyunder, 1969 and Silin, 1973). The growth rates for a weak field, when $\rho_{pH} \ll 1$ (see (7.70)) and $s = 1$, are simpler. In this case in the frequency range $\omega^2 \gg \omega_L^2$ (where ω_L is the lower-hybrid frequency), when the resonance branches $\omega_1(\theta)$ and $\omega_2(\theta)$ are described by formula (4.49) with a $\pm$ sign in front of the kernel, we have

$$\gamma_{max} \simeq -\frac{\rho_{pH}}{4}\frac{\Omega_0^2 \tan\theta}{[\omega_0\omega_H|\cos\theta|]^{1/2}}. \tag{7.75}$$

With an increase in the angle θ, the frequency $\omega_2(\theta)$ is described by the formula (4.52), and for $\theta = \pi/2$ the growth rate for this wave branch is

$$\gamma_{max} \simeq -\frac{\rho_{pH}}{4}\frac{\Omega_0^2}{\Omega_H(1 + \Omega_0^2/\Omega_H^2)^{1/4}}. \tag{7.76}$$

In an isotropic plasma the maximum growth rate for the parametric excitation of high-frequency waves lies close to the Langmuir frequency of the electrons, when

$$s\omega_p \simeq \omega_0\left[1 + (0.6\text{–}0.8)\left(\frac{m}{M}\right)^{1/3} J_s^{2/3}(\rho_p)\right]. \tag{7.77}$$

In this case

$$\gamma_{max} = -(0.6\text{–}0.8)\left(\frac{m}{M}\right)^{1/3} J_s^{2/3}(\rho_p)\omega_0. \tag{7.78}$$

For parametric excitation of low-frequency waves, the limitations imposed upon the frequency of the external source in an isotropic plasma are less strict. For example, for the case when $\rho_p \ll 1$, i.e., $r_e/\lambda \ll 1$, it is only necessary that $\omega_p < \omega_0$, for which

$$\gamma = -\frac{1}{2^{1/2}}\left(\frac{m}{M}\right)^{1/2} \frac{\omega_p}{\left[1-\left(\frac{\omega_p}{\omega_0}\right)^2\right]^{1/2}}. \tag{7.79}$$

For arbitrary values of ρ_p, the quantity γ is given by the formula

$$\gamma = -[\phi_x(\rho_p) - 1]^{1/2}\Omega_0, \tag{7.80}$$

and, for growth of the oscillations to occur, the function

$$\phi_x(\rho_p) = \frac{\pi x}{\sin \pi x} J_x(\rho_p) J_{-x}(\rho_p), \tag{7.81}$$

(where $x = \omega_0/\omega_p$) must be greater than unity.

It is very important to determine the minimum, or *threshold*, value of the field intensity $E_{0,min}$, for parametric instability. For the high-frequency branches $\omega_1(\theta)$ and $\omega_2(\theta)$, it was found that

$$E_{0,min}^2 = 16\pi N_0 k(T_e + T_i)\left(\frac{\omega_p}{\omega_0}\right)^4 \frac{\varepsilon''(\omega_p, K, \theta_r)}{f_{pH}^2(\theta_r)} \tag{7.82}$$

where ε'' is the imaginary part of the permittivity of the longitudinal waves being considered, and $f_{pH}(\theta_r)$ is the maximum value of the function f_{pH} representing the measure of the strength of an RF field in a magnetoplasma (see (7.70); Andreev & Kirii, 1971). For the HF resonance branches under review here,

$$\left.\begin{aligned}\varepsilon''(\omega_p, K, \theta_r) = {} & \left(\frac{\omega_0}{\omega_p}\right)^2 \frac{\nu_{ei}}{\omega_p}\left\{\cos^2\theta + \sin^2\theta \frac{\omega_p^2(\omega_p^2 + \omega_H^2)}{(\omega_p^2 - \omega_H^2)^2}\right\} \\ & + \left(\frac{\pi}{2}\right)^{1/2} \frac{\omega_0^2\omega_p}{(Kv_e)^3} \frac{1}{|\cos\theta|}\left\{\exp\left[-\left(\frac{\omega_p}{Kv_e\cos\theta}\right)^2\right]\right. \\ & \left. + \left(\frac{Kv_e \sin\theta}{\omega_p}\right)^2 \exp\left[-\left(\frac{\omega_H - \omega_p}{Kv_e \cos\theta}\right)^2\right]\right\},\end{aligned}\right\} \tag{7.83}$$

while for $\omega_p > \omega_H$ or $\omega_p < \omega_H$, respectively, we have

$$[f^2_{max}(\theta_r)]_\pm = \left(\cos\theta_r \cos\chi_0 \pm \frac{\omega_p^2}{\omega_p^2 - \omega_H^2} \sin\theta_r \sin\chi_0\right)^2, \tag{7.84}$$

or

$$f^2_{max}(\theta_r) = \frac{\omega_H^2}{\omega_H^2 - \omega_p^2}[f^2_{max}]_-, \quad f^2_{max}(\theta_r) = [f^2_{max}]_\pm, \tag{7.85}$$

where in (7.84) and in the second formula of (7.85) the $\pm$ signs correspond to the signs of $\cos\phi_{max} = \cot\theta \cot\chi$. The angles χ, ϕ and θ_r were defined above (see (7.70*a*) and (7.72)). If the losses in the plasma are mainly due to collisions, then

$$E^2_{0,min} \simeq 16\pi N_0 \kappa(T_e + T_i)\left(\frac{\omega_p}{\omega_0}\right)^3 \frac{\nu_{ei}}{\omega_0}\left[1 + \frac{2(\omega_p^2 - \omega_0^2)}{\omega_p^2 - \omega_H^2}\right]. \tag{7.86}$$

In formulas (7.82)–(7.86) it is assumed that the plasma is quasi-neutral: $N_e \simeq N_i = N_0$. With a decrease in $\omega_2(\theta)$ when $\omega_2 \to \omega_L$ and $\theta \to \pi/2$, the threshold value of the field $E_{0,min}$ decreases as well.

Parametric instability of the plasma with regard to low-frequency waves manifests itself most strongly if the condition

$$\omega_p = \omega_{12}(\theta) + \omega_s \tag{7.87}$$

is satisfied, where $\omega_{12}(\theta)$ are the HF resonance branches and $\omega_s = Kv_s \gg \Omega_H$ is the frequency of the ion-acoustic waves in a nonisothermal plasma (see (5.59) and (5.64)). In this case in the field of a strong wave the frequency of an excited ion wave is $\omega = \omega_s + \Delta\omega_s$; in the absence of a field, $\omega = \omega_s$. The expression for $\Delta\omega_s$ will not be given here, however, because of the cumbersomeness of the corresponding formula (see Andreev & Kirii, 1971, and Silin, 1973). When (7.87) is satisfied, the growth rate is a maximum, while the threshold value of the field is a minimum. Fig. 7.36 shows the calculated ratios $\frac{\dot{\mathbf{r}}_e}{v_e} = \frac{eE_{0,min}}{m\omega v_e}$, characterizing the threshold value of the field strength $E_{0,min}$ of the pump wave, as a function of the ratio ω_H/ω_P, for different values of T_e/T_i (Andreev, 1969, 1971). Naturally, when considering plasma instability with regard to ion-acoustic waves, we assume that the condition (5.58) is satisfied, namely that $v_i \ll v_e/\cos\theta \ll v_e$. Inspection of Fig. 7.36 shows that, as the nonisothermality of the plasma increases, the threshold field decreases, because the attenuation coefficient of the ion-acoustic oscillations is reduced accordingly. Note that the curves in

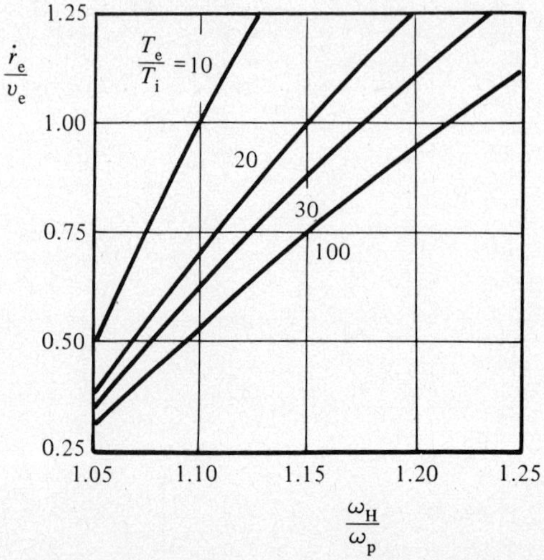

Fig. 7.36. Ratio of the threshold velocity $\dot{r}_e$ of electron oscillations to the thermal velocity v_e, as a function of ω_H/ω_p (for ω_p see (7.87)). The ratio $\dot{r}_e/v_e$ defines the threshold field $E_{0,\min}$ for parametric instability of the plasma (Andreev, 1969).

Fig. 7.36 were calculated for a plasma consisting of electrons and hydrogen ions, when $\mathbf{E}_0$ is perpendicular to $\mathbf{H}_0$.

7.3.3. Parametric excitation of transverse waves

Next let us describe some cases, which are of interest within the framework of the topics being considered, of the parametric excitation of nonpotential (transverse) waves in a plasma, when the pump wave is quite strong: $\rho_p = r_e/\lambda \gg 1$.

In a magnetically active plasma, when the electric field $\mathbf{E}$ of the generated oscillations is perpendicular to $\mathbf{H}_0$ but their wave vector $\mathbf{K}$ is parallel to $\mathbf{H}_0$, the most effective resonant excitation of HF waves will be that satisfying the conditions

$$\omega_p = \omega_0 - \Delta\omega, \quad \Delta\omega \ll \omega_0. \tag{7.88}$$

Under these conditions, the maximum growth rate corresponds to a frequency of the excited waves

$$\omega = (\omega_0^2 + \Omega_0^2)^{1/2} + \Delta\omega. \tag{7.89}$$

In the case where the condition $\Delta\omega \ll \omega_H$ is satisfied as well, we have

$$\left.\begin{aligned} K^2c^2 &= (\Delta\omega)^2 - \omega_0^2\Delta\omega/\omega_H, \\ (\Delta\omega)^2 &= -\frac{1}{16}\left(\frac{eE_0}{mc}\right)^2 \frac{\omega_0 K^2c^2}{(\omega_p - \omega_H)^2(2\Delta\omega - \omega_0^2/\omega_H^2)} \end{aligned}\right\} \tag{7.90}$$

(Gorbunov & Silin, 1969). The maximum growth rate of these oscillations is

$$\gamma_{\max} = -\tfrac{1}{8}\frac{eE_0}{mc}\frac{\omega_0^2}{(\omega_0^2 - \omega_H^2)}. \tag{7.91}$$

if the frequency of the external field

$$\omega_p = \left[1 - \left(\frac{eE_0}{2mc}\right)^2 \frac{(\omega_0^2 + \omega_H^2)}{(\omega_0^2 - \omega_H^2)^2}\right]\omega_0. \tag{7.92}$$

It is evident from (7.91) and (7.92) that the equality $\omega_0 = \omega_H$ is forbidden.

Another case of parametric resonance, one which involves a decay of the pump wave into two waves, corresponds to the condition (see Andreev, 1969)

$$(\omega_p - \omega_A)^2 = \omega_0^2 + K^2c^2 \tag{7.93}$$

where

$$\omega_A = KV_A/[1 + (V_A/c)^2] \tag{7.94}$$

is the frequency of the Alfvén waves (V_A is the Alfvén velocity, see (5.56)). If the resonance condition (7.93) is satisfied, two waves with the same wavelength are excited, one of which is an HF transverse wave with a frequency $\omega = (\omega_0^2 + K^2c^2)^{1/2}$, while the other is an Alfvén wave. The growth rate is

$$\gamma_A = -\frac{1}{4}\left(\frac{eE_0}{mc}\right)\frac{\omega_0^2}{\omega_H\omega_p}\left[\frac{(1 + (2\Delta\omega/\omega)(c/V_A)^2)^{1/2} - 1}{1 + (c/V_A)^2 - (1 + (2\Delta\omega/\omega)(c/V_A)^2)^{1/2}}\right]^{1/2} \tag{7.95}$$

If the condition

$$\omega_0 < \omega_p < \omega_H \tag{7.96}$$

is satisfied, then decay instability appears for a threshold value of the pump field satisfying the inequality

$$\frac{\dot{r}_e}{v_e} = \frac{eE_0}{m\omega}\frac{1}{v_e} \gtrsim 4(2\pi)^{1/2}\left(\frac{m}{M}\right)\frac{\nu_{ei}}{\omega_p}. \tag{7.97}$$

Here an HF transverse wave is excited, as well as a low-frequency longitudinal ion wave, propagated along $\mathbf{H}_0$; the wave vectors of these waves satisfy the condition

$$K^2 = \frac{\omega_p^2}{c^2} + \frac{\omega_0^2}{c^2}\frac{\omega_p}{\omega_H - \omega_p}. \tag{7.98}$$

Instances of parametric instability relative to transverse waves in the field of a strong wave, when $r_e/\lambda \gg 1$, were described above. However, the excitation of nonpotential plasma oscillations, when $r_e \lesssim \lambda$, is also possible (see Gorbunov & Silin, 1969; Andreev, 1969; Silin, 1973). As an illustration, let us cite only the following effects of the appropriate type. In a nonisothermal plasma ($T_e \gg T_i$) when $\omega_p > \omega_0$, a transverse pump wave excites two waves having frequencies $\omega_p - \omega_s$ and $\omega_p + \omega_s$, where ω_s is the frequency of the ion-acoustic wave (see (5.59)). This is the phenomenon known as Mandel'shtam – Brillouin combination scattering. In a nonisothermal plasma the decay conversion of a transverse wave into a low-frequency transverse wave and a Langmuir longitudinal wave is also possible. Here the growth rate is a maximum when $2\omega_0 < \omega_p < \omega_0 c/v_e$.

In an isotropic plasma the excitation of HF transverse waves of frequency $\omega \simeq \omega_0$ is possible when the wave vector

$$K = (\omega_p/c)(1 - 2\omega_0/\omega_p)^{1/2}. \tag{7.99}$$

and $\omega_p > 2\omega_0$ (Montgomery & Alexeff, 1966). Another case of parametric excitation of HF nonpotential oscillations for $\omega_p = \omega_0 - \Delta\omega$ can take place for $r_e/\lambda \gg 1$. Then,

$$\omega_p = \omega_0\left[1 - \left(\frac{\dot{r}_e}{c}\right)^2\right], \quad \gamma_{max} = \tfrac{1}{8}\left(\frac{\dot{r}_e}{c}\right)^2 \omega_0 \tag{7.100}$$

and it is assumed that $\omega_p \gg (v_e/c)\omega_0$.

The excitation of low-frequency transverse oscillations in an isotropic plasma is possible at the frequencies

$$\omega = \frac{v_\phi}{c}(\omega_p^2 - \omega_0^2 - \Omega_0^2)^{1/2}. \tag{7.101}$$

These have the following growth rates:

$$\left.\begin{aligned} \gamma &= -\frac{1}{2(2)^{1/2}}(Kr_e)\Omega_0, \quad \Omega_0 r_e > v_e, \\ \gamma &= -\frac{1}{2(2)^{1/2}}\frac{1 - 8D_e^2/r_e^2}{1 + (KD_e)^2}(Kr_e)\Omega_0, \quad v_i < \Omega_0 r_e < v_e, \end{aligned}\right\} \tag{7.102}$$

if $r_e^2 > 8D_e^2$, ($r_e = eE_0/m\omega^2$). Various cases of excitation of low-frequency oscillations in an isotropic plasma are described in Silin, 1973.

7.4. Nonlinear waves in a plasma. Changes in plasma properties in a nonuniform RF field

In the foregoing sections of this chapter various nonlinear phenomena observed in a plasma under the influence of an external source have been described. The electric field of a wave propagated through a plasma changes its fundamental parameters, and a nonlinear interaction of the wave with the plasma occurs, together with an inverse effect of the plasma on the structure of the wave field: the wave becomes nonlinear. An important aspect of this process is the appearance of a spatially nonuniform distribution of the field of the wave $\mathbf{E} = \mathbf{E}_0(\mathbf{r})$. As a consequence, in a quite strong field a perceptible pressure is exerted on the charged particles (in practice, just on the electrons). Thus, their distribution function and permittivity are ultimately altered. This phenomenon will be described briefly below, in the rest of this chapter. In general, the nonlinear processes taking place in a plasma are very diverse, and the study of them requires the self-consistent solution of some rather complex problems. Consequently, on the whole, only by solving a specific type of problem, for definite specified conditions, will it be possible not only to explain in greater detail and to describe theoretically the results of various experiments, but also often to grasp the general physical essence of the nonlinear phenomena. Some results of solving problems of the corresponding type were described above, in Sections 7.1 and 7.2.

However, as mentioned previously, nonlinearity lies at the very basis of the physical nature of the plasma. Therefore, despite the fact that the linear theory of plasma oscillations and waves makes it possible to explain and describe quantitatively many real plasma processes and experimental results, still it is frequently impossible using this theory to trace out the development even of small-amplitude disturbances in an equilibrium plasma. The nonlinear description of these, on the other hand, reveals some qualitatively new phenomena, which play a major role in the plasma. As an illustration, let us examine here very briefly some such phenomena, mainly of the slightly nonlinear type, i.e., involving plasma disturbances with relatively small amplitudes. The purpose of this study will be to get some insight into nonlinear processes of this type. They apparently play a major role in the plasma regions of interest to us, which very often can be assumed to be collisionless. Some of their features will now be described.

It is very important that these phenomena take place in a *dispersive medium.* Therefore, since nonlinear processes are involved, in the course of time the shape of the oscillations becomes distorted. Actually, together with the initial wave $(\omega, \mathbf{K})$, in the plasma higher harmonics of it will gradually make their appearance $(2\omega, 2\mathbf{K}, \ldots)$: i.e., new components with shorter wavelengths are generated. Let us consider two cases of this: when the initial and the new waves are propagated at the same speed and when they have different speeds. In the former case strong coupling exists between harmonics; they will be in resonance. This leads to a transfer of energy from one harmonic to another. Short waves tend to distort the wave form

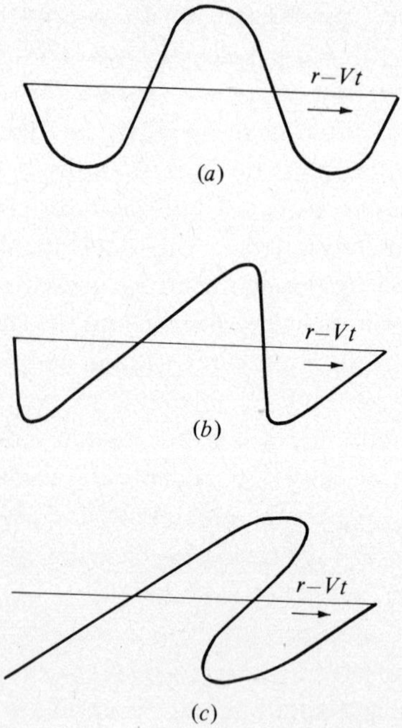

Fig. 7.37. Schematic representation of the nonlinear process of 'steepening' of a wave front (variation of its shape) in a dispersive medium.

in the direction of wave propagation. The leading edge of each wave becomes steeper (see Fig. 7.37).

Ultimately the wave front becomes vertical, and its derivative with respect to $r - Vt$ goes to infinity. Here r is the spatial coordinate in the direction of propagation of the wave, V is its velocity, and t is time. During the further development of this process, the wave front becomes inverted, i.e., the function describing the wave becomes many-valued relative to $r - Vt$ (Fig. 7.37*c*). At the moment of appearance of a vertical front in a collisional medium, a shock wave forms, and then decays as it transfers energy to the medium.

In the second case, when all the harmonics are propagated in the dispersive medium at different velocities, the coupling between waves is greatly weakened; the above phenomenon of energy transfer, leading to steepening and inversion of the wave front, is absent, and the nonlinear development of the process is different. Depending on the typical parameters of the medium, the disturbance of the plasma may lead to the excitation in it of *periodic waves* or *solitary waves* (*solitons*) and to the appearance in it of *collisionless shock waves*: an example is the Earth's bow shock. Some properties of these waves will be described below.

It is also important to note that these nonlinear waves may appear in the presence of *steady motions* in a quasi-equilibrium plasma. The wave processes are thus also steady-state; this is possible only if the linear dimensions of the disturbances are comparable to or less than the Debye length D_e of the electrons. In this case the dependences of the propagation velocity and amplitude of the wave process on the wave number K may be mutually compensating, paving the way for the appearance of a steady-state process. If $KD_e \ll 1$, i.e., if the plasma disturbances have long wavelengths, then they will occasion the appearance of unsteady waves (waves whose form varies with time).

Note, too, that the frequencies of the branches describing the natural nonlinear plasma oscillations, as compared with their frequencies in the linear approximation (for instance, the Langmuir frequency of the electrons ω_0 or the oscillation frequencies $\omega_1(\theta), \omega_2(\theta)$, and $\omega_3(\theta)$), vary only if the thermal motion of the electrons and ions is taken into account, i.e., if the problems are viewed kinetically. In a cold plasma the frequency of, for instance, the nonlinear Langmuir oscillations does not vary; although in themselves the oscillations may be nonlinear, their potential depends on the velocity of wave propagation and the velocity of the thermal motion of the particles (Akhiezer *et al.*, 1974).

7.4.1. Nonlinear wave processes

As an illustration, let us see how a disturbance of the electron density develops in a nonisothermal ($T_e \gg T_i$) isotropic plasma. Here it is assumed that external fields are absent and that the waves are one-dimensional. In this case the characteristic plasma parameters are the velocity of nonisothermal sound $v_s = (kT_e/M)^{1/2}$ and the Debye length of the electrons D_e. In the linear approximation, the solution of the corresponding problem shows that ion-acoustic waves may exist in the plasma (the properties of these waves are discussed in Chapter 5 (see formulas (5.59)–(5.62)). A nonlinear approach to small plasma oscillations, on the other hand, gives a pattern of development of a disturbance which is quite different: more interesting and more complex. Depending on the value of some characteristic dimensionless quantity σ_s, to be defined below (see formulas (7.103)–(7.105)), any one of the three above-mentioned types of steady nonlinear waves can be excited. The dispersion equation describing these waves will naturally be nonlinear. In the linear approximation it was possible, when considering this problem, to use the linearized dispersion equation

$$\omega(K) = v_s K - \beta K^3, \tag{7.103}$$

which contains only the nonlinear dependence of ω on K, but is not itself nonlinear. It was shown that the cubic term in the expansion of $\omega(K)$ is important for ascertaining the basic properties of these waves. Thus the coefficient β can be called the nonlinearity parameter. As $\beta \to 0$, equation (7.103) defines the phase velocity of the ion-acoustic waves

$$v_s = \omega/K. \tag{7.104}$$

In order to examine the nonlinear development of a wave process, the problem is posed for a weakly dispersive medium by assuming a specified initial disturbance. The posing of this problem, together with the results of its solution, will now be outlined schematically.

Let us assume that at a time $t = 0$ a disturbance of the plasma electron density $(N_e - N_{e0})/N_{e0}$ with a linear size L_{n0} has a velocity v_{n0} directed along coordinate r. Here, in accordance with the foregoing, L_{n0} is taken to be of the order of, or less than, D_e. A nonlinear solution of the corresponding one-dimensional problem for a nonisothermal plasma indicates that the development of this disturbance is determined by the following dimensionless quantity:

$$\sigma_s = \frac{L_s}{2^{1/2}}\left(\frac{v_{n0}}{6\beta}\right)^{1/2} = \frac{1}{(12)^{1/2}}\left(\frac{L_s}{d_s}\right)\left(\frac{v_{n0}}{V}\right)^{1/2} \tag{7.105}$$

where L_s is the characteristic length of the process, V is its velocity of propagation, and d_s is the so-called *dispersion length*, to be defined below (see, for instance, Karpman, 1973). An analysis of the complete nonlinear system of equations, including the Poisson equation and the continuity equation, shows that for $\sigma_s \ll 1$ a periodic solution of this problem exists. Nonlinear steady *periodic waves*, with a potential

$$\phi(r - Vt) = \bar{\phi} + \phi_{\max}[a_1 \cos(r - Vt) + a_2 \cos 2(r - Vt) + \ldots] \tag{7.106}$$

appear in the plasma (Fig. 7.38*a*), where $a_1 \simeq 1$ and $a_1 \gg a_2$. The wavelength of these oscillations

$$\lambda \simeq a_e D_e, \tag{7.107}$$

where a_e is a coefficient that varies with $\phi_{\max}$ and v_{n0} from approximately $a_e \simeq 1$ to $a_e \lesssim 3$ to 5. Their velocity is

$$V = v_s - \beta K^2 + \bar{V} \tag{7.108}$$

where $\bar{V}$ is the mean velocity. Consequently, if $\bar{V} = 0$, the disturbance is propagated through the plasma with the phase velocity $v_\phi = \omega/K$ of a linear ion-acoustic wave (see (7.103) and (7.104)).

With an increase in σ_s the shape of the periodic wave is seen to change considerably, since the effect of higher harmonics is enhanced in it

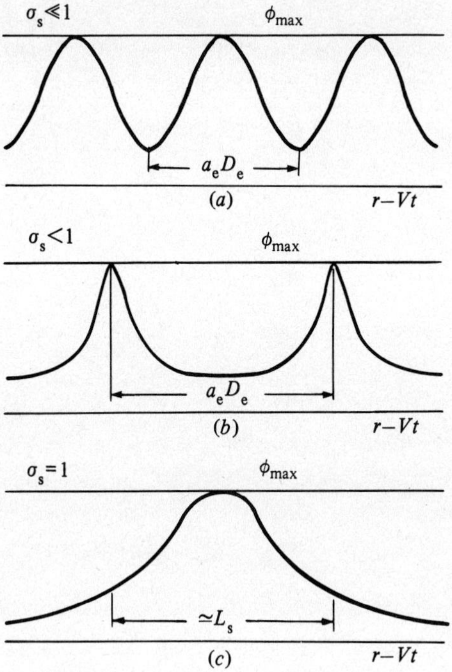

Fig. 7.38. Nonlinear variation of the wave shape in a dispersive plasma with increase in the dimensionless nonlinearity parameter σ_s (see (7.105)); the ordinate is the potential ϕ.

(Fig. 7.38*b*). Finally, when $\sigma_s \simeq 1$, the periodic solution disappears completely, and the development of the disturbance reduces to the appearance in the plasma of a steady *solitary wave*, or *soliton*, propagated through the plasma at a supersonic speed (see Fig. 7.38*c*). In this case the dispersion equation (7.103) has a different meaning: $\omega = 0, \lambda \to \infty$. The concept of a wave number disappears, and equation (7.103) defines the nonlinearity parameter

$$\beta = v_s d_s^2 \tag{7.109}$$

(see (7.105)). The meaning of the length d_s becomes clear if the parameter β is substituted into the dispersion equation (7.103), describing the periodic steady waves. The phase velocity

$$\omega/K = V = v_s[1 - (2\pi d_s/\lambda)^2] \tag{7.110}$$

differs considerably from the velocity of the ion-acoustic waves, when the length d_s becomes commensurable with $\lambda/2\pi$.

The potential of a solitary wave

$$\phi = \frac{\phi_{\max}}{ch^2\left[\dfrac{x - Vt}{2D_e}\left(1 - \left(\dfrac{v_s}{V}\right)^2\right)^{1/2}\right]}, \quad \phi_{\max} = \frac{3}{2}\frac{KT}{e}\left(1 - \left(\frac{v_s}{V}\right)^2\right)^{1/2} \tag{7.111}$$

(see Akhiezer *et al.*, 1974, and Karpman, 1973). Far away from its maximum, we have

$$\phi \simeq \exp\left\{-\frac{x - Vt}{D_e}\left[1 - \left(\frac{v_s}{V}\right)^2\right]^{1/2}\right\}. \tag{7.112}$$

Hence, the order of magnitude of the wavelength in the vicinity of $\phi = \exp(-1)\phi_{\max}$ is

$$L_s = \frac{2D_e}{1 - \left(\dfrac{v_s}{V}\right)^2}. \tag{7.113}$$

The velocity of propagation of a soliton, as shown above and as is evident from (7.108), satisfies the condition $V > v_s$, i.e., it is greater than the velocity of sound. It is defined as

$$V = \frac{v_s}{2^{1/2}} \frac{\exp(e\phi_{\max}/\kappa T) - 1}{\{\exp(e\phi_{\max}/\kappa T) - (e\phi_{\max}/\kappa T) - 1\}^{1/2}}. \tag{7.114}$$

A soliton cannot have an arbitrarily high amplitude. Actually, from (7.114) and the condition $e\phi_{\max} \simeq \frac{1}{2}MV^2$, signifying that at the crest of the solitary wave the release of the kinetic energy of the ions is complete, it follows that

$$e\phi_{\max} = 1.3KT_e, \quad V_{\max} \simeq 1.6v_s. \tag{7.115}$$

The foregoing results for steady nonlinear waves correspond to a quite small nonlinear perturbation: $\sigma_s \lesssim 1$. Studies of *severe perturbations*, for which $\sigma_s \gg 1$, are also of considerable interest. In such cases the nonlinear effects start to influence the development of the perturbation earlier and more intensively than do the dispersion effects of the medium. If this condition is satisfied, depending on whether $\sigma_s > 1$ or $\gg 1$ the soliton begins to disintegrate into individual wave groups, forming several solitons (Zabusky & Kruskal, 1965; Berezin & Karpman, 1964, 1966; Zabusky, 1967. Gardner *et al.*, 1967).

Fig. 7.39 shows the calculated results for the formation of solitons (numerical solutions of Korteweg–de Vries equation) corresponding to $\sigma_s = 1.9, 2.83$, and 16.9. These data were taken from Berezin & Karpman,

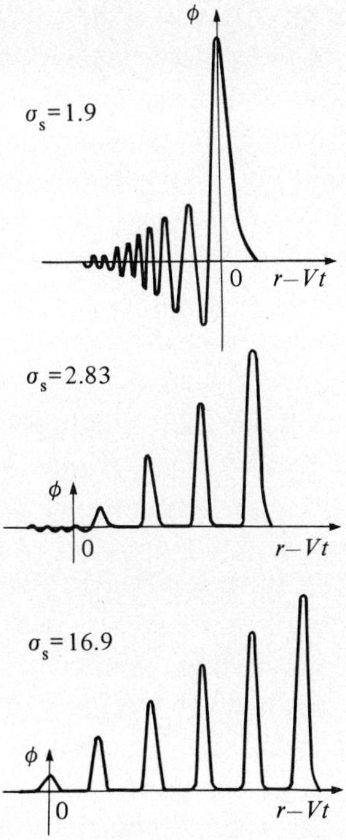

Fig. 7.39. Diagram illustrating the formation of solitons for different values of σ_s (see (7.105) and Berezin & Karpman, 1964).

1964. For $\sigma_s = 1.9$ the development of a perturbation, in the case considered in Berezin & Karpman, 1964 and 1966, reduces to the formation of a single soliton, in the tail of which extends a long wave packet of quite large amplitude. In the second case ($\sigma_s = 2.83$) four solitons are formed, and a wave packet of small amplitude is still perceptible in the tail. For $\sigma_s = 16.9$, the disturbance is transformed into six solitons. It should be noted that the velocities of the solitons are proportional to their amplitudes, since $V = v_{n0}/3$. Thus the vertexes of the solitary waves lie along a straight line. In Zabusky & Kruskal, 1965, numerical methods were also used to study the development of a sinusoidal perturbation specified at the time $t = 0$. First there is an increase in the steepness of the leading edge of the disturbance, the position of which is close to the state of inversion (see Fig. 7.37). However, here dispersion delays the inversion of the front, and soon the first soliton separates from it. Then the second, third, and other solitons gradually appear, the number of these in the example of Zabusky & Kruskal being eight.

The foregoing description of the development of plasma perturbations, which in the last stage of this process involves the formation of steady waves with linear dimensions of the order of D_e, shows that nonlinear waves may play a major role in the regions of the near-Earth plasma of interest to us, especially at quite large distances from the Earth, where the medium is practically collisionless. Let us briefly illustrate this conclusion with some qualitative considerations concerning the effects to be expected.

Assume that a wave packet of small amplitude is being propagated in a plasma. This packet, excited by an external source, has a wavelength $\lambda \ll D_e$, so that its frequency is considerably higher than the frequency of ion-acoustic waves. If this wave packet passes through a plasma region in which periodic or solitary nonlinear waves are propagating, then the two processes will interact with each other. Under certain conditions the packet may *be trapped* by these waves. Then each periodic or solitary nonlinear wave, with this high-frequency internal structure, may be propagated further into the medium. Apparently this effect can occur if there is a strong interaction between the two processes.

A weaker effect of nonlinear waves may manifest itself as follows: if the width of a wave packet from some external source exceeds the spatial period of the nonlinear periodic or solitary waves $T_s = \Omega_s^{-1} = (v_s/D_e)^{-1}$, then the amplitude of the wave packet may become modulated with the frequency $\Omega_s = (D_e/v_s)^{-1}$ or else broken up into individual packets, if the modulation index is high enough. Another type of effect may consist, for example, of the following. As a result of the pressure of the field of the external source, or, so to speak, of the particle flux, a cloud-like inhomogeneity may form in some region of the plasma. Under certain conditions the development of the cloud should be accompanied by the excitation of nonlinear waves of the type described above, as well as by high-frequency (for example, Langmuir or cyclotron) plasma oscillations, since an external source is acting. The final stage of this process is the excitation of ion-acoustic waves with high-frequency internal structure, which can be recorded using the appropriate receiving devices. Naturally, however, these qualitative considerations cannot be given any reliable basis without a theoretical solution of the corresponding self-consistent problems and an analysis of the experimental data. On the other hand, the solution of such problems involves some serious difficulties. Theoretical problems whose solutions indicate the possibility of the effects mentioned have indeed been considered (see, for example, Karpman, 1971; Istomin & Karpman, 1972;

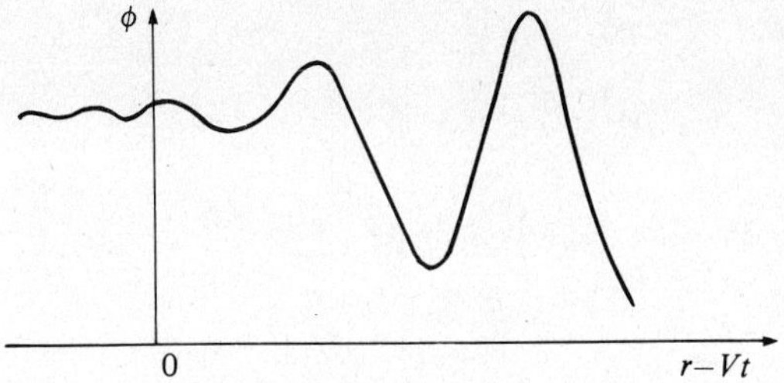

Fig. 7.40. Schematic representation of the shape of a nonlinear bow shock in a plasma.

Helliwell & Crystal, 1973). Experimental data also exist which may be explained by such effects (see, for instance, Bell & Helliwell, 1971).

In conclusion, let us just mention that in some cases the symmetry of the spatial distribution of the electric field or the particle density is disturbed, relative to the direction of propagation of the disturbance. Such asymmetry may, for instance, cause ions to be reflected from a potential barrier formed by the wave. Under such conditions a nonlinear wave which is usually called a *collisionless shock wave*, or *bow shock*, appears in the plasma. Its leading edge in an isotropic plasma has a width of the order of the Debye length D_e. The bow shock divides the plasma into two regions: an undisturbed region ahead of the wave front and a disturbed region behind the front in which there are plasma oscillations of decreasing amplitude (see Fig. 7.40). This kind of wave is apparently formed in the near-Earth plasma, where the solar wind begins to feel the influence of the Earth's magnetosphere, i.e., several Earth radii upstream from the magnetopause (see Fig. 2.9).

7.4.2. Pressure on the electrons in a nonuniform RF electric field

In an alternating field $\mathbf{E}_0 \cos \omega t$, an electron oscillates at a rate $\dot{\mathbf{r}}_\mathrm{E} = (e\mathbf{E}_0/m\omega) \sin \omega t$ and the energy of its motion, averaged over the period, is

$$W_\mathrm{E} = (m/2)\overline{(\dot{\mathbf{r}}_\mathrm{E})^2} = e^2 E_0^2/4m\omega^2. \tag{7.116}$$

Therefore, if the field is nonuniform, $\mathbf{E} = \mathbf{E}_0(\mathbf{r}) \cos \omega t$, it exerts a pressure on the electrons, corresponding to a force per electron

$$\mathbf{F} = -\operatorname{grad} W_\mathrm{E} = -(e^2/4m\omega^2) \operatorname{grad} E_0^2(\mathbf{r}). \tag{7.117}$$

This expression gives the *pressure force* on an electron in an isotropic plasma. It can also be derived from the general formula for $\mathbf{F}$ in a magnetoplasma:

$$\left.\begin{aligned}\mathbf{F} &= (-e^2/4m\omega^2)\operatorname{grad}\sum_{s,p}^{3} a_{sp}E_s^* E_p \\ &= -(e^2/4m\omega^2)\operatorname{grad}(a_{11}E_x^* E_x + a_{12}E_x^* E_y + a_{21}E_y^* E_x \\ &\quad + a_{22}E_y^* E_y + a_{33}E_z^* E_z)\end{aligned}\right\} \tag{7.118}$$

where E_x, E_y, E_z and E_x^*, E_y^*, E_z^* are the field components and their complex conjugates, and the coefficients a_{sp}, which were obtained from a calculation of the force $\mathbf{F}$ using the tensor elements (4.1), are

$$a_{11} = a_{22} = \frac{\omega^2}{\omega^2 - \omega_H^2}, \quad a_{12} = -a_{21} = -\frac{\omega\omega_H}{\omega^2 - \omega_H^2},$$

$$a_{33} = 1, \quad a_{13} = a_{31} = a_{23} = a_{32} = 0 \tag{7.119}$$

(Pitaevskii, 1959). In an isotropic plasma, where $\mathbf{H}_0 = 0$, we have $a_{11} = a_{22} = a_{33} = 1$, and $a_{12} = 0$, so the formula (7.117) follows from (7.118).

The presence of a pressure force $\mathbf{F}$ will obviously necessitate the solution of nonlinear equations. Actually, taking the force $\mathbf{F}$ into account, the Boltzmann distribution for the electrons is

$$N = N_0 \exp\left(\frac{e\phi}{kT} - \frac{e^2}{4m\omega^2}\sum_{s,p}^{3} a_{sp}E_s^* E_p\right). \tag{7.120}$$

Thus the Poisson equation has the form

$$\begin{aligned}\Delta\phi &= \operatorname{div}(\operatorname{grad}\phi) = \operatorname{div}\mathbf{E} = -\frac{e}{\varepsilon_0}\Bigg[\int f_i d^3v \\ &\quad - N_0 \exp\left(\frac{e\phi}{KT} - \frac{1}{kT}\frac{e^2}{4m\omega^2}\sum_{s,p}^{3} q_s a_p E_s^* E_p\right)\Bigg],\end{aligned} \tag{7.121}$$

and the tensor elements (4.1) are replaced by the expression

$$\varepsilon_{s,p}(\mathbf{r}) = \delta_{sp} - \frac{\omega_0^2}{\omega^2}\exp\left[-\frac{e^2}{8m\omega^2}\frac{1}{kT}\sum_{l,m}^{3} E_l^* E_m\right] a_{sp}, \tag{7.122}$$

where the coefficients a_{sp} are defined in (7.119), and δ_{sp} are the Kronecker numbers: $\delta_{sp} = 1$ when $s = p$, and $\delta_{sp} = 0$ when $s \neq p$.

In an isotropic plasma, where $\mathbf{H}_0 = 0$, we have

$$\varepsilon(\mathbf{r}) = n^2(\mathbf{r}) = 1 - \frac{\omega_0^2}{\omega^2}\exp\left[-\left(\frac{e^2}{8m\omega^2}\frac{E_0^2(\mathbf{r})}{kT}\right)\right] \tag{7.123}$$

and, if $(e^2/8m\omega^2)(E_0^2/kT) \ll 1$, i.e., if the field $E(\mathbf{r})$ is weak enough, it follows that

$$n^2(\mathbf{r}) = 1 - \frac{\omega_0^2}{\omega^2}\left(1 - \frac{e^2}{8m\omega^2}\frac{E_0^2(\mathbf{r})}{kT} + \dots\right) = n_0^2 + \delta n^2(E_0(\mathbf{r})) \quad (7.124)$$

and the nonlinear disturbance of the refractive index $\delta n^2[E_0(\mathbf{r})] = (e^2\omega_0^2/8m\omega^4)(E^2/kT)$.

Consequently, the wave phenomena in a plasma are described in terms of the effect of a pressure force $F(\mathbf{r})$ by a nonlinear refractive index. The nonlinear effects arising in this case are usually referred to as being of the *striction type.* Naturally, the tensor elements (7.122) or (7.123) have to be used only when the effect of the nonuniformity of the field $E(\mathbf{r})$ predominates over the other factors making necessary the solution of nonlinear problems.

8

Group velocity, trajectories, and trapping of electromagnetic waves in a magnetoplasma

An important property of any wave process is the transfer of its energy. But it is usually a very complicated task to arrive at a sufficiently complete theoretical explanation of such phenomena. The structural characteristics of the medium, as well as its wave properties, have to be considered in detail. In an inhomogeneous magnetoplasma, the corresponding problems are especially complex when the plasma consists of several species of ions and its state is described by a large number of independent parameters. Consequently, we often limit ourselves just to individual aspects of the process, and in many instances this provides, as will be demonstrated below, not only an understanding of the nature of the phenomena taking place, but also a quantitative interpretation of the experimental data.

A favourable circumstance that aids considerably in solving such problems is the fact that in a collisionless plasma the direction of the energy flux of a quasi-monochromatic packet of quasi-plane waves, i.e., the direction of the Poynting vector **S**, coincides with the direction of the group-velocity vector

$$\mathbf{u} = \partial\omega/\partial\mathbf{K} \tag{8.1}$$

(Al'pert, 1946, 1948; Rytov, 1947; Bremmer, 1949; Stix, 1962). This theorem stating the colinearity of the vectors **S** and **u** makes it possible to consider separately the various aspects of the travel of a wave packet. For instance, the trajectory of the packet can be considered on the basis of the ray theory, in the approximation of geometrical optics, using the quantity $\partial\omega/\partial\mathbf{K}$. The corresponding results have been given by, for example, Booker, 1939; Maeda & Kimura, 1956; Haselgrove 1955, 1957; Haselgrove & Haselgrove, 1960; Yabroff, 1961; Kimura, 1966; Walter, 1969; Aikyo & Ondoh, 1971; Edgar, 1976a,b. Here, of course, it is not necessary to determine the field components and the Poynting vector, which, mathematically speaking, usually constitutes an incomparably more complex problem. The possibility of a quite wide application of theorem (8.1) is furthered by the fact that, in a collisional plasma, when the wave number **K** is complex but the wave attenuation is not particularly high ($\kappa \ll n$), this theorem is general in the sense that the vector

$$\mathbf{u} = \mathrm{Re}\left\{\frac{\partial \omega}{\partial \mathbf{K}}\right\} \tag{8.2}$$

is always parallel to the vector **S** while the vector $\{\partial\omega/\partial\mathbf{K}\}$ does not affect the propagation velocity of a quasi-monochromatic wave group (Suchy, 1972a,b). Similarly, a number of important features of such phenomena as the trapping and ducting of electromagnetic waves by a magnetic field or by inhomogeneous formations can also be explained solely on the basis of an analysis of the properties of the group velocity (see Smith *et al.*, 1960; Smith, 1961; Booker, 1962, 1975; Copal Rao & Booker, 1963; Al'pert, 1967).

Naturally, however, to arrive at a sufficiently complete description and understanding of the trapping and guiding of electromagnetic waves and the phenomena associated with these, the structure of the wave field has to be analysed carefully. Moreover, there are a number of attendant phenomena which cannot be studied without taking into account the thermal motion of the particles and the resonance properties of the plasma, i.e., they call for a kinetic treatment of the corresponding problems. Consequently, we are in some cases confronted by the need for a self-consistent solution of the kinetic and wave equations. Such problems are known to be quite complex, and with regard to the topic being considered here the solutions obtained are by no means as comprehensive as they ought to be. Some investigators have nevertheless made use of a careful study of the field equations to analyse specific problems of this sort. The corresponding results can be found in: Adachi, 1965, 1966; Aubry *et al.*, 1970; Bitoun *et al.*, 1970; Scarabucci & Smith, 1971; Walker, 1971; Suchy, 1972; Al'pert & Kaufman, 1973, 1974; Laird & Nunn, 1975; Matsumoto & Kimura, 1974; Kimura & Kawai, 1976; Moiseev, 1976; Kaufman, 1977; Laird, 1977; Al'pert & Moiseev, 1980.

The theory of ray tracing for electromagnetic waves in a magnetoplasma has been developed in recent years by Budden and others (Budden & Daniell, 1965, Budden & Terry, 1971, Terry, 1971, Budden, 1980, Budden & Stott, 1980, Herring, 1980). The methods used in this series of papers offer the possibility of a more detailed theoretical investigation and understanding of ray propagation, i.e., of the transfer of the electromagnetic energy of waves, particularly in the ionosphere and magnetosphere.

In this chapter we will consider the properties of the group-velocity vector and we will present the results of calculations of its direction throughout the entire frequency range. Trajectories of electromagnetic waves in the ionosphere and the magnetosphere will be examined. Finally, the findings of a wave study of certain phenomena of 'trapping' of electromagnetic waves by a static magnetic field $\mathbf{H}_0$ and by inhomogeneities will be presented, since these illustrate a number of important properties of this process.

8.1. Group velocity u in a homogeneous magnetoplasma

The group velocity $\mathbf{u}$ characterizes the velocity of propagation of the envelope of a quasi-monochromatic wave packet, or in other words, the velocity $\mathbf{u}_\varepsilon$ of transfer of the signal energy. The quantity $\mathbf{u}$ plays a role both in the analysis of various cases of wave propagation or excitation and in the analysis of a number of processes in the vicinity of a body moving in a plasma.

The general expression for the velocity of energy transfer in a dispersive medium is

$$\mathbf{u}_\varepsilon = \mathbf{S}_u / W, \tag{8.3}$$

where $\mathbf{S}_u$ is the mean energy-flux density (energy per unit time, averaged over a wave period, per unit surface area perpendicular to the direction in which this flux is greatest); in a cold magnetoplasma (but not in a warm one) $\mathbf{S}_u$ is equal to the mean Poynting vector

$$\mathbf{S} = \overline{\mathbf{E} \times \mathbf{H}} \tag{8.3a}$$

where the bar denotes a time average. The term W is the mean energy density (energy per unit volume, averaged over a wave period). These properties refer to an infinite plane wave of fixed frequency. In dispersive nonabsorbing media, and in particular in the plasma regions of interest to us here, the velocity of energy transfer is equal, in magnitude and in direction, to the group velocity, the expression for which is

$$\mathbf{u} = \frac{\partial \omega}{\partial \mathbf{K}} = \frac{\partial \omega}{\partial K_x}\hat{\mathbf{x}} + \frac{\partial \omega}{\partial K_y}\hat{\mathbf{y}} + \frac{\partial \omega}{\partial K_z}\hat{\mathbf{z}} \tag{8.4}$$

where $\hat{x}, \hat{y}$ and $\hat{z}$ are the unit axial vectors. From (8.4) it follows that

$$|\mathbf{u}| = \left|\frac{\partial \omega}{\partial \mathbf{K}}\right| = \left[\left(\frac{\partial \omega}{\partial K_x}\right)^2 + \left(\frac{\partial \omega}{\partial K_y}\right)^2 + \left(\frac{\partial \omega}{\partial K_z}\right)^2\right]^{1/2}. \tag{8.4a}$$

For simplicity, we take coordinates with the z-axis parallel to $\mathbf{H}_0$, and with the x-axis in the plane defined by $\mathbf{H}_0$ and $\mathbf{K}$, so that $K_y = 0$. Then

u also lies in this plane, and thus we have

$$\left.\begin{aligned} u &= \frac{c}{n+\omega\dfrac{\partial n}{\partial\omega}}\frac{1}{\cos\psi} = \frac{c}{n_{\mathrm{g}}}[1+\tan^2\psi]^{1/2}, \\ \tan\psi &= \frac{\sin\theta}{n}\frac{\partial n}{\partial\cos\theta} = \frac{\sin\theta}{2n}\frac{\partial n^2}{\partial\cos\theta}, \\ \tan\alpha = \tan(\theta-\psi) &= \frac{\tan\theta\dfrac{\partial(n\cos\theta)}{\partial\cos\theta}}{\dfrac{\partial(n\cos\theta)}{\partial\cos\theta}-\dfrac{1}{\cos\theta}\dfrac{\partial n}{\partial\cos\theta}}, \end{aligned}\right\} \tag{8.5}$$

where ψ and α are the angles between the vector **u** and, respectively, the vectors **K** and $\mathbf{H}_0$ (Al'pert, 1948, 1967). Note that

$$\left.\begin{aligned} \cos\psi &= \left[1+\frac{\sin^2\theta}{n^2}\left(\frac{\partial n}{\partial\cos\theta}\right)^2\right]^{-1/2}, \\ n_{\mathrm{g}} &= \frac{\partial(\omega n)}{\partial\omega} = n+\omega\frac{\partial n}{\partial\omega}. \end{aligned}\right\} \tag{8.5a}$$

The quantity n_{g} is called the group refractive index, although it is not equal to c/u but to $c/(u\cos\psi)$, as may be seen from the first of the equations (8.5). In other words, it corresponds, not to the magnitude of the group velocity, but to the component of the group velocity along the wave vector. There are historical reasons for this somewhat inconsistent terminology.

In general the magnitude and direction of the group velocity are found from the dispersion equation (3.1). If the dispersion equation is expressed in the form $D = D(n, \omega, \theta)$, then it can be shown (Sturrock, 1965) that the components of the group-velocity vector parallel and perpendicular to the wave vector **K** are, respectively,

$$\left.\begin{aligned} u_{\parallel} &= c\frac{\dfrac{\partial D}{\partial n}}{n\dfrac{\partial D}{\partial n}-\omega\dfrac{\partial D}{\partial\omega}}, \\ u_{\perp} &= c\frac{\dfrac{\partial D}{\partial\theta}}{n\dfrac{\partial D}{\partial n}-\omega\dfrac{\partial D}{\partial\omega}}. \end{aligned}\right\} \tag{8.5b}$$

From (8.5*b*) we then get the following expression for the angle between **u** and **K**:

$$\tan\psi = \frac{1}{n}\frac{\partial D}{\partial \theta}\left(\frac{\partial D}{\partial n}\right)^{-1}. \tag{8.5c}$$

It should be noted that formulas (8.5*b*) and (8.5*c*) define the group velocity of either a cold ($T = 0$) or warm ($T \neq 0$) plasma, depending on whether the thermal motion of the particles is taken into account.

Formulas (8.5) are convenient for determining the modulus of the group velocity **u** and the angle α between the energy-flux vector **u** and the magnetic-field vector $\mathbf{H}_0$. To do this, the derivatives $\partial n_{12}/\partial \cos\theta$ and $\partial n_{12}/\partial\omega$ have to be calculated. It was shown above that in the plasma regions of interest to us here the refractive index of a two-component plasma is defined over the entire frequency range by formulas (4.13) and (4.17), of which the ranges of validity overlap one another. Using (4.13), we find for the frequency range $0 \ll \omega \ll \omega_H$ that

$$\frac{\partial n_{12}}{\partial \cos\theta} = -\frac{(n_{12}^2-1)^2}{\left(\dfrac{\Omega_0}{\Omega_H}\right)^2 n_{12}}\cos\theta$$
$$\times\frac{\left(\dfrac{\Omega_0}{\Omega_H}\right)^2\left(\dfrac{\omega^2}{\Omega_0^2}+1\right)+(n_{12}^2-1)\left[2\left(\dfrac{\omega}{\Omega_0}\right)\sin^2\theta+\left(\dfrac{\omega}{\Omega_H}\right)^2+\left(\dfrac{\omega^2}{\Omega_0^2}+1\right)\left(\dfrac{\omega^2}{\omega_H\Omega_H}-1\right)\right]}{2\left(\dfrac{\Omega_0}{\Omega_H}\right)^2+(n_{12}^2-1)\left[2\left(\dfrac{\omega^2}{\omega_H\Omega_H}-1\right)+\left(\dfrac{\omega^2}{\Omega_0^2}+1\right)\sin^2\theta\right]} \tag{8.6}$$

$$\frac{\partial n_{12}}{\partial\omega} = \frac{\omega(n_{12}^2-1)^2}{\left(\dfrac{\Omega_0}{\Omega_H}\right)^2 n_{12}}\left[\frac{1}{\omega_H\Omega_H}-\frac{\sin^2\theta}{2\Omega_0^2}\right.$$
$$\left.\mp\frac{-\dfrac{\sin^4\theta}{2\Omega_0^2}\left(1-\dfrac{\omega^2}{\Omega_0^2}\right)+\dfrac{\cos^2\theta}{\Omega_H^2}}{\left[\left(1-\dfrac{\omega^2}{\Omega_0^2}\right)^2\sin^4\theta+4\left(\dfrac{\omega}{\Omega_H}\right)^2\cos^2\theta\right]^{1/2}}\right] \tag{8.7}$$

For the range $\omega_L \ll \omega \to \infty$, on the other hand, it follows from (4.17) that

$$\left.\begin{aligned}\frac{\partial n_{12}}{\partial\cos\theta} &= \left[n_{12}(1-n_{12}^2)^2\left(\frac{\omega_H}{\omega}\right)^2\cos\theta\right]\left[2\left(1-\frac{\omega_0^2}{\omega^2}-n_{12}^2\right)\left(\frac{\omega_0^2}{\omega^2}-1\right)\right.\\ &\quad\left.+(1-n_{12}^2)(1-\cos^2\theta)\left(\frac{\omega_H}{\omega}\right)^2\right]^{-1}\end{aligned}\right\} \tag{8.8}$$

$$\frac{\partial n_{12}}{\partial \omega}=\frac{(1-n_{12}^2)^2}{\omega(1-\omega_0^2/\omega^2)n_{12}}\left[1-\left(\frac{\omega_H^4}{\omega^4}\sin^4\theta+4\frac{\omega_H^2(\omega^2-\omega_0^2)^2}{\omega^6}\cos^2\theta\right)\right.$$
$$\times\left\{\frac{1-n_{12}^2}{2\omega_0^2/\omega^2}\left[2\left(2-\frac{\omega_0^2}{\omega^2}\right)-\frac{\omega_H^2}{\omega^2}\sin^2\theta\right]\right.$$
$$\left.\left.-\left(\frac{\omega_H}{\omega}\right)^2\left[\left(\frac{\omega_H}{\omega}\right)^2\sin^4\theta-8\left(1-\frac{\omega_0^2}{\omega^2}\right)\cos^2\theta\right]\right\}. \tag{8.9}$$

8.1.1. Angle α between the group-velocity vector u and the magnetic field $\mathbf{H}_0$

The propagation directions of all the above-mentioned types of waves observed in a plasma (see Fig. 4.4), relative to the vector $\mathbf{H}_0$, can be determined on the basis of calculations of $\tan\psi$ (see (8.5), (8.6), and (8.8)). The results of the corresponding calculations over the whole frequency range, and in particular for ELF waves ($0\leqslant\omega\simeq\Omega_H$) and for a plasma consisting of three species of ions (0^+, He^+, and H^+), will be presented in this section. For a multicomponent plasma, it is more convenient to use the general formulas

$$\left.\begin{aligned}\frac{\partial n_{12}}{\partial\cos\theta}&=\frac{n_{12}\cos\theta[(n_{12}^2-\varepsilon_1)(\varepsilon_1-\varepsilon_3)+\varepsilon_2^2]}{\xi_1[2n_{12}^2-\varepsilon_3-(\varepsilon_1^2-\varepsilon_2^2)/\varepsilon_1]-[(2n_{12}^2-\varepsilon_1)(\varepsilon_1-\varepsilon_3)+\varepsilon_2^2]\cos^2\theta},\\ \tan\alpha_{12}&=\tan\theta\frac{\varepsilon_2^2n_{12}^2\cos^2\theta+(\varepsilon_2^2/\varepsilon_3)[n_{12}^2-(\varepsilon_1^2-\varepsilon_2^2)/\varepsilon_1]^2}{\varepsilon_2^2n_{12}^2\cos^2\theta+\varepsilon_1(n_{12}^2-\varepsilon_1)[n_{12}^2-(\varepsilon_1^2-\varepsilon_2^2)/\varepsilon_1]},\end{aligned}\right\} \tag{8.10}$$

obtained from (4.1) and (8.4), where the tensor elements $\varepsilon_1, \varepsilon_2$, and ε_3 were described in (4.3) (Al'pert, 1980a, 1980b).

Let us show here that overall studies of the direction of the Poynting vector $\mathbf{S}(\alpha)$ are conveniently made using the well-known fact that it is perpendicular to the wave surface. If we use the coordinate system defined above, such that the wave vector $\mathbf{K}=(K_x, 0, K_z)$ and the z-axis is parallel to $\mathbf{H}_0$, which does not spoil the generality, then we get the following formula for α:

$$\tan\alpha=-\tan\theta\frac{\partial K_z^2}{\partial K_x^2}. \tag{8.11}$$

Thus the angle α is related to the slope of the tangent to the dispersion curve in the (K_x^2, K_z^2) plane. The dispersion equation in this plane has the form

$$K_x^4 + \frac{K_x^2}{\varepsilon_1}\left[\left(K_z^2 - \frac{\omega^2}{c^2}\varepsilon_1\right)(\varepsilon_1+\varepsilon_3) + \frac{\omega^2}{c^2}\varepsilon_2^2\right]$$

$$+\frac{\varepsilon_3}{\varepsilon_1}\left[\left(K_z^2 - \frac{\omega^2}{c^2}\varepsilon_1\right)^2 - \frac{\omega^4}{c^4}\varepsilon_1^2\right] = 0 \qquad (8.12)$$

Solving this equation for the variables K_x^2 and K_z^2, which proves to be useful for a number of studies, we get

$$K_{x12}^2 = -\frac{1}{2\varepsilon_1}\left[\left(K_z^2 - \frac{\omega^2}{c^2}\varepsilon_1\right)(\varepsilon_1+\varepsilon_3) + \frac{\omega^2}{c^2}\varepsilon_2^2\right]$$

$$\pm\left\{\frac{1}{4\varepsilon_1^2}\left[\left(K_z^2 - \frac{\omega^2}{c^2}\varepsilon_1\right)(\varepsilon_1+\varepsilon_3) + \frac{\omega^2}{c^2}\varepsilon_2^2\right]^2\right.$$

$$\left. - \frac{\varepsilon_3}{\varepsilon_1}\left[\left(K_z^2 - \frac{\omega^2}{c^2}\varepsilon_1\right)^2 - \frac{\omega^4}{c^4}\varepsilon_2^2\right]\right\}^{1/2},$$

$$K_{z12}^2 = \frac{\omega^2}{c^2}\varepsilon_1 - \frac{\varepsilon_1+\varepsilon_3}{2\varepsilon_3}K_{x12}^2$$

$$\pm\left\{\frac{K_{x12}^4}{4\varepsilon_3^2}(\varepsilon_1-\varepsilon_3)^2 - \frac{\omega^2}{c^2}\frac{\varepsilon_2^2}{\varepsilon_3^2}\left(K_{x12}^2 - \frac{\omega^2}{c^2}\varepsilon_3\right)\right\}^{1/2}.$$

Taking into account collisions between particles does not have much effect on the behavior of the angle α, provided that the collision frequency $\nu \ll \omega$ and the attenuation is small (Moiseyev, 1977). However, the influence of the spatial dispersion on α leads to a number of singularities, mainly in the resonance regions of the plasma (see Section 8.1.1, *f*, below).

(*a*) *Alfvén wave and ion-whistler ELF wave* ($0 \leqslant \omega \leqslant \Omega_H$). Fig. 8.1 shows curves for the angle $\alpha(\theta)$ between the group-velocity vector $\partial\omega/\partial\mathbf{K}$ and $\mathbf{H}_0$. These curves characterize the direction of propagation of the energy of an ELF ion wave. This branch of the waves is strongly guided by the magnetic field, especially for $(\omega/\Omega_H)^2 \ll 1$. In this frequency region, provided that the condition for quasi-longitudinal propagation is satisfied, namely

$$(\omega/\Omega_H)^2 \ll \sin^4\theta/4\cos^2\theta,$$

the quantities $\tan\alpha_1$ and n_1^2 are described by the simple formulas

$$\tan\alpha_1 = \tan\theta\frac{1-(m/M)\tan^4\theta}{1+(\Omega_H/\omega)^2\tan^4\theta}$$

$$n_1^2 = \frac{\Omega_0^2}{\Omega_H^2\cos^2\theta\{1-(m/M)[(\omega/\Omega_H)\tan\theta]^2\}}. \qquad (8.13)$$

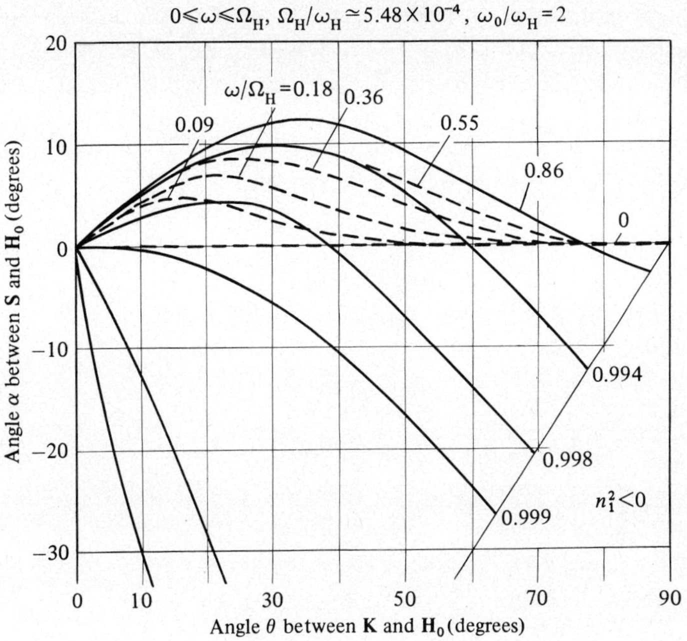

Fig. 8.1. Angle α characterizing the direction of the group-velocity vector $\partial\omega/\partial\mathbf{K}$ of ELF ion waves (n_1, $0 \leqslant \omega \leqslant \Omega_H$) as a function of the direction of the wave vector $\mathbf{K}$ (angle θ) in a plasma consisting of electrons and protons. The numbers on the curves indicate the ratio of the wave frequency ω to the proton gyrofrequency $\Omega_H(H^+)$ (Al'pert 1980b).

Inspection of Fig. 8.1 shows that $\alpha < \theta$ or $\alpha \ll \theta$ right up to $\omega/\Omega_H \simeq 0.999$. Therefore, in a cold plasma the ion mode $n_1(\omega, \theta)$ (Fig. 4.3) is guided effectively by the magnetic field up to the resonance region $\omega \simeq \Omega_H$. The maximum angle of the *trapping cone* of these waves, the axis of which is the magnetic-field vector $\mathbf{H}_0$, depends little on the frequency, being defined by the relation $\tan\alpha_n = 0.22$, $\alpha_n \simeq 12°$; it corresponds to the values $\theta_M \simeq 36°$ and $\omega/\Omega_H \simeq 0.85$ to 0.86.

For $\omega \ll \Omega_H$, when $\alpha \to 0$ the modulus of the group-velocity vectors of both ELF waves is

$$|\mathbf{u}_{12}| = \frac{c}{n_A}\frac{2(1 \mp \omega/\Omega_H)^{3/2}}{(2 \mp \omega/\Omega_H)}, \tag{8.14}$$

while for $\omega \leqslant \Omega_H$ for an ion wave

$$\left.\begin{aligned} |\mathbf{u}_1| &= \frac{c}{n_A}\frac{2(1-\omega/\Omega_H)^{3/2}}{(2-\omega/\Omega_H)}\left\{\frac{2(1+2\cos^4\theta+\cos^6\theta)}{(1+\cos^2\theta)^3}\right\}^{1/2}, \\ \tan\psi_1 &= -\frac{\tan\theta}{1+\cos^2\theta}, \quad \tan\alpha_1 = \tan\theta\frac{\cos^4\theta}{1+\cos^4\theta}. \end{aligned}\right\} \tag{8.15}$$

The resonance values of the functions $\alpha(\theta)$, i.e., the angles α_∞ and θ_∞ corresponding to $n_1^2(\omega,\theta)\to\infty$, are found from the following formulas implied by (8.10), these being applicable to the whole frequency range:

$$\tan\alpha_\infty = -(\varepsilon_1/\varepsilon_3)^{1/2}, \quad \tan\theta_\infty = 1/\tan\alpha_\infty. \tag{8.16}$$

From (8.16) it follows that for ELF waves

$$\tan\alpha_\infty = -\left(\frac{\omega}{\Omega_H}\right)\frac{m}{M}\left[\frac{1}{1-(\omega/\Omega_H)^2}\right]^{1/2}. \tag{8.16a}$$

The more precise formula for $\tan\alpha_1$, in the region $\omega\simeq\Omega_H$, when the condition

$$1-(\omega/\Omega_H)^2 \ll 1 \tag{8.17}$$

is satisfied, has the following form:

$$\tan\alpha_1 = \tan\theta\frac{(\cos^2\theta-\delta^2)-\delta(1+\delta)}{1-(\cos^2\theta-\delta)^2-\delta^2}, \quad \delta = \frac{m}{M}\left(1-\frac{\omega^2}{\Omega_H^2}-\frac{m}{M}\right)^{-1}. \tag{8.18}$$

Then, if the condition (8.17) is satisfied, we have

$$n_1^2 = \frac{\Omega_0^2(1+\cos^2\theta)}{\Omega_1^2\cos^2\theta[1-(\omega^2/\Omega_H^2)^2-(m/M)\tan^2\theta]}. \tag{8.19}$$

(*b*) *Fast magnetoacoustic wave, electron whistler* ($0\leqslant\omega\leqslant\omega_H$) Depending on the angle θ, this branch of the 'electron' wave, which is also known as the *whistler mode*, is continuous and real ($n_2^2>0$) in the frequency range from $\omega=0$ to $\omega\simeq\omega_H$. This range includes the lower-hybrid frequency ω_L ($\theta=\pi/2$). For $\omega_H>\omega_0$ this branch also includes the region of Larmor resonance $\omega=\omega_0$ (see Fig. 4.5). The families of $\alpha(\omega,\theta)$ curves calculated for $\omega_0/\omega_H=2$ and $\omega_0/\omega_H=0.3$ are shown in Figs. 8.2 and 8.3. When considering these, we must note the transition through the region $\omega\simeq\omega_L$. The lower-hybrid frequency ω_L is characteristic for this wave mode, in the sense that it divides $\alpha(\omega,\theta)$ into two families of curves. Since no sufficiently complete calculations of $\alpha(\omega)$ were available in the literature, taking the effect of ions into account, the $\alpha(\omega_L,\theta)$ relation was calculated for $\omega=0$. Also for $\omega=0$, calculations were made of the maximum value of the angle of the *trapping cone of whistler waves* $\alpha_M=19.5°$ ($\tan^2\alpha_M=1/8$) and the corresponding value $\theta_M=54.7°$ ($\cos^2\theta_M=1/3$) (Storey, 1953).

In the quasi-longitudinal approximation,

$$\sin^4\theta/4\cos^2\theta \ll |(\omega^2-\omega_0^2)/\omega\omega_H|, \tag{8.20}$$

if we use the formula giving the refractive index for electron-whistler LF

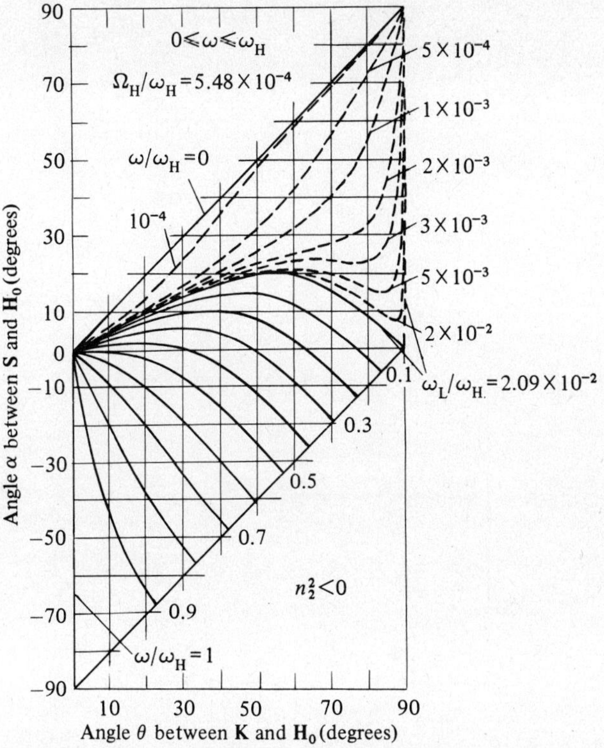

Fig. 8.2. Angle α for the whistler mode n_2, (fast magnetoacoustic wave), in the frequency range $0 \leqslant \omega \leqslant \omega_H$, as a function of the angle $\theta, \omega_0 = 2\omega_H$. The line separating the two families of curves corresponds to the lower-hybrid frequency ω_L (Al'pert, 1980b).

waves ($\omega_L \ll \omega \lesssim \omega_H$):

$$n_2^2 = \frac{\omega_0^2}{\omega(\omega_H \cos\theta - \omega)}, \tag{8.21}$$

then from (8.4) we have

$$\left.\begin{aligned} |u_2| &= \frac{c}{n_g}\left[1 + \frac{\sin^2\theta}{4}\frac{\omega_H^2\omega_0^4}{(\omega_H\cos\theta - \omega)^2(\omega_0^2 - \omega^2)^2}\right]^{1/2}, \\ \tan\psi &= -\frac{\sin\theta}{2}\frac{\omega_H}{(\omega_H\cos\theta - \omega)}\frac{\omega_0^2}{(\omega_0^2 - \omega^2)}, \end{aligned}\right\} \tag{8.22}$$

where

$$n_g = \frac{n_2}{2}\frac{\omega_H\cos\theta}{\omega_H\cos\theta - \omega}.$$

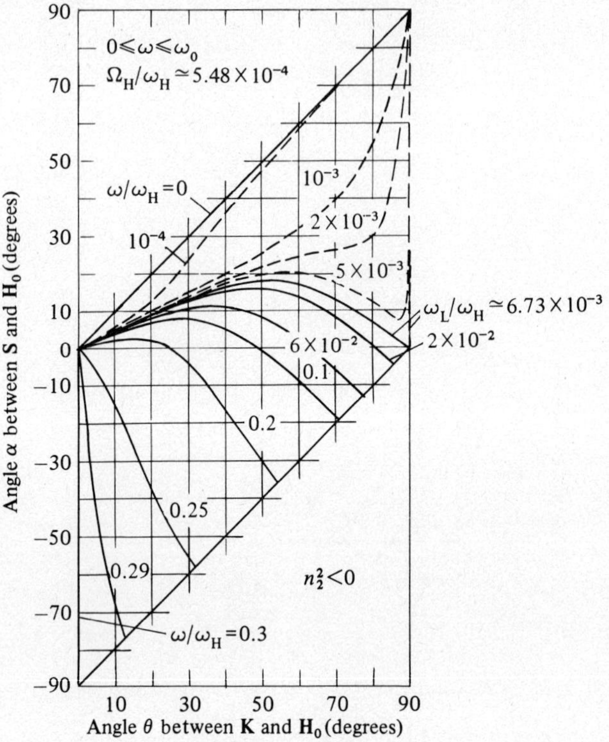

Fig. 8.3. As for Fig. 8.2, but for $\omega_0 = 0.3\omega_H$

When $\omega_0^2 \gg \omega_H^2$, we have

$$\tan\alpha = \sin\theta \frac{\omega_H \cos\theta - 2\omega}{\omega_H \cos^2\theta - 2\omega\cos\theta + \omega_H},$$

$$\tan\alpha_M = \frac{1}{2(2)^{1/2}}\left[1 - \frac{3^{1/2}\omega/\omega_H}{(1-\omega^2/\omega_H^2)^{1/2}}\right]^{3/2}, \quad (8.23)$$

$$\tan\alpha_\infty = \frac{\omega}{\omega_H}\left(1 - \frac{\omega^2}{\omega_H^2}\right)^{-1/2}.$$

It is seen from Figs. 8.2 and 8.3 that whistler LF waves are guided effectively by the magnetic field ($\alpha < \theta$, $\tan\psi < 0$) right down to $\omega \simeq 0.3\,\omega_H$ if $\omega_0/\omega_H > 1$, and down to $\omega \simeq 0.1\omega_H$ if $\omega_0 < \omega_H$. In the two cases, $\alpha = \pm\pi/2$, respectively, as $\omega \to \omega_H$ and $\omega \to \omega_0$. In the range $\omega < \omega_L$ the direction of propagation of VLF waves gradually becomes isotropic. In the vicinity of the lower-hybrid resonance, when $\omega < \omega_L$ and $\theta \lesssim \pi/2$, the angle $\alpha \to \pi/2$. In the ELF range ($0 \leqslant \omega < \Omega_H$), the propagation of the

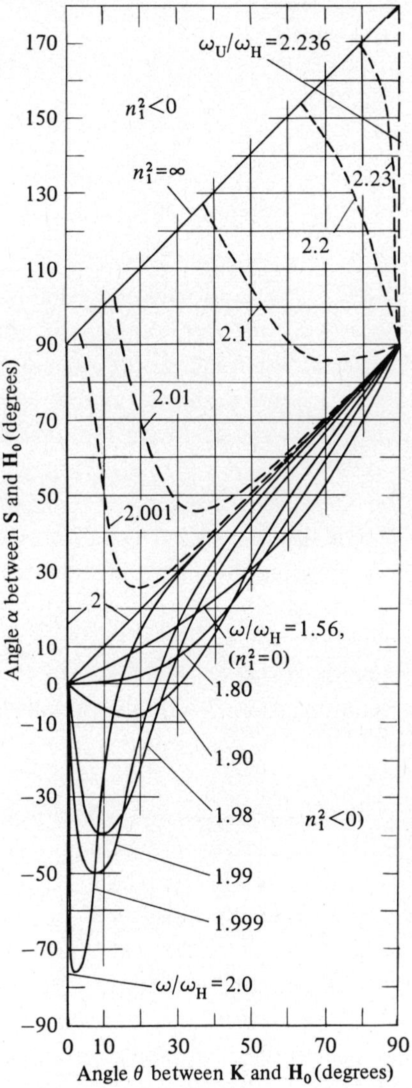

Fig. 8.4. Angle α for the slow extraordinary HF wave n_1, in the frequency range $\omega_0 \leqslant \omega \leqslant \omega_U$ (see Fig. 4.3) as a function of the angle θ, for $\omega_0 = 2\omega_H$. The line separating the two families of curves corresponds to the frequency $\omega = \omega_0$ (Al'pert 1980b).

electron wave is quasi-isotropic. For $\omega \simeq \Omega_H$,

$$\left.\begin{aligned} |u_2| &= \frac{c}{n_A}\left(\frac{1+3\cos^2\theta}{1+\cos^2\theta}\right)^{1/2}, \\ \tan\psi &= -\tan\theta\frac{\cos^2\theta}{1+\cos^2\theta}, \end{aligned}\right\} \tag{8.24}$$

while at the limit, for $\omega \ll \Omega_H$, we have $\partial n_2/\partial\cos\theta \to 0$, that is, $\psi \to 0$; the direction of the modified Alfvén wave coincides with the direction of the wave vector **K**.

(c) *Extraordinary slow wave* ($\omega_- \leqslant \omega \leqslant \omega_U$). This wave branch is real in the frequency range lying between the zero of the refractive index $n_1^2(\omega_-) = 0$ and the upper-hybrid frequency ω_U, for which in a collisionless cold plasma $n_1^2(\omega_U) = \infty$. Fig. 8.4 shows families of $\alpha(\omega, \theta)$ curves for various values of ω/ω_H, with $\omega_0/\omega_H = 2$. In this case the limiting values of the frequency are, respectively, $\omega_- = 1.562\omega_H$ and $\omega_U = 2.236\omega_H$, so the overlapping range of frequencies includes the Langmuir resonance frequency $\omega_0 = 2\omega_H$. This makes the $\alpha(\omega, \theta)$ variation rather complex, and, depending on the frequency, the $\alpha(\theta)$ curves include two families. One $\alpha(\theta)$ family corresponds to the range $\omega_- \leqslant \omega \leqslant \omega_0$, and it is characterized by the following properties. For values of $\theta \leqslant 10$ to $15°$, the angle $\alpha < 0$, and depending on the frequency, it ranges from $\alpha = 0$ to $\alpha \lesssim -\pi/2$; the absolute value $|\alpha(\theta)|$ has a maximum. Here $\alpha = 0$ when $\theta = 0$. For $\theta \geqslant 20$ to $30°$, when $\alpha > 0$, the angle $\alpha \lesssim \theta$, i.e., waves of this mode are little guided by the magnetic field; their propagation is almost isotropic.

The other family of $\alpha(\omega, \theta)$ curves corresponds to the range $\omega_0 \leqslant \omega \leqslant \omega_U$. It is characterized by the fact that, in certain intervals of the frequency ω and of the angle θ, the values of α lie between $\pi/2$ and π. This means that in such cases the energy flux of the waves being considered is guided in directions $\alpha > \pi/2$; when $\omega \to \omega_U$ and $\theta \to \pi/2$, the angle $\alpha \to \pi$, i.e., the direction of the flux **S** becomes opposite to the direction of the magnetic field $\mathbf{H}_0$ and normal to the wave vector **K**. Naturally, in the resonance region ($\omega_0 \leqslant \omega \leqslant \omega_U$) an accurate determination of $\alpha(\theta)$ can be made on the basis of a kinetic study.

In the region $\omega_0 < \omega < \omega_U$, when $\omega_0^2 \gg \omega_H^2$, the function $\alpha(\theta)$ can be calculated with the aid of the following asymptotic formulas:

$$\tan\alpha = \tan\theta\frac{\sin^2\theta - \delta(1-\delta\tan^2\theta)}{\sin^2\theta - 2\delta}, \quad \tan\alpha_\infty = \left(\frac{1-\delta}{\delta}\right)^{1/2}, \tag{8.25}$$

$$n_1^2 = \frac{\omega_H^2\sin^2\theta}{\omega_H^2\sin^2\theta - \omega^2 + \omega_0^2} = \frac{\sin^2\theta}{\sin^2\theta - \delta}, \tag{8.26}$$

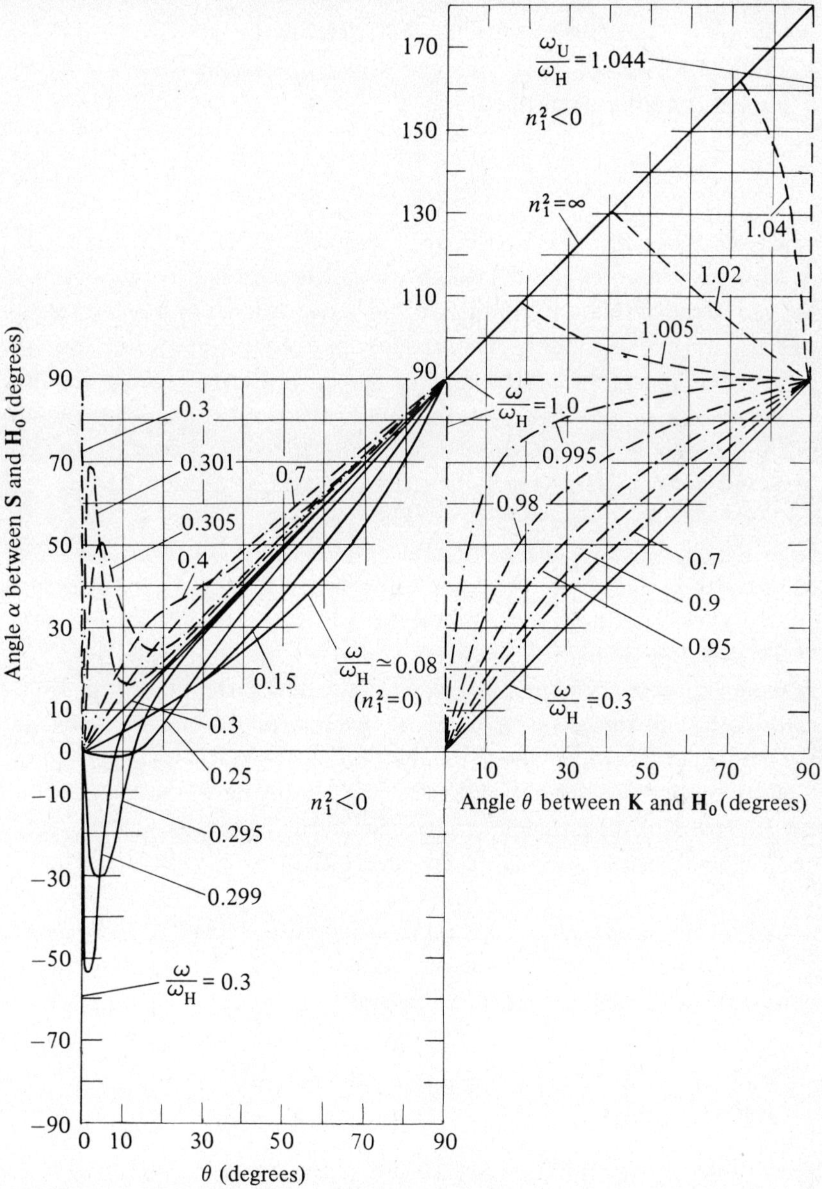

Fig. 8.5. As for Fig. 8.4, but for $\omega_0 = 0.3\omega_H$. The lines separating the three families of curves correspond to the frequencies $\omega = \omega_0$ and $\omega = \omega_H$.

where $\delta = (\omega^2 - \omega_0^2)/\omega_H^2$. These formulas imply that the values of θ for which $\alpha \to \pi/2$ are given by

$$\sin\theta(\alpha = \pi/2) = [2(\omega^2 - \omega_0^2)/\omega_H^2]^{1/2}. \tag{8.27}$$

Formulas (8.25)–(8.27) can be used with considerable accuracy if the following condition is satisfied:

$$\frac{\sin^4\theta}{4\cos^2\theta} \geqslant \left|\frac{\omega^2 - \omega_0^2}{\omega\omega_H}\right|^2, \tag{8.20a}$$

which describes the quasi-transverse propagation of HF waves. The $\alpha(\omega, \theta)$ variation is more complicated when $\omega_0 < \omega_H$, because the frequency range $\omega_- < \omega < \omega_H$ of this wave branch includes two resonance frequencies ω_0 and ω_H. In this case there are three families of $\alpha(\theta)$ curves (or possibly four, depending on the nature of the variation; see Fig. 8.5). The properties of the $\alpha(\omega, \theta)$ curves of this wave branch in the various frequency ranges exhibit the above-considered singularities of the wave branches, as well as of the ordinary HF wave to be examined next, in Section 8.1.1,*d*.

(*d*) *Ordinary* ($\omega \gtrsim \omega_0$) *and extraordinary* ($\omega \gtrsim \omega_+$) *fast HF waves.* The $\alpha(\omega, \theta)$ curves for both of these wave branches for $\omega_0/\omega_H = 2$ and $\omega_0/\omega_H = 0.3$ (Figs. 8.6 and 8.7) indicate that in the specified ranges HF waves are guided very little by a static magnetic field $\mathbf{H}_0$. The direction of the group velocity of an extraordinary fast HF wave differs little from the direction of the wave vector $\mathbf{K}$. At frequencies $\omega \geqslant \omega_+$, where ω_+ is the frequency at which the refractive index $n_1^2(\omega) = 0$, we have $\alpha < \theta$. Conversely, for the ordinary HF wave we always have $\alpha > \theta$, and, with the approach to the Langmuir frequency, when $\omega \to \omega_0$, we have $\alpha \to \pi/2$, i.e., the energy flux is normal to the vector $\mathbf{H}_0$.

(*e*) *Multicomponent plasma, ELF waves.* Figures 8.8 and 8.9 show $\alpha(\omega, \theta)$ curves calculated with the aid of formula (8.10). These pertain to a quasi-neutral plasma containing three species of ions:

$$\left.\begin{array}{l} \text{atomic oxygen } O^+\text{, for which } m/M_1 = 3.425 \times 10^{-5}, \\ \text{helium } He^+\text{, for which } m/M_2 = 1.37 \times 10^{-4}, \\ \text{protons } H^+\text{, for which } m/M_3 = 5.48 \times 10^{4}; \end{array}\right\}$$

$$\omega_{res1} = \Omega_{H1}(O^+) = 3.425 \times 10^{-5}\omega_H = 0.0625\Omega_{H3}(H^+),$$
$$\omega_{res2} = \Omega_{H2}(He^+) = 1.37 \times 10^{-4}\omega_H = 0.25\,\Omega_{H3}(H^+),$$
$$\omega_{res3} = \Omega_{H3}(H^+) = 5.48 \times 10^{-4}\omega_H,$$

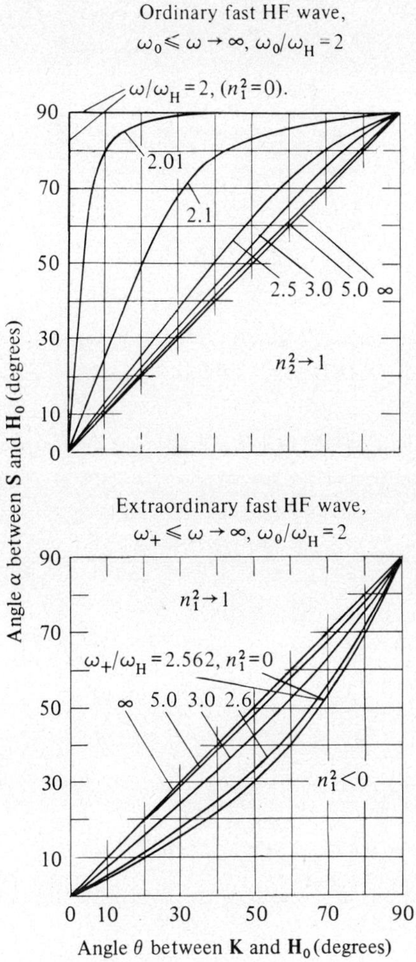

Fig. 8.6. Angle α for the fast ordinary and extraordinary HF waves ($\omega_0 \leqslant \omega \to \infty$, $\omega_+ \leqslant \omega \to \infty$), as a function of the angle θ, for $\omega_0 = 2\omega_H$ (Al'pert, 1980b).

with relative values of the ion densities as follows:

$$\alpha_1 = N_1/N_0 = 0.4, \quad \alpha_2 = N_2/N_0 = 0.1, \quad \alpha_3 = N_3/N_0 = 0.5$$

where $\alpha_1 + \alpha_2 + \alpha_3 = 1$ and N_0 is the electron density.

On approaching the resonance frequencies $\omega(\theta)$, the $\alpha(\omega, \theta)$ curves naturally have the same form as in a plasma with a single species of ion (Fig. 8.1). The nature of the $\alpha(\omega, \theta)$ curves is generally the same as in Figs. 8.1. and 8.2. These relationships become interesting close to the ion–ion

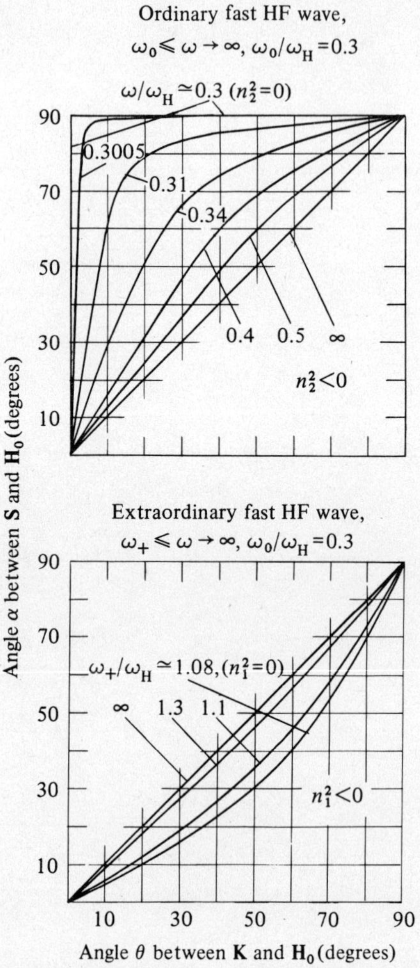

Fig. 8.7. As for Fig. 8.6, but for $\omega_0 = 0.3\omega_H$.

hybrid frequencies $\omega_{L12} = 1 \times 10^{-4}\omega_H$ and $\omega_{L23} = 2.1 \times 10^{-4}\omega_H$. In the vicinity of these, the variation of $\alpha(\theta)$ with frequency is similar to its variation for the electron-whistler mode ω_L in the vicinity of the lower-hybrid frequency (compare Figs. 8.8 and 8.9 with Fig. 8.2). The function $\alpha(\omega, \theta)$ also has some important properties between the cutoff frequencies ω_{c1}, ω_{c2} and the resonance frequencies Ω_{H2}, Ω_{H3}, i.e., in the frequency ranges

$$(\omega_{c1} - \omega_{res2}) \quad \text{and} \quad (\omega_{c2} - \omega_{res3}),$$

which include the crossover frequencies ω_{cr12} and ω_{cr23}. At the crossover

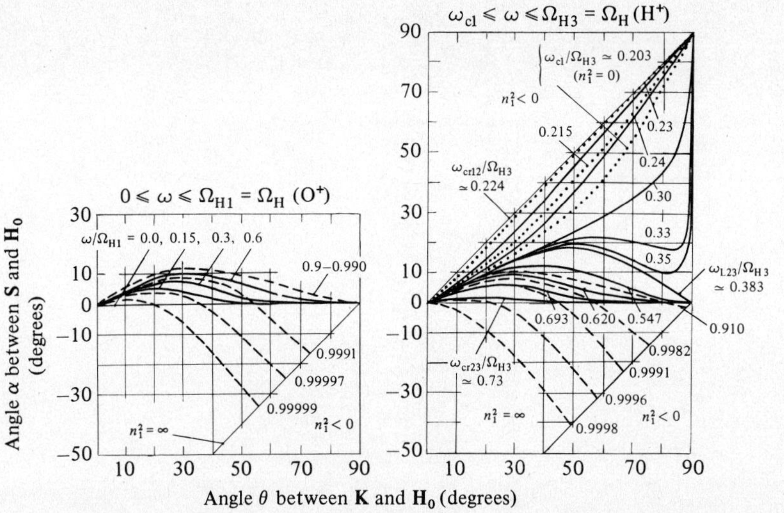

Fig. 8.8 Angle α for the two branches of ELF waves n_1, as a function of the angle θ, in a plasma containing three ion species: $1, 2, 3(0^+, He^+, H^+)$. The following notation is used: ω_{c1} and ω_{c2} are the cutoff frequencies, at which $n_1 = 0$ and $n_2 = 0$ respectively ω_{cr12} and ω_{cr13} are the crossover frequencies, and ω_{L12} and ω_{L23} are the ion–ion hybrid frequencies; $\Omega_{H1}/\Omega_{H3} = \frac{1}{16}$, $\Omega_{H2}/\Omega_{H3} = \frac{1}{4}$, $\Omega_{H3}/\omega_H = 5.48 \times 10^{-4}$ (Al'pert 1980b).

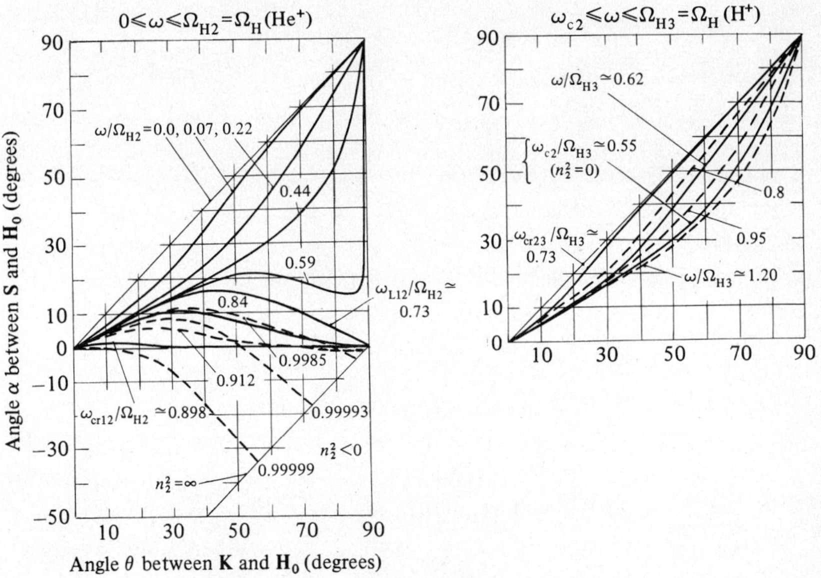

Fig. 8.9. As for Fig. 8.8, but for the other two ELF wave branches n_2; $\Omega_{H2}/\Omega_{H3} = \frac{1}{4}$, $\Omega_{H3}/\omega_H = 5.48 \times 10^{-4}$

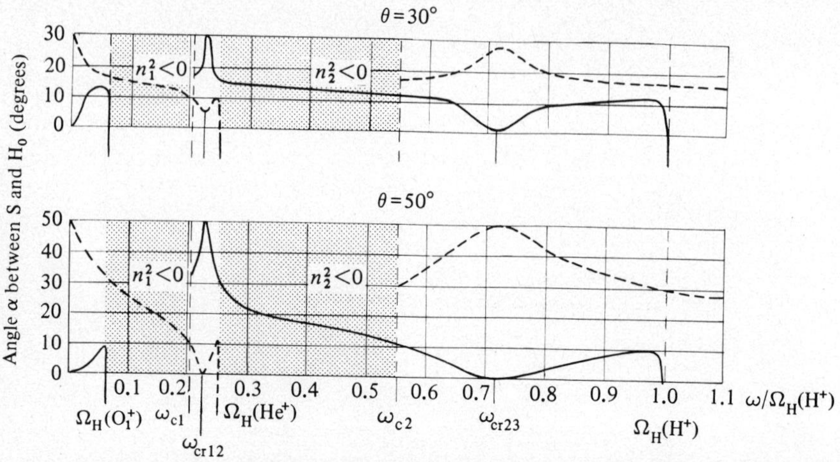

Fig. 8.10. Direction of the group velocity $\partial\omega/\partial\mathbf{K}$ (angle α) as a function of the ratio $\omega/\Omega_H(H^+)$ in a plasma containing the three ion species O^+, He^+, H^+ for $\theta = 30°$ and $50°$; the solid curve corresponds to the branch n_1^2, and the dashed curve to the branch n_2^2 (Al'pert, 1980b).

frequencies, the angle $\alpha(\theta) = 0$. These properties of the angle $\alpha(\omega, \theta)$ manifest themselves most clearly in the specified frequency ranges if we consider the relation $\alpha(\omega, \theta = \text{constant})$ (Fig. 8.10).

In the vicinity of the ion–ion hybrid frequencies $\omega = \omega_{L12}$ and $\omega = {}_{L23}$, the angle $\alpha(\omega, \theta)$ varies little with the frequency:

$$\tan\alpha = \tan\theta \frac{\cos^2\theta}{1 + \cos^2\theta}, \quad \tan\alpha_M = \frac{1}{2(2)^{1/2}}, \quad \alpha_M = 19.5°. \tag{8.28}$$

Thus the maximum angle of the 'trapping' cone of the waves, i.e., the angle between $\partial\omega/\partial\mathbf{K}$ and $\mathbf{H}_0$, is equal to the angle α_M of the whistler mode (see formulas (8.23)); it corresponds to the ω values equal to the ion–ion hybrid frequencies. The crossover frequencies ω_{cr}, which are found from the condition $\varepsilon_2 = 0$ (see Section 4.4 above), satisfy the inequalities

$$\omega_{L12} < \omega_{cr12} < \Omega_{H2}, \quad \omega_{L23} < \omega_{cr23} < \Omega_{H3}.$$

As can be seen from Fig. 8.9, the wave branches $n(\omega, \theta)$ lying in these frequency ranges are everywhere guided effectively by the magnetic field: $\alpha <$ or $\ll \theta$, whereas at frequencies ω_{cr}, as we have seen, the vector $\partial\omega/\partial\mathbf{K}$ is in general colinear with $\mathbf{H}_0$. However, the other wave branches, also including the crossover frequencies lying in the frequency intervals

$$\omega_{c1} < \omega_{cr12} < \omega_{L23}, \quad \omega_{c2} < \omega_{cr23} < \Omega_{H3},$$

are guided little or not at all by the magnetic field, while at frequencies

$\omega = \omega_{\text{cr1 2}}, \omega_{\text{cr23}}$ for these wave branches we have $\alpha(\omega, \theta) = \theta$ (see Figs. 8.9 and 8.10).

An analysis of the quantity $\alpha(\omega, \theta)$ shows that the nature of its frequency dependence does not change with variations in the relative ion densities α_1, α_2 and α_3. With a decrease in α_1 and α_2, when in the plasma a single species of ion begins to predominate, the widths of the regions of opaqueness $n_1^2 < 0$ and $n_2^2 < 0$, and the frequency bands $(\omega_{c1} - \Omega_{H2})$ and $(\omega_{c2} - \Omega_{H3})$ naturally become narrower. Accordingly, although they become narrower, the regions of strong ducting of one of the waves along vector $\mathbf{H}_0$ remain. Thus the presence in the plasma of even a trace amount of an 'impurity', in the form of an ion of another species, alters considerably the properties of the energy-flux vector of the ELF waves propagated in it. At the limit, when $\alpha_1 = \alpha_2 = 0$ and $\alpha_3 = 1$, the $\alpha(\omega, \theta)$ curves for both waves naturally coincide with those obtained from Figs. 8.1 and 8.2 for $\omega_0/\omega_H = 2$.

(*f*) *Effect of the thermal motion of the particles on the group velocity.* So far, the influence of spatial dispersion on the behavior of the group velocity $u(\omega, \theta)$ has been little studied. A detailed kinetic study of this subject over the entire frequency range, similar to the study which has been carried out neglecting the effect of thermal motion in a multicomponent plasma, would certainly be worthwhile. There is reason to assume that such a study would reveal new properties of the transfer of wave packets in a plasma, familiarity with which could be a key toward understanding the processes observed. Let us illustrate this point briefly with the aid of some examples.

It was mentioned above (Section 8.1.1,*e*) that in a multicomponent plasma containing, in particular, three species of ions the energy flux of one of the wave branches is guided completely by the magnetic field in the vicinity of the crossover frequencies ω_{cr}. To make some of the relations more accurate in this frequency range, for instance the relations for the crossover angles θ_{cr}, it is apparently vital to clarify the role of electron-electron collisions and of collisions between ions of the same species, and this can be done only in the kinetic approximation (see Section 4.4). On the other hand, at the crossover frequencies a transformation of the various wave branches takes place; under certain conditions they can interact considerably with each other. Naturally, this process must be accompanied by a transformation of the direction of the Poynting vector for these waves as well, a comprehensive study of which is also impossible without taking into account the thermal motion of particles. Consequently, it is clear that for a multicomponent plasma there are a number of problems which have to be examined with the aid of the kinetic theory of a plasma.

The already obtained, albeit meagre, results of such studies present another example. For instance, it was shown (Gershman, 1960b) that the group velocity of the longitudinal HF plasma waves described by refractive index n_3^2 (see (5.5)) is under certain conditions directed opposite to the wave vector $\mathbf{K}$. This means that the angle α between $\mathbf{K}$ and $\mathbf{u}$ can be obtuse, and may reach π. Note that α for the ordinary and extraordinary transverse waves (n_1^2 and n_2^2) is always acute, even when the effect of thermal motion of the particles is neglected. In the vicinity of the resonance frequencies $\omega_1(\theta)$ and $\omega_2(\theta)$ (see (4.49)–(4.51)), angle $\alpha \simeq \pi/2$ over the entire angle interval $0 < \theta < \pi/2$, i.e., including the region of the upper-hybrid resonance of HF waves (Stepanov, 1962; Kitsenko & Stepanov, 1963). In the region of low-frequency (LF) plasma resonances $\omega_2(\theta)$ and $\omega_3(\theta)$ (see (4.51)–(4.53)), and in particular at the lower-hybrid frequency ω_L and the gyrofrequencies Ω_H and $2\Omega_H$, under some conditions a phenomenon is also observed which is analogous to the above-mentioned one for HF waves, namely oppositely directed vectors of the phase velocity ($\mathbf{K}$) and the group velocity ($\mathbf{u}$). Thus, in a magnetoplasma, cases are possible in which the direction of the energy flux is opposite to the direction of wave propagation (i.e. the wave-normal direction); since the energy always flows outwards from the source, the wavefronts move in towards the source in such cases. Consequently, these phenomena have to be investigated over the whole frequency range, taking the effect of thermal motion of the particles into account, if we are to arrive at a complete understanding of the ducting of electromagnetic waves by a magnetic field.

8.1.2. Langmuir, ion-acoustic, and magnetoacoustic waves

For longitudinal HF electron Langmuir waves, from (5.5) we get

$$\frac{\partial\omega}{\partial\mathbf{K}} = \left(\frac{3}{2}\right)^{1/2}\left(\frac{\omega^2-\omega_0^2}{\omega^2}\right)^{1/2} v_e = \left(1-\frac{\omega_0^2}{\omega^2}\right)v_\phi = u, \tag{8.29}$$

from which it follows that $u \ll v_\phi$, since $(1-\omega_0^2/\omega^2) \ll 1$. The modulus of the group velocity of ion-acoustic waves (ion-Langmuir waves) is

$$\left|\frac{\partial\omega}{\partial\mathbf{K}}\right| = \frac{c^3}{v_s^2}\frac{1}{n^3} = \left(1-\frac{\omega^2}{\Omega_0^2}\right)v_\phi. \tag{8.30}$$

It has a magnitude close to that of the phase velocity v_ϕ when $\omega \ll \Omega_0$; i.e., when the waves are quite long, $Kv_s \ll \Omega_0$ (see Section 5.6 above). For short waves in the upper part of the frequency range, when $\omega \to \Omega_0$

and the ion-acoustic waves are gradually transformed into longitudinal Langmuir ion waves, the group velocity $u \ll v_\phi$. At the limit, of course, $u \to 0$, and the frequency no longer depends on the wave vector. For fast and slow ion-acoustic waves in a magnetoplasma ($\mathbf{H}_0 \neq 0$), we then obtain from (5.63):

$$|\mathbf{u}_{12}| = \frac{c^3}{v_s^2 n_{12}^3}\left[\frac{(\omega_{12}^2 - \Omega_H^2)^2 + (v_s/c)^4 n^4 \Omega_H^4 \sin^2\theta\cos^2\theta}{\omega_{12}^2(\omega_{12}^2 - \Omega_H^2\cos^2\theta) - \Omega_H^2(\omega_{12}^2 - \Omega_H^2)}\right]^{1/2},$$

$$(\tan\alpha)_{12} = \tan\theta\left[1 - \frac{\Omega_0^2\Omega_H^2}{\omega_{12}^2(\Omega_0^2 + \Omega_H^2 - \omega_{01}^2)}\right]$$
$$= \tan\theta\left[1 - \frac{\Omega_H^2}{\omega_{12}^2}\left(1 + \frac{\Omega_H^2}{\Omega_0^2} - \frac{\omega_{01}^2}{\Omega_0^2}\right)^{-1}\right] \tag{8.31}$$

(Al'pert, 1967).

Inspection of (8.31) indicates that the vector **u** of a fast wave, which exists in the range $\Omega_H < \omega_1 < (\Omega_H^2 + \Omega_0^2)^{1/2}$, is aligned with $\mathbf{H}_0$ when its frequency ω_1 is at the bottom of this range. However, with increasing frequency the situation changes, and when $\omega_1 \to \Omega_0$ the vector **u** becomes aligned with **K**. The direction of the group-velocity vector of a slow wave ($0 < \omega_2 < \Omega_H$) remains close to that of $\mathbf{H}_0$ right down to values of α close to $\pi/2$.

8.1.3 Magnetoacoustic transverse waves in a nonisothermal plasma

Here we will also present formulas giving the group velocity for magneto-acoustic transverse ELF waves. The latter are formed in a nonisothermal plasma as a result of the transformation of a modified fast Alfven wave. For $T_e \gg T_i$, three wave branches appear instead of two waves. One of these is the ordinary slow Alfven wave:

$$\omega_1 = K_1 V_A \cos\theta. \tag{8.32}$$

However, because of the effect of the motion of electrons, two fast waves are formed:

$$V_{23}^2 = \frac{V_A^2 + v_s^2}{2} \pm \tfrac{1}{2}[(V_A^2 + v_s^2)^2 - 4V_A^2 v_s^2 \cos^2\theta]^{1/2}. \tag{8.32a}$$

Their group velocities are

$$|u_{23}| = \frac{c}{n_{23}}\left\{1 + \frac{4V_A^2 v_s^2 \cos^2\theta \sin^2\theta}{V_{23}^4[(V_A^2 + v_s^2)^2 - 4V_A^2 v_s^2\cos^2\theta]}\right\}^{1/2},$$

$$\tan\alpha_{23} = \tan\theta\frac{V_{23}^2[(V_A^2 + v_s^2) - 4V_A^2 v_s^2\cos^2\theta]^{1/2} + 2V_A^2 v_s^2\cos^2\theta}{V_{23}^2[(V_A^2 + v_s^2) - 4V_A^2 v_s^2\cos^2\theta]^{1/2} - 2V_A^2 v_s^2\cos^2\theta}. \tag{8.33}$$

It is clear from (8.33) that for both waves the vector **u** is not aligned with $\mathbf{H}_0$, but rather it deviates from $\mathbf{H}_0$ by an angle $\alpha > \theta$.

8.2. Trajectories of electromagnetic waves in a smoothly varying magnetoplasma

The ray method of calculating the trajectories of electromagnetic waves in a smoothly varying anisotropic medium is based on the use of Fermat's principle:

$$\delta\phi = \delta\int n\cos\psi\,\mathrm{d}s = 0. \tag{8.34}$$

In (8.34) the variable n is the real part of the refractive index, ψ is the angle between the ray direction and the wave vector, and ds is an element of distance along the wave trajectory. By virtue of the mathematical analogy between Fermat's principle and Hamilton's principle for a moving mechanical system, it is possible to arrive at a system of differential equations similar to the Euler–Lagrange system, which describes the ray trajectory and is suitable for numerical integration with a computer. These differential equations were also first used to compute the trajectories of short radio waves in the ionosphere (Haselgrove, 1955, 1957; Haselgrove & Haselgrove, 1960). The method worked out on the basis of these equations was used to calculate the trajectories of electromagnetic waves in the plasmasphere and magnetosphere. They are suitable both for a cold plasma ($T = 0$) (Yabroff, 1961; Kimura, 1966; Walter, 1969; Aikyo & Ondoh, 1971) and for a hot plasma ($T \neq 0$) (Aubry *et al.*, 1970; Bitoun *et al.*, 1970; Hashimoto *et al.*, 1977).

In a spherical coordinate system (R, θ, χ), assuming that the ray does not leave the plane of incidence (R, θ), the wave trajectory is a plane curve described by the equations

$$\left.\begin{aligned}
\frac{\mathrm{d}R}{\mathrm{d}t} &= \frac{1}{n^2}\left(n_\mathrm{R} - n\frac{\partial n}{\partial n_\mathrm{R}}\right),\\
\frac{\mathrm{d}\theta}{\mathrm{d}t} &= \frac{1}{Rn^2}\left(n_\theta - n\frac{\partial n}{\partial n_\theta}\right),\\
\frac{\mathrm{d}n_\mathrm{R}}{\mathrm{d}t} &= \frac{1}{n}\frac{\partial n}{\partial R} + n_\theta\sin\theta\frac{\mathrm{d}\theta}{\mathrm{d}t},\\
\frac{\mathrm{d}n_\theta}{\mathrm{d}t} &= \frac{1}{R}\left(\frac{1}{n}\frac{\partial n}{\partial \theta} - n_\theta\frac{\mathrm{d}R}{\mathrm{d}t}\right),
\end{aligned}\right\} \tag{8.35}$$

where R and θ are the geocentric distance and angle, n_θ and n_R are the components of a vector directed along the wave vector **K**, the modulus of which is equal to the real part of the refractive index ($n^2 = n_\theta^2 + n_R^2$), and t is the phase time of wave propagation along the ray (Yabroff, 1961).

Equations (8.35) have been integrated numerically in various studies of ionospheric models that are inhomogeneous vertically, and are close approximations to the actual height distribution of plasma. These calculations were carried out for an altitude interval including the boundary of the outer ionosphere (plasmapause), for a plasma consisting of one or several species of ions. They revealed the features of whistler-mode trajectories of electromagnetic waves in a wide frequency range. The main results of these calculations will now be presented.

8.2.1. Trajectories guided by the magnetic field H_0

If the effects of the ions are neglected, then a basic trait of the electromagnetic waves in the frequency range corresponding to the whistler mode n_2 (see Figs. 4.3, 4.5, 4.6) is the fact that they are guided by a static magnetic field $\mathbf{H}_0$ and that the ray trajectory is asymmetrical with respect to the

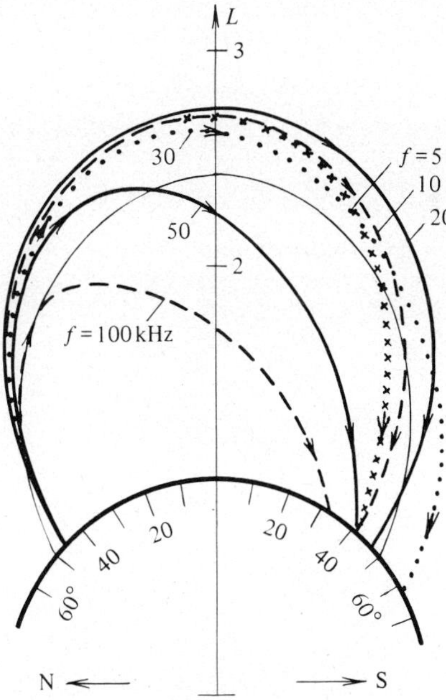

Fig. 8.11. Ray trajectories for the whistler mode in the outer ionosphere at various frequencies, when the effect of the ions is neglected or when $\omega \gg \omega_L$. The initial wave-normal angle, at the point of entry into the ionosphere, is $\theta_0 = 0$ (Yabroff, 1961).

geomagnetic equator and to the magnetic force lines. The degree of asymmetry of the trajectory varies considerably with the initial wave-normal angle θ_0, i.e., the angle between the wave vector and the direction of a magnetic force line $\mathbf{H}_0(R) = \mathbf{H}_0(R, \phi)$ at the point where the ray enters the base of the ionosphere. The asymmetry also varies with the frequency ω, the geomagnetic latitude ϕ_0, the initial point of the trajectory, and the distribution of the electron density $N(\mathbf{R})$. Nevertheless, for a given distribution $N(\mathbf{R})$, in certain intervals of ϕ_0 and ω there exist certain ray trajectories that are quite symmetrical about the magnetic force lines.

Moreover, for the ionospheric model used in Yabroff, 1961, at latitudes $\phi_0 = 40$ to $50°$ and frequencies $f = \omega/2\pi = 5$ to $20\,\text{kHz}$, the ray trajectories are quite symmetrical about the geomagnetic equator. The general character of their variation with frequency f and geomagnetic latitude ϕ is shown by Figs. 8.11 and 8.12. At frequencies close to the electron gyrofrequency, in the vicinity of the apogee, the ray trajectories deviate greatly from the magnetic force lines, towards latitudes $\phi > +\phi_0$, i.e., toward latitudes higher than those of their origin at the Earth's surface, and they lie generally on the side of the geomagnetic equator corresponding to

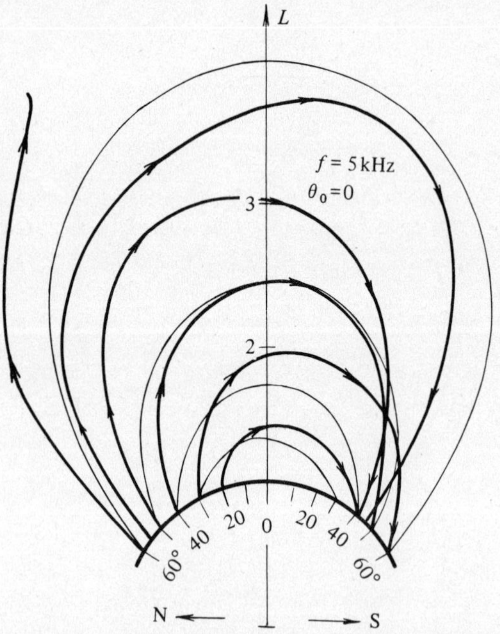

Fig. 8.12. Ray trajectories in the outer ionosphere at the frequency $f = 5\,\text{kHz}$, for ray origins at different geomagnetic latitudes ϕ_0; ion effects are neglected (Yabroff, 1961).

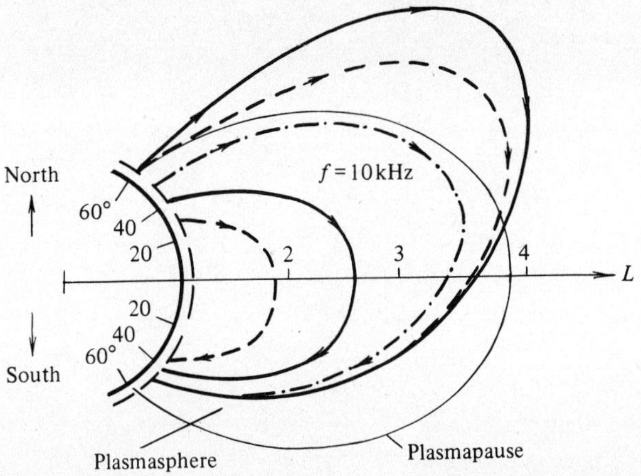

Fig. 8.13. Ray trajectories at the frequency $f = 10$ kHz, which is higher than the lower-hybrid frequency $f_L = \omega_L/2\pi$ throughout the entire region of wave propagation; the effects of the ions are taken into account (Aikyo & Ondoh, 1971).

positive values of ϕ_0 (Fig. 8.13). The ray trajectories even become focused to some extent in the magnetically conjugate latitude region $(-\phi_0)$.

It is important here to note that the curves presented in Fig. 8.13 have been calculated for a plasma containing three species of ions (H^+, He^+, O^+) (Aikyo & Ondoh, 1971). However, for the ionospheric model that these authors used, the frequency $f = 10$ kHz is everywhere higher than the lower-hybrid frequency f_L, so the nature of the trajectory would be the same if the effect of the ions were neglected. With the approach to f_L, at frequencies $f \lesssim f_L$, the nature of a ray trajectory becomes greatly altered, and the rays acquire a novel character. The corresponding data are presented in the next section.

8.2.2. Unguided ray trajectories ($\omega \leqslant \omega_L$)

The trajectories of wave propagation in the outer ionosphere, the apogees of which reach the plasmapause or lie beyond it (these were considered in the previous section), describe the guided propagation of electromagnetic waves: they are guided in the sense that, over the entire path of wave propagation, the ray closely follows a magnetic force line, the end of which is located at a magnetically conjugate point. In some cases the ray trajectories are quite symmetrical about the geomagnetic equator, or even follow very closely along the magnetic force line passing through their starting point. The latter situation is most likely to be realized if the electron density along the force line is higher than in the surround-

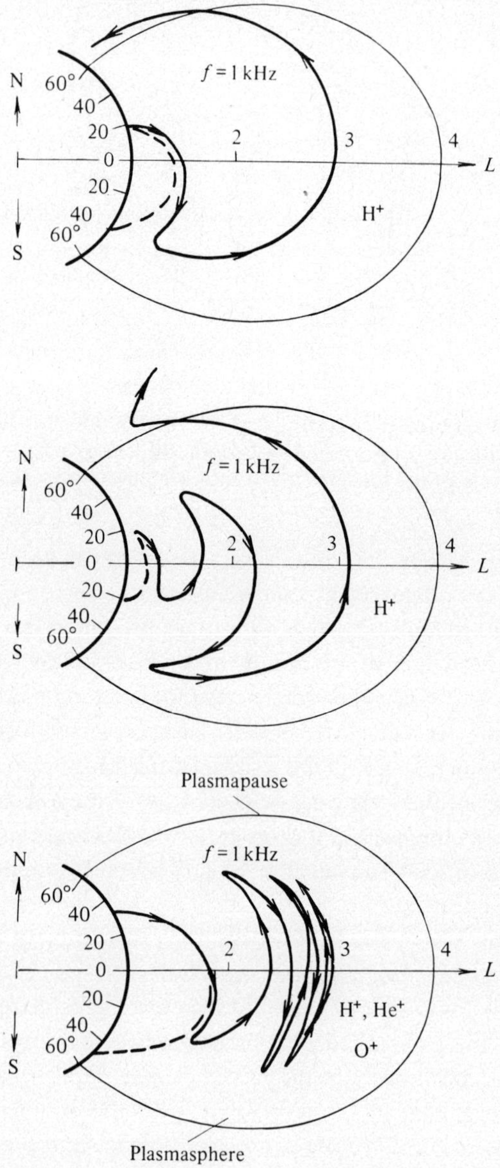

Fig. 8.14. Ray trajectories at the frequency $f = 1$ kHz, in the case where $f < f_L$ over the entire outer ionosphere. The initial wave-normal angle $\theta_0 = 0$, and the height of the start of the trajectory $z = 300$ km. The dashed curves were calculated neglecting the effect of the ions (Kimura, 1966).

ing plasma, i.e., if over a sufficiently long segment of the force line elongated inhomogeneous formations are present (ionized clouds). Calculations of the corresponding wave trajectories confirm this conclusion (Yabroff, 1961).

However, these properties are possessed only by LF waves in the whistler mode, the frequencies of which are higher than the local values of the lower-hybrid frequency f_L. At frequencies lower than f_L, the propagation of VLF waves assumes a distinctly unguided character. Their trajectories start to be loop-shaped, and they do not terminate in a latitude region that is magnetically conjugate to their starting point. Under some conditions the trajectories appear to oscillate about the geomagnetic equator (Fig. 8.14, Kimura, 1966). Ultimately the waves may become trapped in some region of the outer ionosphere (plasmasphere). This property of the trajectories of VLF waves explains the so-called magnetospherically reflected whistling atmospherics, trapped in the plasmasphere, which were discovered with artificial satellites. These atmospherics are discussed in Volume 2. Some calculated results illustrating the nature of the trajectories of VLF waves of the whistler mode are presented in the figures of the present section.

As mentioned above, a novel feature of the wave trajectories is that they do not follow a magnetic force line: the rays have magnetospheric reversal points, or reflection points. This is because, for $f < f_L$, the energy flux of VLF waves may have an arbitrary direction relative to $\mathbf{H}_0$; the group-velocity vector may thus be directed at a right angle to $\mathbf{H}_0$. This property of the trajectories was described above, in Section 8.1 (see Figs. 8.2 and 8.3). However, the nature of the trajectory varies considerably if several species of ions are present in the plasma. The latitudes of the reversal points of the ray naturally will affect the shape of the trajectory, too (see Fig. 8.14). When the effect of ions is neglected, the ray trajectories become smooth curves of the guided type (dashed curves in Fig. 8.14). If the wave source is situated in the equatorial plane close to the plasmapause, then the shape of the wave trajectory will depend considerably on the initial direction of the wave normal relative to $\mathbf{H}_0$ and on the altitude of the source (Figs. 8.15 and 8.16; Aikyo & Ondoh, 1971).

8.3. Trapping of electromagnetic waves by the magnetic field $\mathbf{H}_0$

The following question is of considerable interest: how does the trapping of electromagnetic waves by a magnetic field $\mathbf{H}_0$ take place, i.e., what is the characteristic width of the zone within which the vector $\mathbf{S}$ turns towards

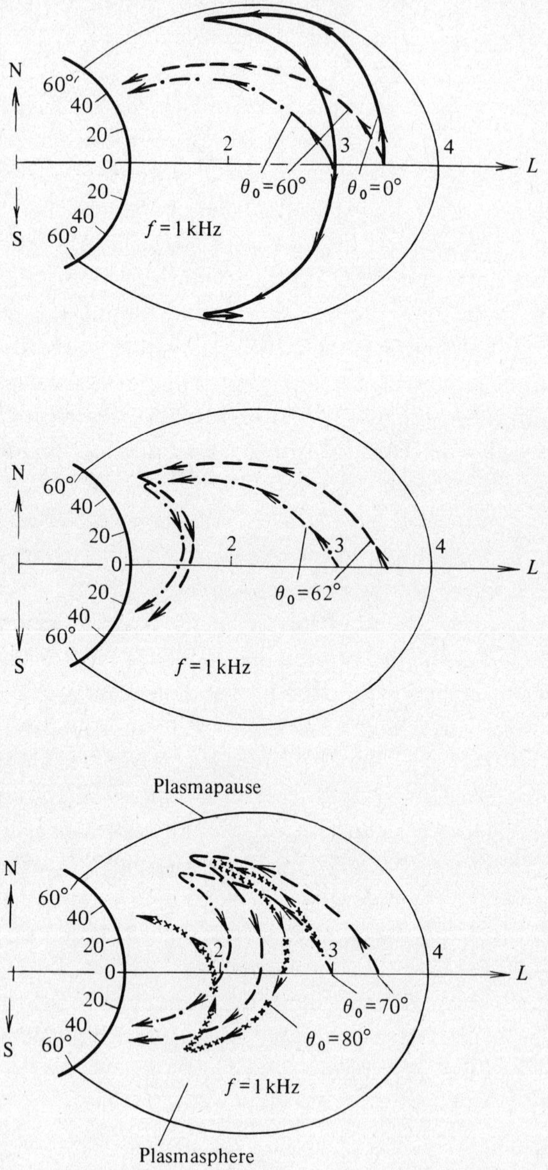

Fig. 8.15. Ray trajectories at the frequency $f = 1\,\mathrm{kHz}\,(< f_L)$, beginning on the geomagnetic equator. These figures illustrate the change in the nature of the trajectories with variation of their initial wave-normal angle θ_0 and of the altitude of the start of their trajectory (quantity L) (Aikyo & Ondoh, 1971).

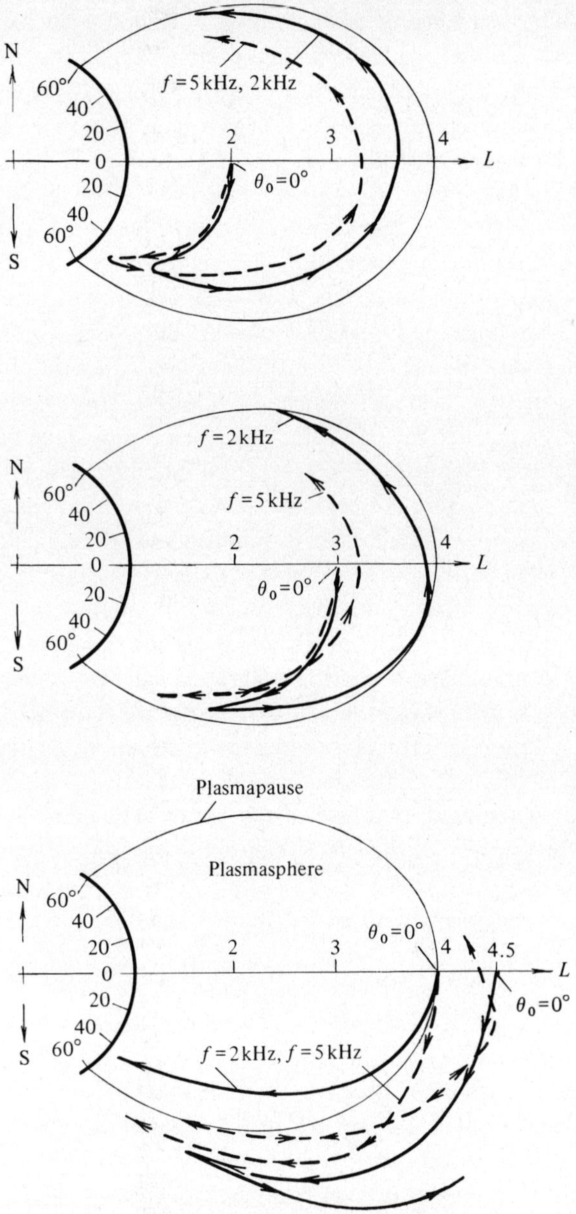

Fig. 8.16. Ray trajectories at the frequencies $f = 2\,\text{kHz}$ and $5\,\text{kHz}$ ($< f_L$), starting on the geomagnetic equator at various altitudes (quantity L). Initial wave-normal angle $\theta_0 = 0$ (Aikyo & Ondoh, 1971).

the direction of $\mathbf{H}_0$? A study of this problem naturally involves a careful analysis of the incidence of a wave upon an inhomogeneous magnetoplasma or else upon an isolated magnetoplasma irregularity. Both of these problems are, in general, difficult to solve. The former has been considered only for a number of particular cases (Al'pert & Kaufman, 1973, 1974). However, the results obtained still provide an adequate general idea of the zone of trapping of waves by a magnetic field (see Section 8.3.1, below). The trapping of electromagnetic waves in an isolated irregularity, on the other hand, is a more complex, many-faceted problem.

To solve this boundary problem, even purely as a wave problem for surfaces of simple shape, is no easy matter. But, when considering this problem, it is not enough just to investigate the structure of the field in the vicinity of and inside irregularities. If the solution is to cover the expected physical phenomena adequately, then it must take into account the effects of the resonance properties of an inhomogeneous structure as a whole, as well as the wave resonance properties of the plasma itself. It can be assumed that the effective trapping of electromagnetic waves by an isolated irregularity has a complex resonant character. Thus there is a definite need for a more complete analysis of the (purely phenomenological) wave solution of this problem, taking into account the effect of the motion of the charged particles. Here the particle distribution may differ greatly from a Maxwellian. Ultimately we have to find a self-consistent solution of the wave and kinetic equations of the problem, having assigned ourselves the following general tasks:

(*a*) to study the field structure of the electromagnetic waves and the conditions favourable for the effective trapping of radiation incident upon an irregularity;

(*b*) to determine the lifetime of the trapped radiation and the conditions of its disappearance;

(*c*) to ascertain the resonance properties of this process.

Note that an isolated irregularity in a magnetoplasma may move and it may be continually, or only for short periods of time, under the influence of an incident electromagnetic wave. At present, not only is an approximate solution of this problem lacking, but also particular aspects of it have not been considered in sufficient detail, especially with regard to isolated irregularities. In a number of works only certain features of the phenomena of interest to us here, in a plane-stratified magnetoplasma layer, have been examined, using both the methods of geometrical optics and with the aid of rigorous wave methods. The results obtained using wave methods characterizing the trapping of the radiation of an incident wave in a plane-stratified layer will be presented in Section 8.3.2 and 8.3.3. They give a quantitative idea of the trapping of radiation by a layer. The results of similar studies, obtained using the methods of geometrical optics, which played a major role in the development of this problem, can be summed up briefly as follows:

(1) guiding of a certain type of wave was found to be possible, if the irregularity is aligned with the magnetic field (Smith *et al.*, 1960; Smith, 1961; Booker, 1962; Copal Rao & Booker, 1963;

(2) for guiding, it is sufficient that the electron density $N = N_0 + \Delta N$ of the irregularity differ only slightly from the value N_0 of the surrounding plasma: $\Delta N/N_0 \simeq 10^{-2}$ to 10^{-1};

(3) guiding is possible for both $\Delta N > 0$ and $\Delta N < 0$;

(4) depending on the sign of ΔN, the properties of this process will be different. For instance, the electron-whistler mode is guided at frequencies $f < f_H/2$, provided that $\Delta N > 0$; here the direction of the wave vector $\mathbf{K}$ must be close to that of $\mathbf{H}_0$.

Note the following very important result of these studies: it was found that cutoff for guided waves of the whistler mode occurs at a frequency $f \simeq f_H/2$, rather than at the frequency f_H given by kinetic theory (Roederer, 1970). A wave study of this problem (see Section 8.3.3, below) indicates that at a frequency $f \simeq f_H/2$ the amplitude of the field in a plane-stratified layer increases considerably, for a given flux of incident radiation; the latter is trapped effectively by the layer (Moiseyev, 1976).

8.3.1. Turning of the Poynting vector of plane waves on entering a plane-stratified inhomogeneous medium

Assume that a plane wave is incident normally from free space upon a magnetoplasma with a specified electron density distribution $N(z)$; the plasma is inhomogeneous in the z direction. The lower boundary of the plasma is in the xy plane, and the static magnetic field $\mathbf{H}_0$ makes an angle θ with the normal (z direction) to the xy plane. In this case the following question is of interest: what is the thickness of the region in which the vector $\mathbf{S}$ of the energy flux assumes a direction quite close to that of the vector $\mathbf{H}_0$ which it has in a homogeneous medium? If, for instance, we consider the whistler mode, when the vector $\mathbf{S}$ is aligned with $\mathbf{H}_0$, then the problem consists of determining the thickness of the zone within which the direction of the vector $\mathbf{S}$ changes completely. Although the problem can be posed as simply as this, it is mathematically difficult to arrive at a general solution of it. Thus we have to limit ourselves to a study of particular instances of the problem. The results obtained thereby apparently still give a quite general idea of the turning of the Poynting vector in an inhomogeneous medium (Al'pert & Kaufman, 1973, 1974).

The calculated results given here correspond to plasma models which, in particular, are fully adequate to describe the altitude dependence of the electron density at the base of the near-Earth plasma (the so-called D region of the ionosphere) at night, namely

$$\left.\begin{aligned} &\text{I.} \quad N(z) = N_{\infty 1}(1 + be^{-\alpha z})(1 - e^{-\alpha z}), \\ &\text{II.} \quad N(z) = N_{\infty 2}(1 - be^{-\alpha z})^2, \end{aligned}\right\} \tag{8.36}$$

where $N_{\infty 1}$ and $N_{\infty 2}$ are the electron densities far away from the plasma boundary ($z \to \infty$). One of the formulas for $N(z)$, namely formula I in (8.36), has a maximum at $z = z_M = \alpha^{-1}\ln[2b/(b-1)]$. The choice of the functions (8.36) was dictated by the fact that they enable an analytical solution of the wave equation to be found. Other limitations in the formulation of this problem making possible an analytical solution of it are as follows. The angle θ between the z-axis (the direction of the wave vector $\mathbf{K}_0$ of the incident wave) and $\mathbf{H}_0$ is so selected as to satisfy the condition of quasi-longitudinal propagation, and two frequency ranges are considered: electron-whistler LF waves ($\omega_L \ll \omega \leqslant \omega_H$), for which the effect of the ions can be neglected; and ELF waves ($0 \leqslant \omega \simeq \Omega_H$, $\omega \gtrsim \Omega_H$). For these two frequency ranges, the conditions of quasi-longitudinality and the refractive index, taking the collision frequency v into account, are given by formulas (4.31)–(4.34). Here it is assumed that $v = $ constant. This assumption, which makes it possible to solve the wave equation, turns out to be valid, since the zone of turning of the vector **S**, as will be seen below, is quite narrow (its direction varies within the limits of just the free-space wavelength $\lambda_0 = 2\pi c/\omega$), so the function $v(z)$ has little effect on the amount and nature of the change in the angle $\psi(z)$, and thus on $\alpha(z)$ as well.

The solution of the problem in the *LF range* $\omega \ll \omega \leqslant \omega_H$ *and* $\omega \gg \Omega_0$ (where Ω_0 is the plasma frequency of the ions), for a completely ionized plasma, leads to the following results. The angle $\psi(z)$ between the wave vector $\mathbf{K}(z)$ and the Poynting vector **S** is found using the formula

$$\tan\psi(z) = -\tfrac{1}{2}\frac{F(z)}{R_e F(z)}\frac{\omega_H\omega_0^2\sin\theta}{|\omega_H\cos\theta - \omega + iv|\,|\omega_0^2 - \omega^2 + iv|}, \tag{8.37}$$

where

$$F(z) = n_\infty^* + i\frac{c}{\omega}f_0^*(z), \tag{8.37a}$$

$$f_0(z) = a(\alpha e^{-\alpha z})\left[1 - \frac{\beta}{\gamma}\frac{\phi(\beta+1, \gamma+1, \xi)}{\phi(\beta, \gamma, \xi)}\right]. \tag{8.37b}$$

and $f^*(z)$ and n^*_∞ are respectively the complex conjugates of the function $f_0(z)$ and of the refractive index

$$n_\infty = \left[1 + \frac{\omega_{0\infty}^2}{\omega(\omega_H \cos\theta + \omega + i\nu)}\right]^{1/2} \tag{8.37c}$$

with $z \to \infty$, when $N = N_\infty$. The expression ϕ (...) in (8.37*b*) is a confluent hypergeometric function (Whittaker & Watson, 1927), the parameters of which are

$$\left.\begin{aligned} \beta &= \tfrac{1}{2} - i\frac{\omega}{c\alpha}\left\{n + \frac{b-1}{b^{1/2}}\left[\frac{\omega_{0\infty}^2}{\omega(\omega - \omega_H \cos\theta - i\nu)}\right]\right\}^{1/2}, \\ \gamma &= 1 - i\frac{2\omega n}{c\alpha}, \quad \xi = 2a\exp(-\alpha z), \end{aligned}\right\} \tag{8.37d}$$

$$a = \frac{\omega}{c\alpha}\left\{\frac{b\omega_{0\infty}^2}{\omega(\omega - \omega_H \cos\theta - i\omega)}\right\}^{1/2}. \tag{8.37e}$$

Fig. 8.17 presents the $\psi(z)$ relations for the values:

$$\theta = 15^\circ, \quad \omega = 2\pi \times 10^5\,\mathrm{s}^{-1}, \quad \nu = 10^6\,\mathrm{s}^{-1}, \quad \alpha = 0.03\,\mathrm{km}^{-1};$$

$$b = 5.8, \text{ model I};$$
$$b = 1, \text{ model II}$$

(see formulas (8.36)). Inspection of the figure shows that, already at distances $z \simeq 0.2\lambda_0$ for model I and $z \simeq 2\lambda_0$ to $3\lambda_0$ for model II, where λ_0 is the wavelength *in vacuo*, the angle ψ is more or less constant and is close to the value in a homogeneous medium, with $N = N_\infty$. It can be shown that, as $z \to \infty$, when $\nu = 0$, the formula (8.37) for $\tan\{\psi(z)\}$ is exactly the same as the corresponding formula for a homogeneous medium (see 8.23)). Note that the dotted curve in Fig. 8.17 indicates the region where the $\psi(z)$ relation has a small maximum, apparently due to the use of the quasi-longitudinal approximation in the calculations.

Therefore, the following general, very important conclusion can be drawn from Fig. 8.17: the Poynting vector **S** of the extraordinary electron-whistler wave very rapidly becomes aligned with the magnetic-field vector $\mathbf{H}_0$ on entering an inhomogeneous medium. The calculations carried out in Al'pert & Kaufman, 1973, for certain values of dN/dz (different values of α) and for certain frequencies indicate that the zone of turning of **S** for both models varies only within the limits $z_t \simeq 0.15\lambda_0$ to $3\lambda_0$, where λ_0 is the wavelength *in vacuo*.

For ELF waves, $0 \ll \omega \ll \omega_L$, a solution of the problem under review

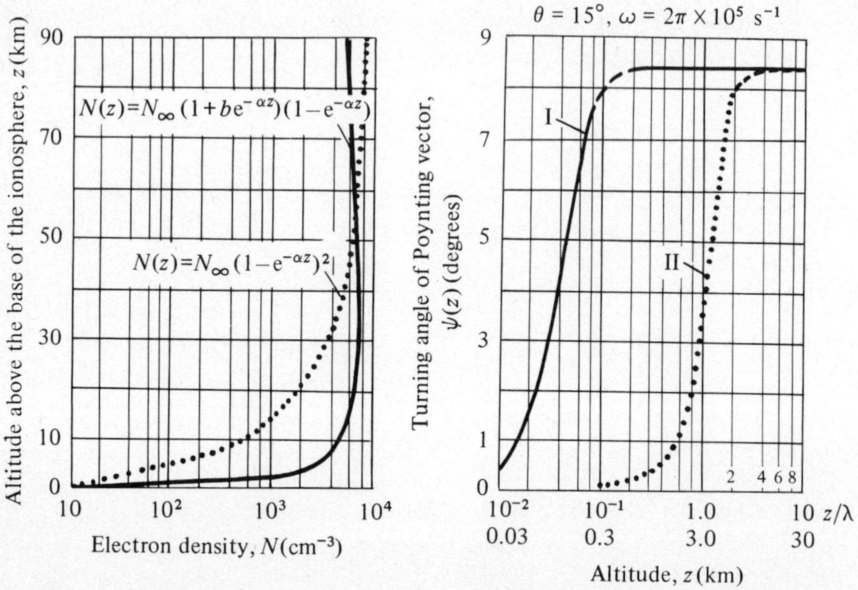

Fig. 8.17. Altitude profiles $N(z)$ of the transition region at the base of the ionosphere and curves of the angle $\psi(z)$ between the Poynting vector **S** and the vertical z (Al'pert & Kaufman, 1973).

was obtained in [27] assuming the additional conditions

$$(m/M)\omega\nu \ll \Omega_0^2(z), \quad \Omega_H^2 \ll \Omega_0^2(z),$$

which necessitate choosing a model of an inhomogeneous plasma in which for $z=0$ we have $N \neq 0$. Ultimately, it is not possible to study how the direction of **S** becomes modified on entering an inhomogeneous medium, beginning from $N=0$. However, in the plasma models used, namely

$$\left.\begin{aligned} &\text{I.} \quad N(z) = N_0 + N_{\infty 1}(1 + be^{-\alpha z})(1 - e^{-\alpha z}), \\ &\text{II.} \quad N(z) = N_0 + N_{\infty 2}(1 - be^{-\alpha z})^2, \end{aligned}\right\} \qquad (8.36a)$$

the quantity N_0 is quite low.

Solution of the wave equation, which in the given case is similar to the equation for LF waves, and subsequent calculations of the Poynting vector, lead to the following formula for the angles between the vector $\mathbf{K} \parallel z$ and $\mathbf{H}_0$ for both ELF waves:

$$\tan\psi_{12}(z) = -\frac{\tan\theta}{2}|f_{12}(\omega)|\frac{F(z)}{\mathrm{Re}\{F(z)\}}. \qquad (8.38)$$

In formula (8.38),

$$f_{12}(\omega) = -\frac{\Omega_H[(m/M)\omega^2[\nu + i(m/M)\omega] + i\Omega_H\cos\theta(\omega^2 - \Omega_H^2 - i\omega\nu)}{\cos\theta\{[\Omega_H^2 + i(m/M)\omega\nu]^2 - \omega^2\Omega_H^2\}}, \quad (8.39)$$

and

$$F(z) = n_{\infty 12} - i\frac{c}{\omega} f(z),$$

where

$$f(z) = \exp(-a_{12}e^{-\alpha z})\phi(\beta_{12}, \nu_{12}, \xi_{12}), \quad (8.42)$$

and

$$n_{12}^2 = \frac{\Omega_0^2}{\Omega_H\cos\theta(\Omega_H\cos\theta \pm \omega) + i(m/M)\nu\omega},$$

with

$$a_{12} = \frac{\omega}{c\alpha}(b)^{1/2} n_{12}(N_1), \quad \beta_{12} = \tfrac{1}{2} - i\frac{\omega}{c\alpha} n_{\infty 12} - \tfrac{1}{2}\frac{b-1}{b} a_{12},$$

and (8.41)

$$\gamma_{12} = 1 - i\frac{2\omega}{c\alpha} n_{\infty 12}, \quad \xi = 2a_{12}e^{-\alpha z}, \quad n_{\infty 12} = n_{12}(N_0 + N_1).$$

Numerical results presented in Al'pert, 1966, for the values

$$\left\{\begin{array}{l} \alpha = 0.03\,\mathrm{km}^{-1}, \quad N_0 = 10\,\mathrm{cm}^{-3}, \quad N_{\infty 1} = 10^4\,\mathrm{cm}^{-3} \\ \nu = 10^6\,\mathrm{s}^{-1}, \quad \Omega_H = 150\,\mathrm{s}^{-1}, \quad \omega/\Omega_H = 0.5; 7.3; 0.91 \text{ and } 1.2 \end{array}\right\}$$

indicate that the thickness of the zone of turning of the Poynting vector for both ELF waves $z_t <$ or $\ll \lambda_0$. At the limit, as $z \to \infty$ the angles ψ_{12} become exactly equal to their values in a homogeneous medium, obtained with the quasi-longitudinal approximation. From (8.38) it follows that, if $\nu = 0$, we have

$$\tan\psi_{12}(z \to \infty) \simeq -(\tan\theta)/2. \quad (8.38a)$$

8.3.2. Trapping of plane electromagnetic waves in a plane-stratified layer of magnetoplasma

As mentioned above, it is of considerable interest to study the trapping of electromagnetic waves in an isolated irregularity, for instance, one that is elongated along $\mathbf{H}_0$. Such irregularities, whose electron densities can be written as $N = N_0 + \Delta N$, with $|\Delta N| \ll N_0$, are produced in the outer ionosphere and magnetosphere; they promote, in particular, the ducting of hydromagnetic and electron whistlers along the Earth's magnetic force lines. Theoretically, such an irregularity could be modelled as an ellipsoid or cylinder with diffuse boundaries, the diameters of which are considerably smaller than their lengths. At present, however, no such approximate solutions of these problems exist. Thus we have to limit ourselves here just to the results of an analysis of two particular model problems, which to some extent give us an idea of the general properties of the phenomena to be expected. They apparently characterize the trapping of waves by individual irregularities not only qualitatively but also, to a certain degree, quantitatively.

(*a*) *Weakly ionized layer.* In an infinite quasi-neutral homogeneous plasma, which can contain several species of ions, there is assumed to be a plane inhomogeneity, with a constant electron density $N_0 + \Delta N$ inside it, and a thickness h_0. The density increment ΔN is such that $0 < \Delta N \ll N_0$, the latter being the density of the plasma on both sides of the plane layer. The static magnetic field $\mathbf{H}_0$ is parallel to the x-axis, which lies in the boundary plane (xy) of the layer, while the z-axis is normal to this plane. A plane wave with a wave vector $\mathbf{K}_i \equiv (K_x, 0, K_z)$ lying in the xz plane is incident upon the boundary plane (xy) from below (where $z < 0$) at arbitrary angles θ_i (see Fig. 8.18).

The Poynting vector $\mathbf{S}$ within the plane layer of thickness h_0 is studied with the aid of the dispersion equation for a homogeneous plasma (see formulas (8.12) and (8.12*a*)), which can be written conveniently as

$$\left.\begin{aligned} K_{z12}^2 = &-\frac{1}{2\varepsilon_1}\left[\left(K_x^2 - \frac{\omega^2}{c^2}\varepsilon_1\right)(\varepsilon_1 + \varepsilon_3) + \frac{\omega^2}{c^2}\varepsilon_2^2\right] \\ &\pm \left\{\frac{1}{4\varepsilon_1^2}\left[\left(K_x^2 - \frac{\omega^2}{c^2}\varepsilon_1\right)(\varepsilon_1 + \varepsilon_3) + \frac{\omega^2}{c^2}\varepsilon_2^2\right]^2 \right. \\ &\left. - \frac{\varepsilon_3}{\varepsilon_1}\left[\left(K_x^2 - \frac{\omega^2}{c^2}\varepsilon_1\right) - \frac{\omega^4}{c^4}\varepsilon_{12}^2\right]\right\}^{1/2} \end{aligned}\right\} \tag{8.42}$$

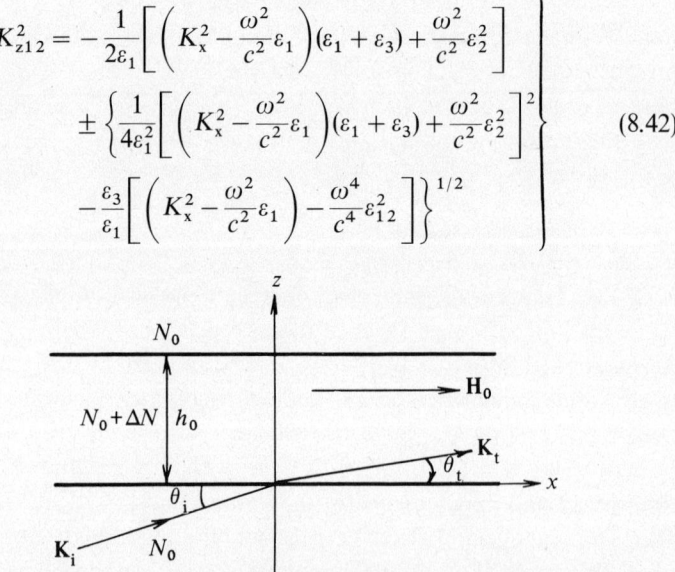

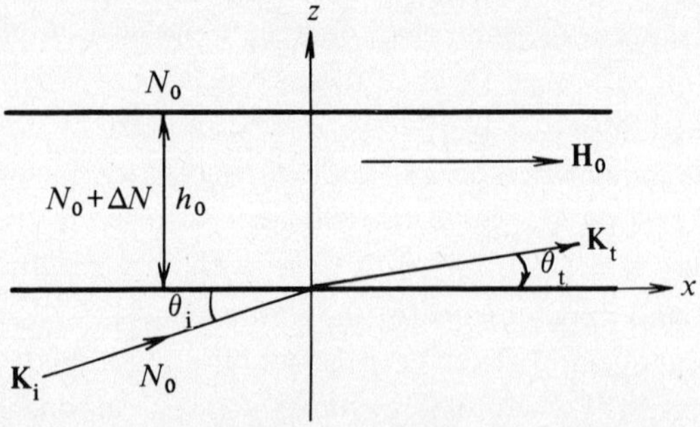

Fig. 8.18. Schematic representation of a plane plasma layer ($\mathbf{H}_0 \neq 0$) upon which a plane wave is incident obliquely ($\Delta N/N_0 \ll 1$).

where the tensor elements were described in (4.3), and $K_x = |K|\cos\theta =$ constant and $K_{z12} = |\mathbf{K}|\sin\theta_{12}$ are, respectively, the components of the wave vector along $\mathbf{H}_0$ and normal to it. The Poynting vector within the layer of thickness h_0 is calculated on the basis of a solution of the wave boundary problem using the perturbation method, which is permissible here, since $\Delta N/N_0 \ll 1$. The field components are thus obtained in the first approximation. This approximation places the following limitation on the layer thickness:

$$|\Delta K_{z12}|h_0 \ll 1, \tag{8.43}$$

and it means that the disturbance of the wave phase introduced by the layer must be small. To the indicated approximations, the component S_{zt} of the energy flux within the layer normal to the boundary equals the corresponding component S_{zi} of the wave outside it. On the other hand, the component S_{xi} naturally is not equal to S_{xt}. It has an interference structure, which disappears if the layer is thin, i.e., if $|K_{z1}|h_0 \ll 1$ and $|K_{z1} + K_{z2}|h_0 \ll 1$. In the opposite case, when the layer is thick, the value $\bar{S}_{xt}$ (i.e., the average of S_{xt} over the layer thickness) has no interference structure. The evaluation of the amount of wave trapping then reduces to an examination of the properties of the energy-flux component S_{xt} along the x-axis (in the $\mathbf{H}_0$ direction).

The solution of this problem shows that the ratio of the mean energy flux $|\overline{\mathbf{S}_{xt}}|$ in the layer (over the spatial period $2\pi/K$) to the energy flux $|\mathbf{S}_{xi}|$ of the incident wave, namely

$$p(\omega, \theta) = \frac{|\overline{S_{xt}}|}{|S_{xi}|} \tag{8.44}$$

can be quite high, i.e., it may be that $p \gg 1$. However, *an enhancement of the field in the layer*, or, in other words, *an effective trapping* of electromagnetic waves by a magnetoplasma layer, takes place in the following ranges of K_{zt}^2.

In the region of the values $K_{zt}^2 \simeq 0$, when the transverse component of the wave vector of the incident waves goes to zero, i.e., when the wave incident upon the layer is quasi-longitudinal and the angle is small. Here the angles θ_i and θ_{t12} (see Fig. 8.18) have to satisfy the condition

$$\sin\theta_i \simeq \sin\theta_{t12} \lesssim (|\Delta N|/N_0)^{1/2} \tag{8.45}$$

and they are close to the *angles of total internal reflection.* In this case the increase in the average value $|\mathbf{S}_x|$ is quite large, namely

$$p = N_0/|\Delta N| \tag{8.46}$$

and, within the framework of the approximations used, the limitation of the layer thickness will be

$$h_0 \geqslant (\lambda_0/2\pi)p \tag{8.47}$$

where $\lambda_0 = 2\pi c/\omega$ is the wavelength *in vacuo*.

In the region of the values $K_{zt1}^2 \simeq K_{zt2}^2$, when

$$\frac{1}{4\varepsilon_1^2}\left[\left(K_x^2 - \frac{\omega^2}{c^2}\varepsilon_1\right)(\varepsilon_1+\varepsilon_3) + \frac{\omega^2}{c^2}\varepsilon_1^2\right]^2 = \frac{\varepsilon_3}{\varepsilon_1}\left[\left(K_x^2 - \frac{\omega^2}{c^2}\varepsilon_1\right) - \frac{\omega^4}{c^4}\varepsilon_2^2\right] \tag{8.48}$$

and

$$K_{z12}^2 = -\frac{1}{2\varepsilon_1}\left[\left(K_x^2 - \frac{\omega^2}{c^2}\varepsilon_1\right)(\varepsilon_1+\varepsilon_3) + \frac{\omega^2}{c^2}\varepsilon_2^2\right]. \tag{8.49}$$

In this case p also satisfies the relation (8.46). An analysis of equations (8.48) and (8.49) indicates that these conditions are satisfied when

$$\frac{n_1^2 - n_2^2}{n^2} \leqslant \left(\frac{\Delta N}{N_0}\right)^{1/2}, \quad \lambda_0 \frac{n^2}{n_1^2 - n_2^2} \leqslant 2\pi h_0, \tag{8.50}$$

where n_1 and n_2 correspond to $K_x =$ constant, and the subscripts 1 and 2 indicate, respectively, a + or − sign in front of the square root in (8.42). Equation (8.49) specifies, for given θ_1 and θ_2, the frequencies at which there is an increase in S_x equal to (8.46). The angles θ_1 and θ_2 may be quite large in this case, but the angles of refraction $((\pi/2) - \theta_t)$ are *close to the angles of total internal reflection* in the layer, which accounts for the high field enhancement.

In the region of values $K_{z1}^2 \simeq K_{z2}^2 \simeq 0$, when, as seen from (8.42),

$$(\varepsilon_1 \pm \varepsilon_2) = -\varepsilon_3. \tag{8.51}$$

In this case the increase in the energy flux along $\mathbf{H}_0$ is quite large:

$$p \simeq (N_0/|\Delta N|)^3. \tag{8.52}$$

Equation (8.51) give the corresponding values of the frequencies and angles for which (8.52) is satisfied. An analysis of equation (8.51) indicates that it has three real roots. One root lies in the region of ELF waves:

$$\omega_1 = \Omega_H(1 - m/M). \tag{8.53}$$

The other two roots are in the HF range. For $\omega_0^2 \gg \omega_H^2$ they are, respectively,

$$\omega_2 \simeq \omega_H/2, \quad \omega_3 \simeq \omega_0 - \omega_H/4. \tag{8.54}$$

At the frequencies ω_1, ω_2, and ω_3, therefore, there will be maximum enhancement of the field in the layer. Then, the angles of incidence must satisfy the condition

$$\tan\theta_{i12} \simeq \sin\theta_{i12} \lesssim (|\Delta N|/N_0)^{1/2} \tag{8.55}$$

and the layer thickness

$$h_0 \gtrsim (\lambda/2\pi)n_{12}\sin\theta_{i12}. \tag{8.56}$$

Just as in case (*a*), intense trapping of electromagnetic waves takes place for quasi-longitudinal incidence upon the layer.

Note here that the field enhancement in the layer is governed, as we have seen, by the dispersion properties of the plasma. Consequently, there is reason to assume that '*trapping of radiation*' of a qualitatively similar nature must take place for other types of inhomogeneities as well. In this sense, the foregoing results can be considered general, and they apparently point to the possibility of considerable field enhancement when $|\Delta N/N_0| \lesssim 1$, too.

(*b*) *Strongly ionized layer.* As in the previous section, let us consider the incidence of a plane wave upon a plane magnetoplasma layer, but without placing any limitations on the electron density N in the layer or on the layer thickness. For simplicity, we assume that on both sides of the layer $n^2 \simeq 1$ and $N_0 \simeq 0$, which in the given case should not affect the nature of the trapping of the incident radiation by the layer. Because of the complexity of the expression for n^2, it will naturally be difficult to calculate the average value (over the layer thickness) of the ratio of the energy flux in the layer parallel to the vector $\mathbf{H}_0$ to the mean energy flux of the incident wave, for arbitrary frequencies and angles. Thus we will examine only a particular problem, the study of which will provide numerical results. It will be seen, however, that this case complements the results obtained in the preceding section for a weakly ionized irregularity and that it provides us to some extent with a more general picture of the trapping of electromagnetic energy at an irregularity.

For a specified value of the Langmuir frequency of the electrons ω_0 in the layer, it is assumed that the angle δ_i of the wave incident upon the layer satisfies the condition of quasi-transverse propagation (see Fig. 8.19):

$$\cos^4\delta_i/4\sin^2\delta_i \gg |(\omega_0^2-\omega^2)/\omega\omega_H|^2, \tag{8.57}$$

and that the frequency of the incident wave lies in the range

$$\omega_- < \omega < \omega_0. \tag{8.58}$$

In this case only the branch of the extraordinary slow wave (see Fig. 4.3) has in the layer a real value of the refractive index $n_1^2 > 0$, the ordinary wave being attenuated ($n_2^2 < 0$). The cutoff frequency $\omega = \omega_-$ corresponds to the zero value of the refractive index $n_1^2(\omega_-) = 0$. To each value of ω in frequency interval (8.58) there corresponds an allowable angle interval

$$0 \leqslant \delta_i \leqslant \delta_{io} \tag{8.59}$$

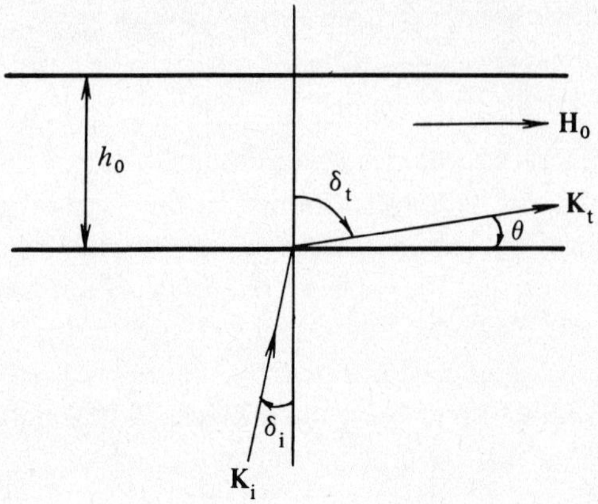

Fig. 8.19. Schematic representation of a plane plasma layer in the case of quasi-transverse incidence of a plane wave.

for the incident wave. The angle δ_{i0} is given by the expression

$$\cos^2 \delta_{i0} = \frac{2\omega_0^2}{\omega^2 + \left[\omega^4 + \dfrac{4\omega^2\omega_H^2\omega_0^2}{\omega_0^2 - \omega_H^2}\right]^{1/2}} \tag{8.60}$$

and it corresponds to the angle of total internal reflection of the wave in the layer, i.e., to the angle $\delta_t = \pi/2$ (here $\theta_t = 0$). It should be noted that the results of the problem under review also lead to a limitation on the value of the refractive index, namely

$$n_1^2 \ll (\omega_0^2 - \omega^2)/2\omega^2. \tag{8.61}$$

This limitation arises from the following factors. The incident wave satisfies, according to the formulation of the problem, the condition for quasi-transverse propagation (8.57). Consequently, its refractive index is

$$n_1^2 = 1 - \frac{\omega_0^2(\omega_0^2 - \omega_H^2)}{\omega^2[(\omega_0^2 - \omega^2) + \omega_H^2\cos^2 \alpha_i]}. \tag{8.61a}$$

On the other hand, it will be seen from the following that intensive trapping of waves takes place for low values of δ_t, when $n_1^2 = [1 - \omega_0^2/(\omega^2 + \omega\omega_H\cos\alpha_t)]$. A comparison of these formulas leads to condition (8.61).

Solution of the problem at hand and calculations of the field and energy flux $\mathbf{S}_t$ in the layer, under the conditions specified, lead to the following results (Kaufman, 1977).

The horizontal components of the flux, S_{xt} and S_{yt}, averaged over the time, oscillate in the layer with a spatial period

$$\Lambda = \tfrac{1}{2}\lambda_0/n_1\cos\theta \tag{8.62}$$

where $\lambda_0 = \omega/2\pi c$. Here, however, we are interested not in the structure of the field in the layer, but rather in the ratio of the mean value (over the layer thickness) of the component $\bar{S}_{xt}$ of the Poynting vector along the $\mathbf{H}_0$ direction to the mean energy density $\bar{S}_{xi}$ of the incident wave (see (8.44)). This ratio is given by the formula

$$p(\omega, \delta_i) = F(\omega, \delta_i)\frac{2\cos^2\delta_i[n_1^2 + \cos^2\delta_i - (\sin^2\xi/2\xi)\phi(\omega, \delta_i)]}{4(n_1^2 - \sin^2\delta_i)\cos^2\delta_i + \sin^2(2\xi)\phi^2(\omega, \delta_i)}, \tag{8.63}$$

where

$$F(\omega, \delta_i) = 1 + \frac{(1-n_1^2)}{n_1^2}\frac{(n_1^2 - \sin^2\delta_i)}{\cos^2\delta_i} + \frac{(\omega_0^2 - \omega^2)}{\omega^2\omega_H n_1^2}\frac{\sin^2\delta_i}{\cos^4\delta_i}, \tag{8.64}$$

$$\phi(\omega, \delta_i) = \frac{\omega_0^2(\omega_0^2 - \omega^2)}{\omega^2[(\omega_0^2 - \omega^2) + \omega_H^2\cos^2\delta_i]}, \quad \xi = 2\pi\frac{h_0}{\lambda_0}(n_1^2 - \sin^2\delta_i)^{1/2}. \tag{8.65}$$

An analysis of the function $p(\omega, \delta_i)$ indicates that at a specified frequency it oscillates about the angle δ_i. The number of its extrema depends on the value of ω, being higher the closer the frequency is to ω_-. The oscillation depth of the function $p(\omega, \delta_i)$ and the number of its maxima are reduced, the further ω is from ω_- (for $\omega \to \omega_0$), and the $p(\omega, \delta_i)$ relation becomes smoothed out. On the other hand, for any allowable values of ω_0, ω_H and h_0, we have $p(\omega, \delta_i) > 1$, i.e., the energy flux along $\mathbf{H}_0$ is enhanced in the

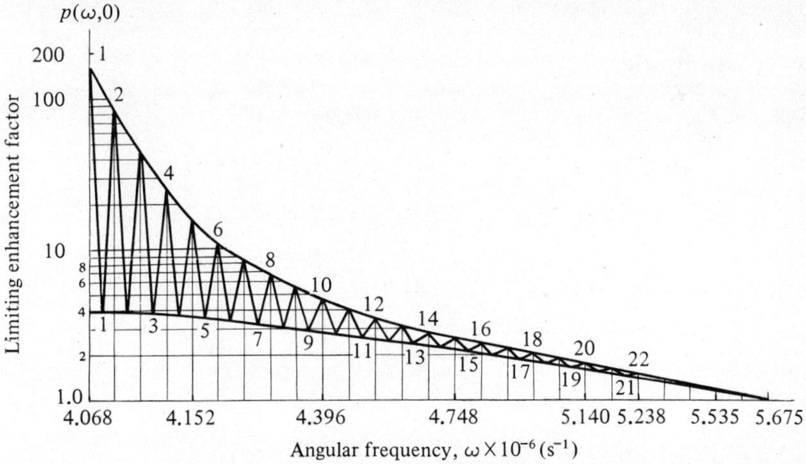

Fig. 8.20. Frequency dependence of $p(\omega, \delta_i = 0)$, which is the ratio of the mean value of the modulus of the Poynting vector $|\mathbf{S}_{xt}|$, directed along the magnetic-field vector $\mathbf{H}_0$ in a plane-parallel layer, to the mean energy density $|\mathbf{S}_{xi}|$ of the incident wave (see (8.63)) (Kaufman, 1977).

layer and may reach values $\gg 1$. The function $p(\omega, \delta_i)$ has its maximum values when $\delta_i \to \delta_{i0}$ and $\omega \to \omega_-$. The general properties of $p(\omega, \delta_i)$ in the interval $\omega_- < \omega < \omega_0$, in which the solution of the given problem applies, are explained by the limit of this function as $\delta_i \to 0$, i.e., the form of the function $p(\omega, 0)$ (Fig. 8.20). It is calculated for the values

$$\omega_0 = 5.675 \times 10^6\,\mathrm{s}^{-1}, \quad \omega_H = 3.86 \times 10^6\,\mathrm{s}^{-1}, \quad h = 4.5\,\mathrm{km}.$$

For these values of ω_0 and ω_H, the cutoff frequency $\omega_- = 4.064 \times 10^6\,\mathrm{s}^{-1}$.

Fig. 8.20 shows the envelopes of function $p(\omega, 0)$, together with its maxima and minima; these are joined arbitrarily by straight lines. Because the oscillation is rapid it is difficult to portray its details. In the allowed frequency interval, the function $p(\omega, 0)$ is seen to have 22 maxima. As the frequency ω approaches ω_0, the function smooths out and its extrema have smaller magnitudes. As ω approaches ω_-, the maxima of $p(\omega, 0)$ become very large (in the given case, the first largest maximum $p_{max} = 184.9$). Figure 8.21 shows the relation $p(\omega = \text{const.}, \delta_i)$ for three fixed values of ω. The $p(\omega, \delta_i)$ curves tend to zero as $\delta_i \to \delta_{i0}$, when the incidence angle approaches the angle of total internal reflection, where the formulas obtained no longer apply. The numbers of the maxima are given near the $p(\omega, \delta_i)$ curves, in accordance with their numbering on the $p(\omega, 0)$ curve. With an increase in frequency, when $\omega \to \omega_0$, the number of maxima is greater but their magnitudes are lower. Thus, as noted previously, the

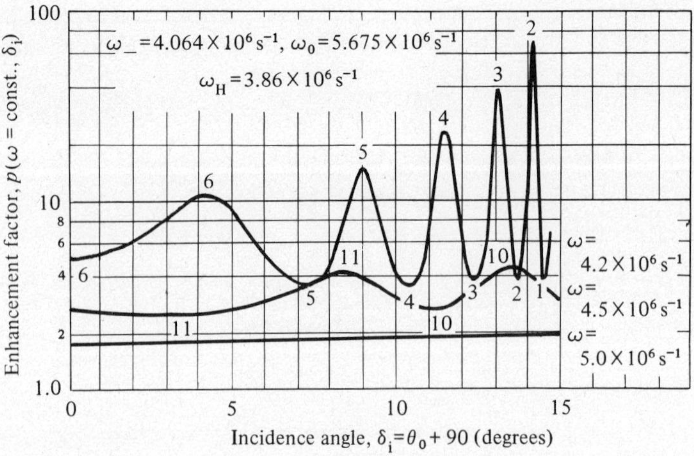

Fig. 8.21. Enhancement factor for the energy flux $p(\omega, \delta_i)$ of a slow extraordinary HF wave in a plane-parallel layer (see (8.63)), as a function of the angle δ_i for three fixed values of the angular frequency ω, expressed in s^{-1}; $\omega_- < \omega < \omega_0$ (Kaufman, 1977).

maximum 'trapping' of the energy flux in the layer corresponds to frequencies close to the cutoff frequency ω_- of the HF extraordinary slow wave and to angles close to the angles of total internal reflection δ_{i0}.

Note that, for the frequencies used in Fig. 8.21, namely

$$\omega = 4.2, 4.5, \text{ and } 5.0 \times 10^6\,\text{s}^{-1},$$

the angles of total internal reflection are

$$\alpha_{i0} = 14°57', 26°45' \text{ and } 38°58'.$$

For all values of ω and δ_{i0}, the condition (8.30) placed on the refractive index is adequately satisfied. The direction of the mean value (over the layer thickness) of the Poynting vector $\mathbf{S}_t$, characterized here by the angles ψ_S between the z-axis and $\mathbf{S}_t$, is closer to the direction of the magnetic field $\mathbf{H}_0$ than is the wave vector $\mathbf{K}$ of the refracted wave (angle δ_t) and its Poynting vector (angle ψ_K). The corresponding formulas will not be presented here, and we will content ourselves with numerical data for a single case. For the layer parameters used to calculate the $p(\omega, \delta_i)$ curves given in Fig. 8.21, at a frequency $\omega = 4.2 \times 10^6\,\text{s}^{-1}$ the following angle values are obtained:

$$\delta_i = 3°,\quad \delta_t = 9°31',\quad \psi_K = 17°42',\quad \psi_S = 29°20',$$
$$\delta_i = 9°,\quad \delta_t = 27°58',\quad \psi_K = 46°11',\quad \psi_S = 65°29',$$
$$\delta_i = 13°,\quad \delta_t = 39°32',\quad \psi_K = 59°15',\quad \psi_S = 79°15'.$$

The $p(\omega, \delta_i)$ ratio considered above describes the enhancement of the component $\bar{S}_{xt}$ (parallel to $\mathbf{H}_0$) of the wave energy flux inside the layer, in relation to the component of the energy flux of the incident wave $\bar{S}_{xi}$ in this same direction. Since in this case the incident wave makes small angles to the vertical (z direction), a more accurate measure of the energy-flux enhancement would probably be the ratio of $\bar{S}_{xt}$ to the total energy of the incident wave. This quantity can be written as

$$p_0(\omega, \delta_i) = p(\omega, \delta_i) \sin \delta_i. \tag{8.66}$$

Obviously, $p_0 < p$. When $\delta_i \to 0$, it follows that $p \to 0$, since $\bar{S}_{xi}$ and $\bar{S}_{xt} \to 0$. Consequently, in the range of angles from 0 to δ_{i0}, the quantity $p_0(\omega, \delta_i)$ increases rapidly with increasing δ_i and its maxima for $\delta_i \simeq \delta_{i0}$ are less than the values shown in Fig. 8.21. For example, for $\omega = 4.2$, 4.5 and $5.0 \times 10^6\,\text{s}^{-1}$ its values are, respectively, $p_0(\omega, \delta_{i0}) = 21.0$, 2.2, and 1.3.

A comparison of the results in this and previous sections, obtained for two cases which are extreme with regard to the physical conditions, reveals a certain generality of the observed phenomena. In both cases 'trapping' occurs, i.e., an enhancement of the energy flux of the electromagnetic waves in the layer in the direction parallel to the magnetic field $\mathbf{H}_0$. The greatest enhancement of the energy flux in the layer takes place close to the angles of total internal reflection. Note, too, that for $\omega_0^2 \gg \omega_H^2$ in the case considered in the previous section, when $|\Delta N| \ll N_0$, the greatest enhancement of the energy flux of HF waves was at $\omega \simeq \omega_0 - \frac{1}{4}\omega_H$, whereas in the case being considered in the present section it was at $\omega_0 - \frac{1}{2}\omega_H < \omega < \omega_0$. The values obtained for $p(\omega, \delta_i)$ in these two cases cannot be compared numerically, however, because the problems were solved using different methods.

8.3.3. Distribution of the field of electromagnetic waves emitted by a point source in directions close to $\mathbf{H}_0$

In Section 8.1 the behavior of the group-velocity vector of electromagnetic waves in a homogeneous magnetoplasma was discussed. The results considered there indicate how the energy flux of a plane wave in a homogeneous plasma changes direction relative to the direction of a static magnetic field $\mathbf{H}_0$ in various frequency ranges. Thus these results characterize to some extent the trapping to be expected of whistler-mode waves in the magnetic-field-aligned ducts.

In order to describe such phenomena more completely, of course, it is necessary to know the spatial distribution of the energy flux of the electromagnetic field of these waves in the field-aligned duct. However, before this problem can be solved, another, more complicated problem must be formulated and solved, namely to investigate within the framework of the wave theory the field structure of the radiation emitted by a finite source located at various distances. The results of solving such a problem will be considered in the present section. General formulas will be presented which enable a determination of the components of the field of electromagnetic waves emitted by a point source in a multicomponent magnetoplasma at various distances from the source in the frequency range $0 \lesssim \omega \lesssim \omega_0$ including the resonance branches of the plasma. Numerical results obtained with the aid of asymptotic formulas describe some properties of the field structure of the whistler mode in the range $\omega_L \lesssim \omega \lesssim \omega_H$. (Al'pert & Moiseyev, 1980).

(*a*) *General theoretical formulas.* Let us consider the radiation of an electric dipole with a moment $\mathbf{I}$ oriented along the vector of a static magnetic field $\mathbf{H}_0$. Analysis shows that the use of a cylindrical system of coordinates (z, r, ϕ), with the z-axis parallel to $\mathbf{H}_0$, enables this problem to be solved in a physically adequate way. The field of the source can then be described in a dimensionless coordinate system $\zeta = (\omega/c)z$,

$\rho = (\omega/c)r$, ϕ with the aid of the following system of Maxwell's equations:

$$\left.\begin{aligned} &-\frac{\partial^2 E_\rho}{\partial \zeta^2} + \frac{\partial^2 E_\zeta}{\partial \zeta \partial \rho} = \varepsilon_1 E_\rho + \mathrm{i}\varepsilon_2 E_\phi, \\ &-\frac{\partial^2 E_\phi}{\partial \zeta^2} - \frac{\partial}{\partial \rho}\left[\frac{1}{\rho}\frac{\partial}{\partial \rho}(\rho E_\phi)\right] = +\,\mathrm{i}\varepsilon_2 E_\rho + \varepsilon_1 E_\phi, \\ &\frac{1}{\rho}\frac{\partial}{\partial \rho}\left[\rho\left(\frac{\partial E_\rho}{\partial \zeta} - \frac{\partial E_\zeta}{\partial \rho}\right)\right] = \varepsilon_3 E_\zeta + 4I\frac{\delta(\rho)\delta(\zeta)}{\rho}. \end{aligned}\right\} \tag{8.67}$$

In (8.67), E_ρ, E_ζ, and E_ϕ are the components of the electric field in the dimensionless cylindrical coordinate system, ε_1, ε_2, and ε_3 are tensor elements (see (4.3), (4.8), and (4.10)); and $\delta(\rho)$ and $\delta(\zeta)$ are Dirac delta functions. The system of equations (8.67) can be solved in the following manner. Fourier–Bessel transforms and Fourier transforms of the variables ρ and ζ, respectively, are used. Equations (8.67) are thereby converted into an algebraic system of equations involving the Fourier components of the electric field $\tilde{E}_\rho$, $\tilde{E}_\zeta$, and $\tilde{E}_\phi$ The solution of these, together with an inverse Fourier transformation, gives the components of the field E_ρ, E_ζ, and E_ϕ in terms of integral expressions. An analysis of the vicinities of their poles makes it possible to determine the field of undamped waves corresponding to the resonance branches in the range $0 \leqslant \omega < \omega_0$, i.e., when $\varepsilon_3 < 0$. These have the following form:

$$\begin{Bmatrix} E_\rho \\ E_\zeta \\ E_\phi \end{Bmatrix} = 2\pi \int_0^\infty \mathrm{d}n_\perp n_\perp^2 F(n_\perp) \exp[-n_\parallel(n_\perp)\zeta] \begin{Bmatrix} K(n_\perp) J_1(n_\perp \rho) \\ B(n_\perp) J_0(n_\perp \rho) \\ J_1(n_\perp \rho) \end{Bmatrix} \tag{8.68}$$

where $J_0(n_\perp \rho)$ and $J_1(n_\perp \rho)$ are Bessel functions, and $n_\parallel = n\cos\theta$ and $n_\perp = n\sin\theta$ are the longitudinal and transverse components of the refractive index:

$$\begin{aligned} n_\parallel(n_\perp) = \Bigg\{ &-\tfrac{1}{2}\left[\left(1 + \frac{\varepsilon_1}{\varepsilon_3}\right) n_\perp^2 - 2\varepsilon_1\right] \\ &\pm \left[\tfrac{1}{4}\left(1 - \frac{\varepsilon_1}{\varepsilon_3}\right)^2 n_\perp^4 - \frac{\varepsilon_2^2}{\varepsilon_3} n_\perp^2 + \varepsilon_2^2\right]^{1/2}\Bigg\}^{1/2} \end{aligned} \tag{8.69}$$

$$\left.\begin{aligned} F(n_\perp) &= \mathrm{i}\,\frac{\varepsilon_2}{2\pi}\,\frac{I}{2n_\parallel^2 \varepsilon_3 + (\varepsilon_1 + \varepsilon_3) n_\perp^2 - 2\varepsilon_1 \varepsilon_3}, \\ B(n_\perp) &= \frac{[n_\parallel^2 + n_\perp^2 - \varepsilon_1][\varepsilon_1 - n_\parallel^2] + \varepsilon_2^2}{\varepsilon_2 n_\parallel n_\perp}, \\ K(n_\perp) &= -\frac{\mathrm{i}}{\varepsilon_2}(n_\parallel^2 + n_\perp^2 - \varepsilon_1). \end{aligned}\right\} \tag{8.70}$$

A further transformation of (8.68) with the aid of integral representations of the Bessel function, together with a consideration of its asymptotic expressions in the vicinity of the stationary points of the integrands, make it possible to arrive at a solution of (8.67) in a form that reveals its physical significance. This solution has the form

$$\left\{\begin{matrix} E_\rho \\ E_\zeta \\ E_\phi \end{matrix}\right\} = \{I_1^{(J)} + I_2^{(J)}\} + \{I_1^{(\varepsilon)} + I_2^{(\varepsilon)}\}, \tag{8.71}$$

where

$$\left.\begin{aligned} I_{12}^{(J)} &= \int_0^\infty dn_\perp \int_\Delta^{\pi-\Delta} d\chi\, n_\perp^2 F(n_\perp) \exp[-i\psi_{12}(n_\perp,\rho,\zeta,\chi)] \left\{\begin{matrix} K(n_\perp \exp \mp i\chi) \\ B(n_\perp) \\ \exp \pm i\chi \end{matrix}\right\} \\ I_{12}^{(\varepsilon)} &= \int_0^\Delta d\chi \int_0^\delta dn_\perp\, A(n_\perp,\chi) \exp[-i\psi_{12}(n_\perp,\rho,\zeta,\chi)] \\ &\quad + \int_0^\infty dn_\perp \int_0^\Delta d\chi\, A(n_\perp,\chi) \exp[-i\psi_{12}(n,\rho,\zeta,\chi)], \end{aligned}\right\} \tag{8.72}$$

$$A(n_\perp,\chi) = 2n_\perp^2 F(n_\perp) \left\{\begin{matrix} \mp iK(n_\perp)\sin\chi \\ B(n_\perp) \\ \mp i\sin\chi \end{matrix}\right\}. \tag{8.73}$$

Formula (8.71) shows that the source field consists of two parts: 1 and 2. Component 1 describes the field inside and outside a cone (1) whose axis is vector $\mathbf{H}_0$ and whose surface generatrix is defined by the angle

$$\alpha_1 = \tan^{-1}\left(\frac{\rho}{\zeta}\right)_1 = \tan^{-1}\left(-\frac{\varepsilon_1}{\varepsilon_3}\right)^{1/2} \tag{8.74}$$

at which resonance of the corresponding wave branch takes place ($n_2^2(\omega,\theta) \to \infty$; see Figs. 8.1 and 8.2). This element of the field structure can be called the *resonance cone*; inside it the field is defined by the integral $I_1^{(J)}$, and outside it by the integral $I_1^{(\varepsilon)}$. The function $I_1^{(J)}$ has in a number of cases (see below) a very pronounced maximum close to the surface of cone 1, while $I_1^{(\varepsilon)} \ll I_1^{(J)}$. Component 2 in (8.71) describe the field inside and outside another cone (2), which can be called the *Storey cone.* The generatrix of this cone is defined by the angle

$$\alpha_2 = \alpha_M = \tan^{-1}\left(\frac{\rho}{\zeta}\right)_2 = \tan^{-1}\left\{-\left[\frac{\partial n_\parallel(n_\perp)}{\partial n_\perp}\right]_{n_\perp^{(2)}}\right\}, \quad \left[\frac{\partial^2 n_\parallel}{\partial n_\perp^2}\right]_{n_\perp^{(2)}} = 0 \tag{8.75}$$

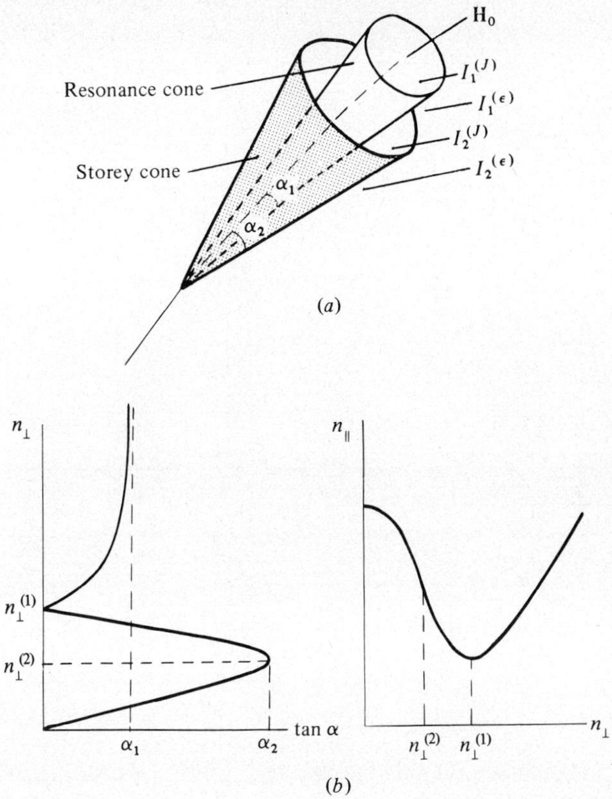

Fig 8.22. (*a*) The two cones within which the flux of electromagnetic waves in the whistler mode is trapped in the frequency range $\omega_L < \omega < \omega_H/2$ $(\omega_0^2 \gg \omega_H^2)$. (*b*) The dispersion branches $n_\perp(\alpha)$, $n_\parallel(n_\perp)$, demonstrating the origin of the two cones.

where the angle α_M corresponds to the maximum of $\alpha(\omega, \theta)$ (see Figs. 8.1–8.3). The amplitude of the field determined by the component I_2 is a maximum close to the surface of cone 2, while $I_2^{(J)} \gg I_2^{(\varepsilon)}$, the field outside the surface of cone 2, also decreases rapidly. One of the cones will obviously be located inside the other (Fig. 8.22), and the total field of the point source drops off rapidly beyond the limit α_1 or α_2, whichever is the greater.

(*b*) *Electric field of the whistler mode* $(\omega_L < \omega \leqslant \omega_H)$. Figures 8.23–8.25 show the results of claculations using the asymptotic formulas (8.71) with the plasma parameters:

$$\left.\begin{aligned} &\omega_0 = 2\omega_H, \quad \omega_H = 7.5 \times 10^6\,\mathrm{s}^{-1}, \quad \lambda_H = 8\pi \times 10\,\mathrm{m}, \\ &m/M(\mathrm{O}^+) = 3.42 \times 10^{-5}, \quad \nu = 10^2\,\mathrm{s}^{-1}. \end{aligned}\right\} \tag{8.76}$$

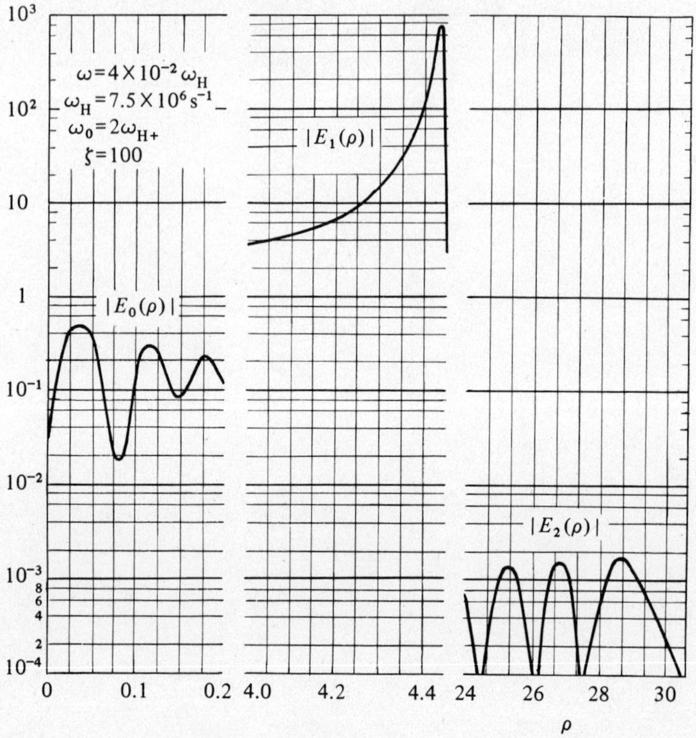

Fig. 8.23. Field amplitude $|\mathbf{E}(\rho)|$ as a function of the normalized horizontal distance $\rho = 2\pi r/\lambda_0$ in vicinities of the three maxima $|\mathbf{E}|_{0\,\max}$, $|\mathbf{E}|_{1\,\max}$, and $|\mathbf{E}|_{2\,\max}$, at a distance from the source such that $\zeta_H = 2\pi z/\lambda_H = 2.5\times10^3$ ($z = 10^5$ m) and at the angular frequency $\omega = 4\times10^{-2}\omega_H$; here $\omega_H = 7.5\times10^6\,s^{-1}$, $v = 100\,s^{-1}$, $\omega_0 = 2\,\omega_H$ (Al'pert & Moiseyev, 1980).

These data indicate that the structure of the electric field has the following features. At a given distance ζ from the source the field amplitude $|\mathbf{E}(\zeta = \text{const.}, \rho)|$ has an oscillatory nature as a function of the normalized horizontal distance ρ. In the frequency range $\Omega_L < \omega < \omega_H/2$ the electric field has three maxima: $|\mathbf{E}|_{0\,\max}$, $|\mathbf{E}|_{1\,\max}$, and $|\mathbf{E}|_{2\,\max}$ (Fig. 8.23). One of these, $|\mathbf{E}|_{0\,\max}$, appears due to the superposition of field components $|\mathbf{E}|_1$ and $|\mathbf{E}|_2$. Most of their energy flux is located, respectively, around the resonance cone 1 and the Storey cone 2, but this maximum lies near the $\mathbf{H}_0$ direction. The other two maxima, $|\mathbf{E}|_{1\,\max}$ and $|\mathbf{E}|_{2\,\max}$, are situated on the inner lobes of the cones, close to their surfaces. The maximum in the resonance cone is very pronounced. Its angular width and amplitude depend on the frequency ω: the closer ω is to the lower-hybrid frequency ω_L, the narrower will be maximum and the higher the value of $|\mathbf{E}|_{1\,\max}$.

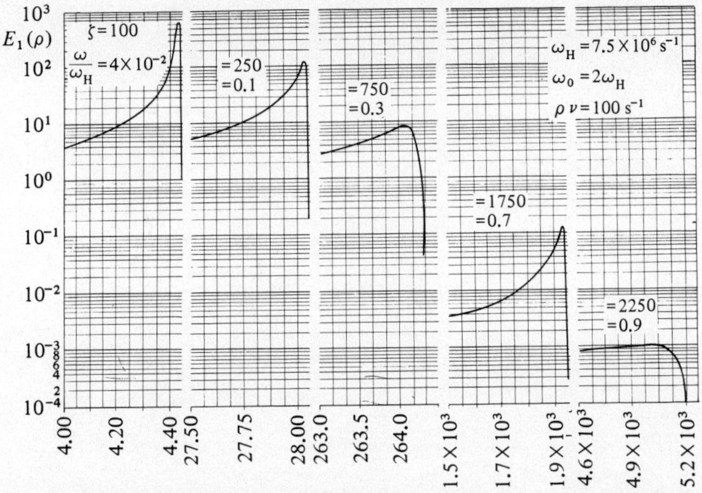

Fig. 8.24. Field amplitude $|\mathbf{E}(\rho)|_{1\,\max}$ close to the surface of the resonance cone 1 at various frequencies: $\omega_H = 7.6 \times 10\,\text{s}^{-1}$, $\omega_0 = 2\omega_H$, $\nu = 100\,\text{s}^{-1}$ (Al'pert & Moiseyev, 1980).

Up to a certain distance from the source, the maximum $|\mathbf{E}|_{1\,\max}$ is very large. Fig. 8.23 shows the relations $|\mathbf{E}_0(\rho)|$, $|\mathbf{E}_1(\rho)|$, and $|\mathbf{E}_2(\rho)|$ for a distance from the source such that $\zeta_H = 2\pi z/\lambda_H = 2.5 \times 10^3$ ($z \simeq 10^5$ m) and a frequency $\omega = 2\omega_L$, and Fig. 8.24 shows $|\mathbf{E}_1(\rho)|$ at various frequencies $\omega_L < \omega < \omega_H$. It is seen that, with an increase in frequency, $|\mathbf{E}|_{1\,\max}$ decreases very rapidly, while the relative width of the resonance maximum widens. When $\omega > \omega_H/2$, or in the case under review here when $\omega_0^2/\omega_H^2 \gg 1$, only the resonance maximum $|\mathbf{E}|_{1\,\max}$ remains, while the maximum $|\mathbf{E}|_{2\,\max}$ disappears, since in this frequency range $\alpha(\omega, \theta)$ does not have an extremum and the cone 2 disappears. With an increase in the distance ζ from the source, the preponderant value of the resonance maximum $|\mathbf{E}|_{1\,\max}$ gradually becomes smoothed out. Beginning at some critical value,

$$\zeta_c \simeq 3/|\gamma| M^{1/2} \tag{8.77}$$

namely, when $\zeta > \zeta_c$, where

$$\gamma = \tfrac{1}{2}\left(-\frac{\varepsilon_1}{\varepsilon_3}\right)^{1/2}\left(\frac{\operatorname{Im}\{\varepsilon_1\}}{|\varepsilon_1|} - \frac{\operatorname{Im}\{\varepsilon_3\}}{|\varepsilon_3|}\right)$$
$$\tag{8.78}$$
$$M = \max\left\{\varepsilon_1, -\varepsilon_3\left[1 + \frac{\varepsilon_2^2}{\varepsilon_1(\varepsilon_1 - \varepsilon_3)}\right], \frac{\varepsilon_2}{[-(\varepsilon_1/\varepsilon_3)(1 - \varepsilon_1/\varepsilon_3)]^{1/2}}\right\}$$

the condition $|\mathbf{E}|_{0\max} > |\mathbf{E}|_{1\max}$ is satisfied, i.e., the field energy begins to be

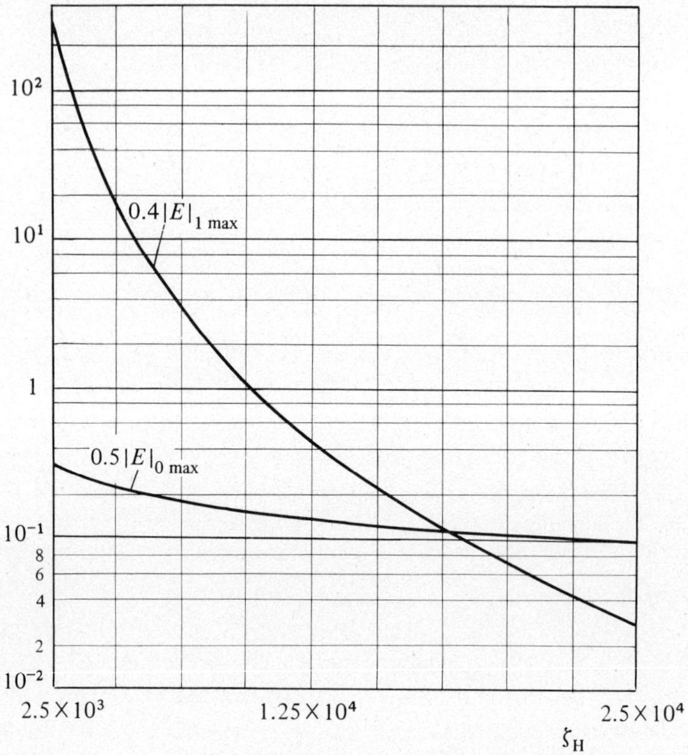

Fig. 8.25. Field amplitudes $|\mathbf{E}|_{1\,\mathrm{max}}$ and $|\mathbf{E}|_{0\,\mathrm{max}}$, as functions of the normalized distance $\zeta_{\mathrm{H}} = 2\pi z \lambda_{\mathrm{H}}$ at the frequency $\omega = 4 \times 10^{-2}\omega_{\mathrm{H}}$; $\lambda_{\mathrm{H}} = 8\pi \times 10\,\mathrm{m}$, $\omega_{\mathrm{H}} = 7.5 \times 10^{6}\,\mathrm{s}^{-1}$, $\omega_0 = 2\omega_{\mathrm{H}}$, $\nu = 100\,\mathrm{s}^{-1}$ (Al'pert & Moiseyev, 1980).

concentrated close to the magnetic force line passing through the source (Fig. 8.25).

The data presented reveal a most important, very interesting property of the field of the whistler-mode waves in the given frequency range: a considerable increase in amplitude compared with the field amplitude *in vacuo*. A dipole field I *in vacuo* is determined in the spherical coordinate system R, θ, ϕ using the formula

$$\mathbf{E}_0 = R^{-1}(\mathbf{I} \times \mathbf{n}_0) \times \mathbf{n}_0 \tag{8.79}$$

where $\mathbf{n}_0$ is the unit vector, normal to the sphere surface, passing through the observation point. In the coordinate system used above, it follows from (8.79) that

$$|\mathbf{E}_0| = (I \sin 2\theta)/2\zeta = (I \sin 2\alpha)/2\zeta \tag{8.80}$$

and for the frequencies ω cited in Figs. 8.23–8.25 the field amplitude $|\mathbf{E}_0|$ in vacuo for $I = 1$, $\rho = 2.5 \times 10^3$ ($z = 10^5$ m), and $\theta = \alpha_1$ (angle of maximum $|\mathbf{E}|_{1\,max}$) has the following values:

ω/ω_H	4×10^{-2}	0.1	0.3	0.7	0.9
$\alpha_1 = \theta$	2.5	6.4	20.6	48.0	67.1
$\|\mathbf{E}_0\|$	4.3×10^{-4}	4.4×10^{-4}	4.2×10^{-4}	2.8×10^{-4}	1.6×10^{-4}
$\|\mathbf{E}\|_{1\,max}/$ $\|\mathbf{E}_0\|$	1.3×10^6	2.5×10^5	2.2×10^4	4.6×10^4	1.9×10

The table (bottom row) also gives the ratios $\rho = |\mathbf{E}_{1\,max}|/|\mathbf{E}_0|$ at these frequencies. The enhancement of the field in the vicinity of the resonance maximum 1 is seen to vary in the range $\rho \simeq 10^6$ to 20. This effect is quite significant and apparently should play a major role in various process taking place in a plasma. However, its importance has not, as far as is known to the author, been taken into account as yet in the various theoretical studies or in the interpretations of experimental results. It should be noted that, with regard to the maximum values of $|\mathbf{E}_0|$, the enhancement of the field in a magnetoplasma varies over a range $\rho \simeq 10^5$ to 10. The $|\mathbf{E}(\alpha)|$ and $|\mathbf{E}_0(\alpha)|$ relations shown in Fig. 8.26 clearly

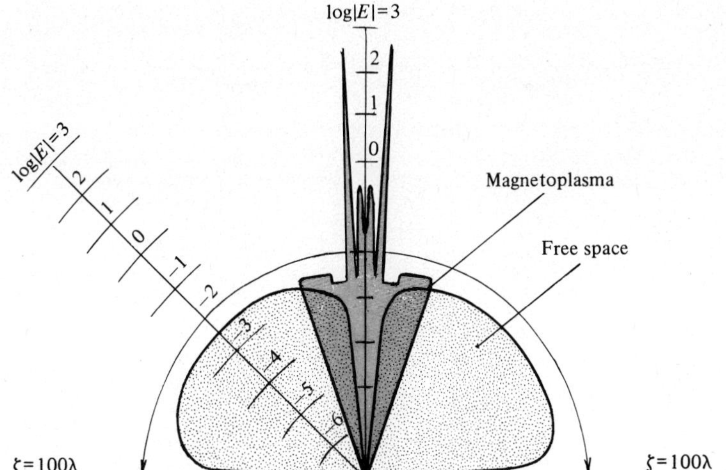

Fig. 8.26. Amplitude of the field excited by a point source with unit dipole moment in a magnetoplasma and *in vacuo*, as functions of the angle $\alpha = \tan^{-1}(\rho/z)$ at the distance $\zeta_H = 2\pi z/\lambda_H = 2.5 \times 10^3$ ($z = 10^5$ m), and at the frequency $\omega = 4 \times 10^{-2}\omega_H \simeq 2\omega_L$ (Al'pert & Moiseyev, 1980).

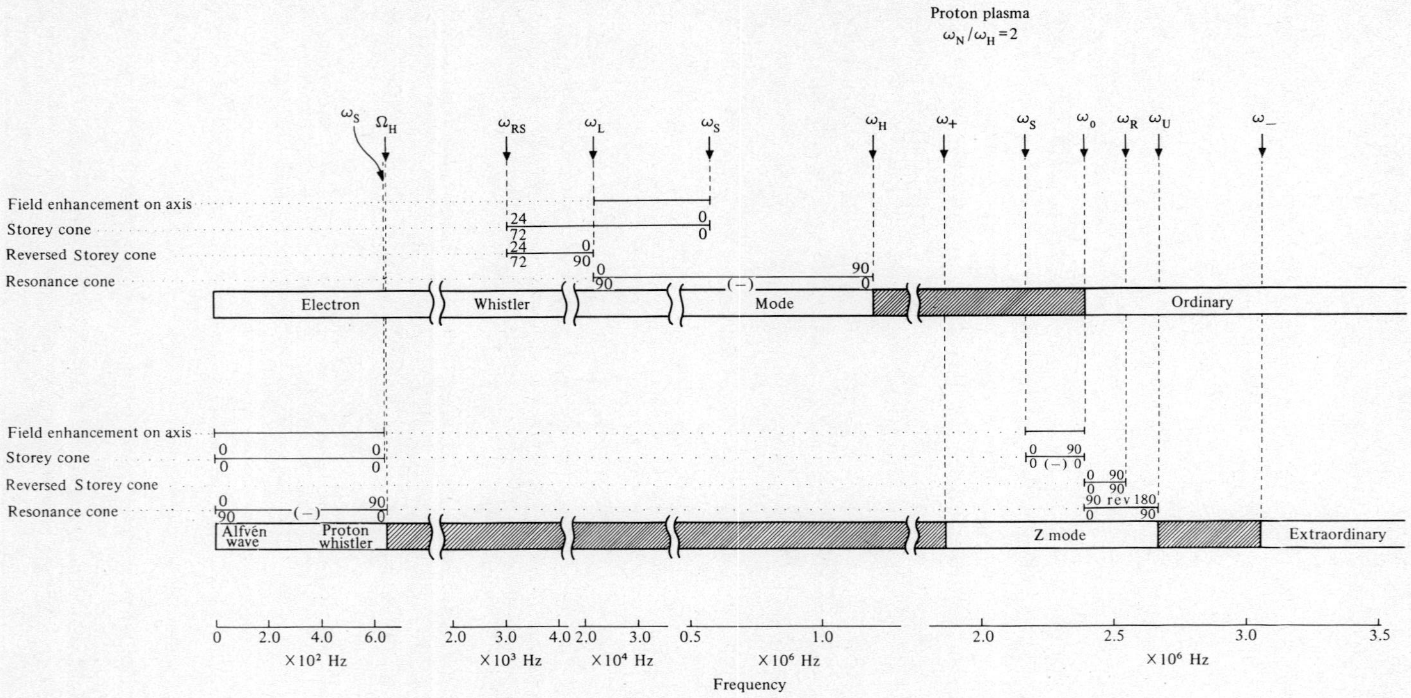

Fig. 8.27. Chart for a cold fully ionized, loss free proton plasma with $\omega_0/\omega_H = 2.0$. The upper and lower strips refer to the Ordinary and Extraordinary waves respectively, and the shaded parts show where these waves are cut off (evanescent) for all real wave normal and ray directions. The transition frequencies marked at the top are explained in the text. The frequency scale is divided into sections and is linear in each section. The horizontal lines show the frequency ranges where the various features of the field, as indicated at the left, are present. The numbers above and below them give the values in degrees of α and θ, respectively, at the ends of the ranges.

illustrate the nature of the variation in the field structure of the whistler-mode waves in a magnetoplasma, compared to the field structure *in vacuo.* Naturally, detailed calculations of the field for various plasma parameters and in different frequency ranges will indicate in which cases a redistribution of the field energy of electromagnetic waves in a magnetoplasma plays some role or other in the various plasma processes.

Recently calculations of the field were done for all the modes in a cold homogeneous magnetoplasma. They covered a full frequency range for a proton plasma and for a plasma with three species of positive ion. The results are discussed in a paper which is in preparation (Al'pert, Budden, Moiseyev & Stott). In addition to the features of the field of the whistler mode described in this section, the further main points of the new results are briefly as follows.

1. For the electron whistler mode there are three transition frequencies, ω_{RS}, ω_L and ω_S in increasing order, and ω_S is very approximately $\frac{1}{2}$ Min (ω_H, ω_0). When $\omega < \omega_{RS}$ there is no real Storey cone or reversed Storey cone and no real resonance cone. When ω increases and passes through ω_{RS}, a real Storey cone and a real reversed Storey cone appear at the same ray direction α. As ω increases further, the ray direction α for the reversed Storey cone decreases and approaches 0° as $\omega \to \omega_L$. When ω passes through ω_L the reversed Storey cone disappears and a resonance cone appears at $\alpha = 0^\circ$. The ray direction α of the Storey cone also decreases and approaches zero as $\omega \to \omega_S$. For $\omega > \omega_S$ the Storey cone is absent.

The field at the maximum of the Storey cone is not very large. The enhancement of the field is large only for the reversed Storey cone, i.e., for $\omega_{RS} < \omega_L$. It is greatest when $\omega \to \omega_L$, and there it is 10^3 to 10^4 times the field at the maximum of the Storey cone.

A resonance cone is present for the frequency range $\omega_L < \omega <$ Min (ω_H, ω_0). Its field enhancement is greatest near the low-frequency end of this range, i.e., for ω slightly greater than ω_L, and here it is about 10^6 times the field in a vacuum.

2. At higher frequencies the extraordinary wave is propagated for the frequency range $\omega_- \lesssim \omega \lesssim \omega_U$, and it is sometimes called the *Z* mode. Here ω_- is the smallest cut-off frequency, $n^2(\omega_-) = 0$, for H.F. waves, and ω_U is the upper hybrid frequency. For parts of this frequency range the *Z* mode shows a Storey cone, a reversed Storey cone and a resonance cone. The behaviour is different for the two cases $\omega_0 < \omega_H$ and $\omega_0 > \omega_H$. The fields at the edges of the Storey cones are of the order 10–20 times greater than the field in a vacuum. For the reversed Storey cones they are 20–40 times greater, and for the resonance cone they are roughly 10^3 times greater.

3. In the ELF range, where $0 \lesssim \omega \lesssim \Omega_{Hs}$ (Ω_{Hs} is the gyrofrequency of ions of the kind $\langle s \rangle$) the enhancement of the field takes place as in the LF and HF ranges. The field has some interesting features in a magnetoplasma with more than one species of positive ion. In such a plasma the behaviour of the field is complicated around the crossover frequencies. But as in the LF range the greatest enhancement of the field takes place around the ion–ion hybrid frequencies ω_{L12}, ω_{L23} etc., at the edges of the reversed Storey cones when $\omega \lesssim \omega_{L12}, \ldots$ and close to the direction of $\mathbf{H}_0$ and of the edge of the resonance cone when $\omega \gtrsim \omega_{L12}, \ldots$. But in these frequency ranges the enhancement factors are less than for the electron whistler mode.

References

Adachi, S., 1965. A study of the guiding mechanism of whistler radio waves, *J. Res. Nat. Bur. Stand.* **69D**, 493.

Adachi, S. 1966. Theory of duct propagation of whistler radio waves, *Radio Science* **1**, 671.

Aikyo, K. & Ondoh, T. 1971. Propagation of nonducted VLF waves in the vicinity of the plasmapause, *J. Radio Res. Lab. Jap.* **18**, 153.

Akhiezer, A.I. & Fainberg, Ya. B. 1949. On the interaction of a charged beam with an electron plasma, *Dokl. Akad. Nauk SSSR* **69**, 555 [In Russian].

Akhiezer, A.I. & Fainberg, Ya. B. 1951a. High-frequency oscillations of an electron plasma, *ZhETF* **21**, 1262 [In Russian].

Akhiezer, A.I. & Fainberg, Ya.B. 1951b. Slow electromagnetic waves, *Usp. Fiz. Nauk* **44**, 321 [In Russian].

Akhiezer, A.I. & Lyubarskii, T.Ya. 1951. On a nonlinear theory of electron oscillations in a plasma, *Dokl. Akad. Nauk SSSR* **80**, 193 [In Russian].

Akhiezer, A.I., Akhiezer, I.A., Polovin, R.V., Sitenko, A.G. & Stepanov, K.N. 1964. *Collective oscillations in a plasma*, Atomizdat, Moscow [In Russian] [1967, Pergamon Press, Oxford, New York].

Akhiezer, A.I., Akhiezer, I.A., Polovin, R.V., Sitenko, A.G. & Stepanov, K.N. 1974. *Plasma electrodynamics*, Nauka, Moscow [In Russian] [1975, Pergamon Press, Oxford, New York].

Akhiezer, A.I., Lapshin, V.I. & Stepanov, K.N. 1976. Theory of damping of magnetohydrodynamic waves in a high-temperature plasma, *ZhETF* **70**, 81 [In Russian].

Aliev, Yu.M. & Silin, V.P. 1965. Theory of plasma oscillations in a high-frequency electromagnetic field, *ZhETF* **48**, 901 [In Russian].

Aliev, Yu.M. & Zyunder, D. 1969. Parametric excitation of lower-hybrid and upper-hybrid resonances, *ZhETF* **57**, 1324 [In Russian].

Allen, E.M., Thome, G.D. & Rao, P.B. 1974. HF-Phased array observations of heater-induced spread-F, *Radio Science* **9**, 905.

Al'pert, Ya.L. 1948. On the trajectories of rays in a magnetoactive ionized medium – the ionosphere, *Izv. Akad. Nauk.SSSR. Ser. Fiz.* **12**, 241, 267 (see also 1946, *Dokl. Akad. Nauk SSSR* **53**, 703) [In Russian].

Al'pert, Ya.L., Gurevich, A.V. & Pitaevskii, L.P. 1963. Effects produced by a satellite rapidly moving in the ionosphere or in interplanetary space, *Usp. Fiz. Nauk* **79**, 23 [In Russian] [1963, *Space Sci. Rev.* **2**, 680].

Al'pert, Ya.L., Gurevich, A. V. & Pitaevskii, L.P. 1965. *Space physics with artificial satellites*, Consultants Bureau, New York [1964, Nauka, Moscow [In Russian]].

Al'pert, Ya.L. 1965. On electromagnetic effects in the neighbourhood of a satellite or a vehicle moving in the ionosphere or interplanetary space, *Geomagnetizm i Aeronomiya* **5**, 3 [In Russian] [1965, *Space Sci. Rev.* **4**, 373].

Al'pert, Ya.L. 1966. On the outer ionosphere and its transition into interplanetary space, *Usp. Fiz. Nauk* **90**, 405 [In Russian] [1967, *Space Sci. Rev.* **6**, 419].

Al'pert, Ja.L. 1967. VLF and ELF waves in the near-earth plasma, *Space Sci. Rev.* **6**, 781.

Al'pert, Ya.L. 1973. *Radio wave propagation and the ionosphere*, Vol. 1 & Vol. 2, Consultants Bureau, New York, [1972, Nauka, Moscow [In Russian]].

Al'pert, Ja.L. & Kaufman, R.N. 1973. The upward propagation of LF waves (electron whistlers) into the ionosphere and the turning of the Poynting vector towards the earth's magnetic field, *J. Atmos. Terr. Phys.* **35**, 1843.

Al'pert, Ja.L. & Kaufman, R.N. 1974. The region in which the Poynting vector of ion ELF waves incident on the ionosphere rotates towards the earth's magnetic field, *J. Atmos. Terr. Phys.* **36**, 1093.

Al'pert, Ya.L. & Fligel', D.S. 1974. A method for estimating the electron concentration altitude dependences and the temperature of the ionosphere using whistlers, *Kosmicheskie Issledovaniya* **12**, 880 [In Russian].

Al'pert, Ja.L. 1975. On an outer ionosphere hydrostatic model, *J. Atmos. Terr. Phys.* **37**, 1375.

Al'pert, Ja.L. & Fligel, D.S. 1977. The locating of hydromagnetic whistler sources and determination of their generating proton spectra, *Planet. Space Sci.* **25**, 487.

Al'pert, Ya.L. 1980a. On elements of the dielectric tensor, the refractive indices n_{12} and attenuation factors k_{12} of a magnetically active collisional plasma, *J. Atmos. Terr. Phys.* **42**, 217.

Al'pert, Ya.L. 1980b. The direction of the group velocity of electromagnetic waves in a multicomponent magnetoactive plasma in the frequency range $0 < \omega \to \infty$, *J. Atmos. Terr. Phys.* **42**, 205.

Al'pert, Ya.L. & Moiseyev, B.S. 1980. On the distribution of the field of electromagnetic waves emitted by a dipole in a homogeneous magnetoactive plasma, *J. Atmos. Terr. Phys.* **42**, 521.

Al'tshul', L.M. & Karpman, V.I., 1965. On the theory of nonlinear oscillations in a collisionless plasma, *ZhETF* **49**, 515 [In Russian].

Andreev, N.E. 1969. On the instability of a magnetoactive plasma in a high-frequency field, *ZhETF* **39**, 1560; 1971, *Radiofizika* **14**, 1160 (see also 1972, *ZhETF* **63**, 1283)[In Russian].

Andreev, N.E. & Kirii, A.Yu. 1971. Parametric instability of a magnetoactive plasma in a high-frequency electric field, *ZhETF* **41**, 1080 [In Russian].

Angerami, J.J. & Carpenter, D.L. 1966. Whistler studies of the plasmapause in the magnetosphere. 2. Electron density and total tube electron content near the knee in magnetospheric ionization, *J. Geophys. Res.* **71**, 711.

Appleton, E.V. 1928. Some notes on wireless methods of investigating the electrical structure of the upper atmosphere, I and II, *Proc. Phys. Soc.* I: **41**, 43; 1930, II: **42**, 321.

Ashour-Abdalla, M. 1972. Amplification of whistler waves in the magnetosphere, *Planet, Space Sci.* **20**, 639.

Aubry, M.P., Bitoun, J. & Graff, P. 1970. Propagation and group velocity in a warm magnetoplasma, *Radio Science* **5**, 635.

Bailey, V.A. & Martin, D.F. 1934. The influence of electric waves on the ionosphere, *Phil. Mag.* **18**, 369.

Bailey, V.A. 1938. On some effects caused in the ionosphere by electric waves, *Phil. Mag.* **26**, 425.

Bailey, V.A. 1959. Some possible effects caused by strong gyrowaves in the ionosphere, I, *J. Atmos. Terr. Phys.* **14**, 299.

Belikovich, V.V., Benediktov, E.A., Getmantsev, G.G., Erukhimov, L.M., Zuikov, N.A., Komrakov, G.P., Korobkov, Yu.S., Kotik., Yu.S., Mityakov, N.A., Rapoport, V.O., Sazonov, V.O., Trakhtengerts, V.Yu., Frolov, V.L. & Cherepovskii, V.A. 1974. Nonlinear phenomena in the upper ionosphere, *Usp. Fiz. Nauk*, **113**, 732 [In Russian].

Belikovich, V.V., Benediktov, E.A., Getmantsev, G.G., Ignat'ev, Yu.A. & Komrakov, G.P. 1976a. Backscatter of radiowaves from the artificial F-spread layer of the ionosphere, *Radiofizika* **19**, 1084 [In Russian].

Belikovich, V.V., Benediktov, E.A., Getmantsev, G.G., Mityakov, N.A., Razin, V.A., Teplykh, A.I., Shavin, P.V. & Tomchinskii, A.M. 1976b. Diffraction of Cassiopeia-A radio emission on artificial inhomogeneities produced in the ionosphere, *Radiofizika* **19**, 1902 [In Russian].

Bell, T.F. & Helliwell, R.A. 1971. Pulsation phenomena observed in long-duration VLF whistler mode signals, *J. Geophys. Res.* **76**, 8414.

Berezin, Yu.A. & Karpman, V.I. 1964. On the theory of nonstationary waves of limited amplitude in a rarefied plasma, *ZhETF* **46**, 1880 [In Russian].

Berezin, Yu.A. & Karpman, V.I. 1966. On the nonlinear evolution of disturbances in a plasma and in other dispersive media, *ZhETF* **51**, 1557 [In Russian].

Bernstein, I.B., Greene, J.M. & Kruskal, M.D., 1957. Exact nonlinear plasma oscillations, *Phys. Rev.* **108**, 546.

Bernstein, I.B. 1958. Waves in a plasma in a magnetic field, *Phys. Rev.* **109**, 10.

Biondi, M.A., Sipler, D.P. & Hake, R.D. 1970. Optical ($\lambda = 6300\,Å$) detection of radio frequency heating of electrons in the F-region, *J. Geophys. Res.* **75**, 6421.

Bitoun, J., Graff, P. & Aubry, M. 1970. Ray tracing in warm magnetoplasma and applications to topside resonances, *Radio Science* **5**, 1341.

Bohm, D. & Gross, E.P. 1949. Theory of plasma oscillations, *Phys. Rev.* **75**, 1851.

Booker, H.G. 1939. Propagation of wave packets incident obliquely upon a stratified doubly refracting ionosphere, *Phil. Trans. Roy. Soc.* **A237**, 411.

Booker, H.G. 1956. A theory of scattering by nonisotropic irregularities with application to radar reflections from the aurora, *J. Atmos. Terr. Phys.* **8**, 204.

Booker, H.G. 1962. Guidance of radio and hydromagnetic waves in the magnetosphere, *J. Geophys. Res.* **67**, 4135.

Booker, H.G. & Dyce, R.B. 1965. Dispersion of waves in a cold magnetoplasma from hydromagnetic to whistler frequencies, *Radio Science* **69D**, 463.

Booker, H.G. 1975. Electromagnetic and hydromagnetic waves in a cold magnetoplasma, *Phil. Trans. Roy. Soc. Lond.* **A280**, 57.

Borisov, N.D., Vas'kov, V.V. & Gurevich, A.V. 1976. Excitation of drift instability under the action of radiowaves at the F-layer of the ionosphere, *Geomagnetizm i Aeronomiya* **16**, 783 [In Russian].

Borisov, N.D., Vas'kov, V.V. & Gurevich, A.V. 1977. Shortwave drift instability, *Fizika Plazmy* **3**, 168 [In Russian].

Bowhill, S.A. 1974. Satellite transmission studies of spread-F produced by artificial heating of the ionosphere, *Radio Science* **9**, 975.

Brace, L.H., and Theis, R.F. 1974. The behavior of the plasmapause at mid-latitudes: Isis 1 Langmuir probe measurements, *J. Geophys. Res.* **79**, 1871.

Bremmer, H. 1949. *Terrestrial radio waves*, Elsevier, London, p. 305.

Brown, L.W. 1973. The galactic radio spectrum between 130 kHz and 2600 kHz, *Atrophys. J.* **180**, 359.

Budden, K.G. 1961. *Radio waves in the ionosphere*, Cambridge University Press, p. 396.

Budden, K.G. Daniell, G.J. 1965. Rays in magnetoionic theory, *J. Atmos. Terr. Phys.* **27**, 395.

Budden, K.G. & Terry, P.D. 1971. Radio ray tracing in complex space, *Proc. Roy. Soc.* A**321**, 275.

Budden, K.G. 1980. The theory of windows in the ionosphere and magnetosphere, *Journ. Atmos. Terr. Phys.* **42**, 287.

Budden, K.G. & Stott, G.F. 1980. Rays in magnetoionic theory II, *J. Atmos. Terr. Phys.* **42**, 791.

Carlson, H.C., Gordon, W.E. & Showen, R.L. 1972. High-frequency induced enhancements of the incoherent scatter spectrum at Arecibo, *J. Geophys. Res.* **77**, 1242 (see also 1971, **76**, 7808).

Carpenter, D.L. 1974. VHF and UHF bistatic observations of a region of the ionosphere modified by a high power radio transmitter, *Radio Science* **9**, 965.

Chapman, S. & Cowling, T.G. 1939. *Mathematical theory of nonuniform gases*, Cambridge University Press.

Chappell, C.R., Harris, K.K. & Sharp, G.W. 1970a. A study of the influence of magnetic activity on the location of the plasmapause as measured by OGO 5, *J. Geophys. Res.* **75**, 50.

Chappell, C.R., Harris, K.K. & Sharp, G.W. 1970b. The morphology of the bulge region of the plasmapause, *J. Geophys. Res.* **75**, 3848.

Chappell, C.R., Harris, K.K. & Sharp, G.W. 1971a. OGO 5 measurements of the plasmapause during observations of stable auroral red arcs, *J. Geophys. Res.* **76**, 2357.

Chappell, C.R., Harris, K.K. & Sharp, G.W. 1971b. The dayside of the plasmasphere, *J. Geophys. Res.* **76**, 7632.

Chappell, C.R. 1972. Recent satellite measurements of the morphology and dynamics of the plasmasphere, *Rev. Geophys. Space Phys.* **10**, 951.

Chappell, C.R. 1974. Detached plasma regions in the magnetosphere, *J. Geophys. Res.* **79**, 1861.

Cohen, R. & Whitehead, J.D. 1970. Radio-reflectivity detection of artificial modifications of the ionospheric F layer, *J. Geophys. Res.* **75**, 6439.

Copal Rao, M.S. & Booker, H.G. 1963. Guiding of electromagnetic waves along a magnetic field in a plasma, *J. Geophys. Res.* **68**, 387.

Cowling, T.G. 1945. The electrical conductivity of an ionized gas in a magnetic field with applications to the solar atmosphere and the ionosphere, *Proc. Roy. Soc. Lond.* **A183**, 453.

Cragin, B.L. & Fejer, J.A. 1974. Generation of large scale field-aligned irregularities in ionospheric modification experiments, *Radio Science* **9**, 1071.

DeForest, S.E. & McIlwain, C.E. 1971. Plasma clouds in the magnetosphere, *J. Geophys. Res.* **76**, 3587.

Domrin, V.I. 1967. On the theory of oscillations in a magnetoactive plasma under the influence of a high-frequency electric field, Preprint FIAN No. 38 [In Russian].

Doyle, P.H. & Neufeld, J., 1959. On the behavior of plasma at ionic resonance, *Phys. Fluids*, **2**, 39.

Dreicer, H. 1959. Electron and ion runaway in a fully ionized gas, *Phys. Rev.* **115**, 238.

DuBois, D.F. & Goldman, M.V. 1965. Radiation-induced instability of electron plasma oscillations, *Phys. Rev. Lett.* **14**, 544.

DuBois, D.F. & Goldman, M.V. 1967. Parametrically excited plasma fluctuations, *Phys. Rev.* **164**, 207.

Edgar, B.C. 1976a. The upper and lower frequency cutoffs of magnetospherically reflected whistlers, *J. Geophys. Res.* **81**, 205.

Edgar, B.C. 1976b. The theory of VLF Doppler signatures and their relation to the magnetospheric density structure, *J. Geophys. Res.* **81**, 3327.

Fadeeva, V.N. & Terent'ev, N.M. 1954. *Tables of the complex probability integral*, Gostekhizdat, Moscow [In Russian].

Fejer, J.A. 1955. Interaction of pulsed radio waves in the ionosphere, *J. Atmos. Terr. Phys.* **7**, 322.

Fialer, P.A. 1974. Field-aligned scattering from a heated region of the ionosphere – Observations at HF and VHF, *Radio Science* **9**, 923.

Försterling, K., 1942. Über die Ausbreitung elektromagnetischer Wellen in einem magnetisierten Medium bei senkrechter Incidenz, *Hochfr. Elek.* **59**, 110.

Frank, L.A. 1967. On the extraterrestrial ring current during geomagnetic storms, *J. Geophys. Res.* **72**, 3753.

Frank, L.A. 1970. Direct detection of asymmetric increases of extraterrestrial 'ring current' proton intensities in the outer radiation zone, *J. Geophys. Res.* **75**, 1263.

Frank, L.A. & Owens, H.D. 1970. Omnidirectional intensity contours of low-energy protons ($0.5 \leqslant E < 50$ keV) in the earth's outer radiation zone at the magnetic equator, *J. Geophys. Res.* **75**, 1269.

Fredricks, R.W., Crook, G.M., Kennel, C.F., Green, I.M., Scarf, F.L., Coleman, P.J. & Russell, C.T. 1970. OGO 5 observations of electrostatic turbulence in bow shock magnetic structures, *J. Geophys. Res.* **75**, 3751.

Fried, B.D. 1959. Mechanism for instability of transverse plasma wave, *Phys. Fluids* **2**, 337.

Fried, B.D. & Conte, S.D. 1961. *The plasma dispersion function*, Academic Press, New York.

Gaponov, A.V. & Miller, M.A. 1958. On the potential for charged particles in a high-frequency electromagnetic field, *ZhETF* **34**, 242 [In Russian].

Gardner, C.S., Green, I.M., Kruskal, M.D. & Miura, R.M. 1967. Method of solving the Korteweg–de Vries equation, *Phys. Rev. Lett.* **19**, 1095.

Georges, T.M. 1970. Amplification of ionospheric heating and triggering of spread-F by natural irregularities, *J. Geophys. Res.* **75**, 6436.

Gershman, B.N., 1953a. Kinetic theory of magnetohydrodynamic waves, *ZhETF* **24**, 453 [In Russian].

Gershman, B.N., 1953b. Propagation of electromagnetic waves in a magnetoactive plasma (kinetic theory), *ZhETF* **24**, 659 [In Russian].

Gershman, B.N., 1960a. Gyroresonance attenuation of electromagnetic waves in a plasma, *ZhETF* **38**, 912 [In Russian].

Gershman, B.N., 1960b. Group velocity of plasma waves in a magnetoactive plasma, *Radiofizika* **3**, 146 [In Russian].

Getmantsev, G.G., Komrakov, N.P., Korobkov, P.P., Mironenko, L.F., Mityakov, N.A, Rapoport, V.O., Trakhtengerts, V.Yu., Frolov, V.L. & Cherepovskii, V.A. 1973. Some results of investigations of nonlinear phenomena in the F-layer of the ionosphere, *ZhETF Lett.* **18**, 621 [In Russian].

Getmantsev, G.G., Zuikov, N.A., Kotik, D.S., Mironenko, L.F., Mityakov, N.A., Rapoport, V.O., Sazonov, Yu.A., Trakhtengerts, V.Yu. & Eidel'man, V.Ya. 1974. Combination frequencies in the interaction between high-power short-wave radiation and ionospheric plasma, *ZhETF Lett.* **20**, 229 [In Russian].

Ginzburg, V.L. 1967. *Propagation of electromagnetic waves in a plasma*, Nauka, Moscow [In Russian] [1970, Pergamon Press, Oxford, New York].

Ginzburg, V.L. & Rukhadze, A.A. 1970. *Waves in a magnetoactive plasma*, Nauka, Moscow [In Russian] [*Handbuch der Physik*, **49/4**, 395].

Gorbunov, L.M. & Silin, V.P. 1969. On the instability of a magnetoactive plasma in a strong high-frequency field, *ZhETF* **39**, 69 [In Russian].

Gordeev, G.V. 1954. Low-frequency plasma oscillations, *ZhETF* **27**, 18 [In Russian].

Gordon, W.E. & LaLonde, L.M. 1961. The design and capabilities of an ionospheric radar probe, *IRE Transactions* **7**, 17.

Gordon, W.E. & Carlson, H.C. 1974. Arecibo heating experiments, *Radio Science* **9**, 1041.

Gross, E.P. 1951. Plasma oscillations in a static magnetic field, *Phys. Rev.* **82**, 232.

Gulel'mi, A.V. & Troitskaya, V.A. 1973. *Geomagnetic pulsations and diagnostics of the magnetosphere*, Nauka, Moscow [In Russian].

Gurevich, A.V. 1955. The distribution function of electrons in a magnetoactive plasma in the presence of an alternating electric field, *Dokl. Akad. Nauk SSSR* **104**, 201 [In Russian].

Gurevich, A.V. 1956a. The influence of radio waves on plasma properties (ionosphere), *ZhETF* **30**, 1112 [In Russian].

Gurevich, A.V. 1956b. The theory of propagation of strong radio waves in a plasma (ionosphere), *Radiotekhnika i Elektronika* **1**, 706 [In Russian].

Gurevich, A.V. 1958a. On the electron temperature in a plasma in an alternating electric field, *ZhETF* **35**, 392 [In Russian].

Gurevich, A.V. 1958b. Distortion of the modulation of strong radio waves in plasmas (ionosphere), *Radiofizika* **1**, No. 4, 21; No. 5–6, 17 [In Russian].

Gurevich, A.V. 1959. Instability and hysteresis of the electron temperature in a plasma in inert gases, *ZhETF* **36**, 624 [In Russian].

Gurevich, A.V. 1960. On certain peculiarities of electron-gas joule heating in a plasma, *ZhETF* **38**, 116 [In Russian].

Gurevich, A.V. 1964. Instability of a disturbed zone in the vicinity of a charge body in a plasma, *Geomagnetizm i Aeronomiya* **4**, 247 [In Russian].

Gurevich, A.V. 1965. Nonlinear effects for powerful radio waves in the ionosphere, *Geomagnetizm i Aeronomiya* **5**, 70 (*see also* 1972, 12, 24) [In Russian].

Gurevich, A.V. 1967. Influence of strong radio waves on the F-layer of the ionosphere, *Geomagnetizm i Aeronomiya* **7**, 291 (see also 1971, **11**, 953) [In Russian].

Gurevich, A.V. 1972. Travelling ionization disturbances in a field of strong electromagnetic waves, *Radiofizika* **15**, No. 1, 11 [In Russian].

Gurevich, A.V. & Shlyuger, I.S. 1975. Investigations of nonlinear phenomena with a powerful radio pulse in the lower ionosphere, *Radiofizika*, **18**, 1237 [In Russian].

Gurevich, A.V. 1978. *Nonlinear phenomena in the ionosphere*, Springer Verlag, Berlin.

Gurnett, D.A., Shawhan, S.D., Brice, N.M. & Smith, R.L. 1965. Ion cyclotron whistlers, *J. Geophys. Res.* **70**, 1665.

Gurnett, D.A. & Frank, L.A. 1974. Thermal and suprathermal plasma densities in the outer magnetosphere, *J. Geophys. Res.* **79**, 2355.

Hall, D.F., Kemp, R.F. & Sellen, J.M. 1964. Plasma-vehicle interaction in a plasma stream, *AIAA Journal* **2**, 1032.

Hartree, D.R. 1931. Propagation of electromagnetic waves in a refracting medium in a magnetic field, *Proc. Camb. Phil. Soc.* **27**, 143.

Hartree, D.R. 1931. Optical and equivalent paths in a stratified medium treated from a wave standpoint, *Proc. Roy. Soc.* **A131**, 427.

Haselgrove, J. 1955. Ray theory and a new method for ray tracing. In: *The Physics of the Ionosphere*, Phys. Soc. Lond., p.355.

Haselgrove, J. 1957. Oblique ray paths in the ionosphere, *Proc. Phys. Soc. Lond.* **B70**, 653.

Haselgrove, C.B. & Haselgrove, J. 1960. Twisted ray paths in the ionosphere, *Proc. Roy. Soc. Lond.* **75**, 357.

Hashimoto, K., Mimura, I. & Kumagai, H., 1977. Estimation of electron temperature by VLF waves propagating in directions near the resonance cone, *Planet. Space Sci.* **25**, 871.

Haslett, J.C. & Megill, L.R. 1974. A model of the enhanced airglow excited by RF radiation, *Radio Science* **9**, 1005.

Helliwell, R.A. & Crystal, T.L. 1973. A feedback model of cyclotron interaction between whistler mode waves and energetic electrons in the magnetosphere, *J. Geophys. Res.* **78**, 7357.

Herring, R.N. 1980. Tracing of *Z*-mode rays in the magnetosphere, *J. Atmos. Terr. Phys.* **42**, 885.

Hess, W.N. 1968. *The radiation belt and magnetosphere*, Blaisdell, USA.

Hultqvist, B. 1975. The ring current and particle precipitation near the plasmapause, *Ann., Geophys.* **31**, 111.

Huxley, L.G.H., Foster, H.G. & Newton, C.C. 1948. Measurements of the interaction of radio waves in the ionosphere, *Proc. Phys. Soc.* **B61**, 134.

Issledovaniya Kosmicheskogo Prostranstva (Studies of Outer Space), 1974. Collection of papers, Izdatel'stvo VNIITI, No. 4 [In Russian].

Istomin, Ya.I. & Karpman, V.I. 1972. Nonlinear evolution of quasi-monochromatic packets of helical waves in plasma, *ZhETF* **63**, 131 (see also 1972, *ZhETF Lett.* **15**, 208) [In Russian].

Jones, D. 1969. The effect of the latitudinal variation of the terrestrial magnetic field strength on ion-cyclotron whistlers, *J. Atmos. Terr. Phys.* **31**, 971.

Jones, D. 1970. The theory of the effect of collisions on ion cyclotron whistlers In: *Plasma waves in space and in the laboratory Vol.* 2, Edinburgh University Press, p. 471.

Jones, D. 1972. Refractive index and attenuation surfaces in the vicinity of the cross-over level, *Planet. Space Sci.* **20**, 1173.

Kantor, I.J. 1974. High-frequency induced enhancements of the incoherent scatter spectrum at Arecibo, 2, *J. Geophys. Res.* **79**, 199.

Karpman, V.I. 1971. High-frequency electromagnetic field in plasma with negative dielectric constant, *Plasma Phys.* **13**, 477.

Karpman, V.I. 1973. *Nonlinear waves in a dispersive medium*, Nauka, Moscow [In Russian] [1975, Pergamon, Oxford].

Kaufman, R.N. 1977. On amplification of high-frequency electromagnetic waves in a plane layer of magnetoactive plasma, *Radiofizika* **20**, 1614 [In Russian].

Kennel, C.F. & Petschek, H.E. 1966. Limit on stably trapped particle fluxes, *J. Geophys. Res.* **71**, 1.

Kimura, I. 1966. Effects of ions on whistler-mode ray tracing, *Radio Science*, **1**, 269.

Kimura, I. & Kawai, M. 1976. Ray path in a stratified absorbing medium, *IEEE Trans.* **AP-24**, 515.

Kitsenko, A.B. & Stepanov, K.N. 1961. On cyclotron instability of an anisotropic plasma, *ZhTF* **31**, 176 [In Russian].

Kitsenko, A.B. & Stepanov, K.N. 1963. Investigation of plasma oscillations in a quasi-hydrodynamic approach, *Plasma Phys. and Thermonuclear Problems* No. 3, 3 [In Russian].

Kovner, M.S. 1960. Theory of the interaction of a beam of charged particles with a magnetoactive plasma, *Radiofizika* **3**, 631, 746 [In Russian].

Laird, M.J. & Nunn, D.A. 1975. Full-wave modes in a cylindrically symmetric enhancement of plasma density, *Planet. Space Sci.* **23**, 1469.

Laird, M.J. 1977. A note on the nonducted propagation of whistlers, *J. Atmos. Terr. Phys.* **39**, 1417.

Landau, L.D. 1946. On electron oscillations in a plasma, *ZhETF* **16**, 574 [In Russian].

Lassen, H. 1927. Die täglichen Schwankungen des Ionisations-Zustandes der Heaviside Schicht, *Electr. Nachr. Tech.* **4**, 174.

Liemohn, H.B. 1967. Cyclotron resonance amplification of VLF and ULF whistlers, *J. Geophys. Res.* **72**, 39.

L'vovich, V.V. 1937. Cross-modulation of the Gorki radio station, *Radiotekhnika* **2**, 5 [In Russian].

Maeda, K. & Kimura, I. 1956. A theoretical investigation of the propagation path of whistling atmospherics, *Rep. Iono. Space Res. Jap.* **10**, 105.

Mandel'shtam, L.I. 1955. Collected Papers, Vol. IV [In Russian].

Mariani, F. & Ness, N.F. 1969. Observations of the geomagnetic tail at 500 earth radii by Pioneer 8, *J. Geophys. Res.* **74**, 5633 (see also 1969, *Rev. Geophys.* **7**, 91).

Matsumoto, H. & Kimura, I. 1974. Propagation velocity of whistler-mode signals in an unstable and abnormally dispersive plasma-beam system, *IEEE Transactions.* **AP-22**, 136.

McIlwain, C.E. 1961. Coordinates for mapping the distribution of magnetically trapped particles, *J. Geophys. Res.* **66**, 3681.

Meltz, G. & LeLevier, R.E. 1970. Heating the F region by deviative absorption of radio waves, *J. Geophys. Res.* **75**, 6406.

Meltz, G., Holway, L.H. & Tomljanovich, N.M. 1974. Ionospheric heating by powerful radio waves, *Radio Science* **9**, 1049.

Mikhailovskii, A.B. 1965. Driven-cyclotron instability of a plasma with hot ions, *Yadernyi Sintez* **5**, 125 [In Russian].

Mikhailovskii, A.B. 1967. Electron temperature instability of an inhomogeneous plasma, *ZhTF* **37**, 1365 [In Russian].

Mikhailovskii, A.B. 1975. *Theory of plasma instabilities*, Vol. 1; 1977, ibid, Vol. 2, Atomizdat, Moscow [In Russian] (see also 1974, *Plasma Instabilities*, Plenum Press, New York).

Miller, M.A. 1960. Candidate's Dissertation, Gorki Univ. [In Russian]; *Proc. Inter. Conf. High-Energy Accel. and Instr., Cern*, p. 661.

Minkoff, J., Kugelman, P & Weissman, I. 1974a. Radio-frequency scattering from a heated ionospheric volume 1, VHF/UHF field-aligned and plasma line backscatter measurements, *Radio Science* **9**, 941.

Minkoff, J., Laviola, M., Abrams, S. & Porter, D. 1974b. Radio-frequency scattering from a heated ionospheric volume 2, Bistatic measurements, *Radio Science* **9**, 957.

Minkoff, J. & Kreppel, R. 1976. Spectral analysis and step response of radio-frequency scattering from a heated ionospheric volume, *J. Geophys. Res.* **81**, 2844.

Mityakov, N.A., Rapoport, V.O. & Trakhtengerts, V.Yu. 1976. Experimental investigations of nonlinear effects in the F-region of the ionosphere, *Radiofizika* **19**, 33 [In Russian].

Mlodnosky, R.F. & Helliwell, R.A. 1962. Graphic data on the earth's main magnetic field in space, *J. Geophys. Res.* **67**, 2207.

Moiseyev, B.S. 1976. On the amplification of electromagnetic waves in a plane inhomogenity of a magnetoactive plasma, *Transactions of IZMIRAN, Ionosphere Physics*, 249 [In Russian].

Moiseyev, B.S. 1977. Angular anisotropy of the energy flux of electromagnetic waves in a multicomponent magnetoactive plasma, *Radiofizika* **20**, 1623 [In Russian].

Montgomery, D. & Alexeff, I. 1966. Parametric excitation of transverse waves in plasma, *Phys. Fluids* **9**, 1363, 2544.

Neufeld, J. & Doyle, P.H. 1961. Electromagnetic interaction of a beam of charged particles with plasma, *Phys. Res.* **121**, 654.

Nicolet, M. 1953. The collision frequency of electrons in the ionosphere, *J. Atmos. Terr. Phys.* **3**, 200.

Nishikawa, K., 1968. Parametric excitation of coupled waves, *J. Phys. Soc. Jap.* **24**, 1152.

Ondoh, T., Tanaka, Y., Nishizaki, R. & Nagayama, M. 1974. VLF emissions and whistlers observed during geomagnetic storms, *J. Rad. Res. Lab. Jap.* **21**, 361.

O'Neil, T. 1965. Collisionless damping of nonlinear plasma oscillations, *Phys. Fluids* **8**, 2255.

Park, C.G. 1977. VLF wave activity during a magnetic storm: a case study of the role of power line radiation, *J. Geophys. Res.* **82**, 3251.

Perkins, W.F., Oberman, C. & Valeo, E.J. 1974. Parametric instabilities and ionospheric modifications, *J. Geophys. Res.* **79**, 1478.

Perkins, W.F. & Valeo, E.J. 1974. Thermal self-focusing of electromagnetic waves in plasmas, *Phys. Rev. Lett.* **32**, 1234.

Pitaevskii, L.P. 1959. Electric forces in a dispersive medium. *ZhETF* **39**, 1450 [In Russian].

Radio Science, 1974. Ionospheric modification by high-power transmitters, **9**, Special Issue No. 11, 881.

Rao, P.B. & Thome, G.D. 1974. A model for RF scattering from field-aligned heater-induced irregularities, *Radio Science* **9**, 987.

Rapoport, V.O. 1960. On the growth rate of electromagnetic waves in a beam moving in a magnetoactive plasma, Radiofizika **3**, 767 [In Russan].

Ratcliffe, J.A. 1959. *The magneto-ionic theory and its applications to the ionosphere*, Cambridge University Press.

Rawer, K. & Suchy, K. 1967. Radio observations of the ionosphere, *Handbuch der Physik*, Springer Verlag, Berlin, **49/2**, 1–534.

Roederer, J.G. 1970. *Dynamics of geomagnetic trapped radiation*, Springer Verlag. Berlin.

Rytov, S.M. 1947. Some theorems of the group velocity of electromagnetic waves, *ZhETF* **17**, 930 [In Russian].

Sagdeev, R.Z. 1964. In: *Voprosy Teorii Plazmy*, **4**, p. 20, Atomizdat, Moscow [In Russian] [1966, *Reviews of Plasma Physics, Consultants Bureau*, New York, **4**, 23].

Scarabucci, R.R. & Smith, R.L. 1971. Study of magnetospheric field-oriented irregularities – the mode theory of bell-shaped ducts, *Radio Science* **6**, 65.

Scarf, F.L. 1970. Microscopic structure of the solar wind, *Space Sci. Rev.* **11**, 234.

Scarf, F.L., Fredricks, R.W., Green, I.M. & Neugebauer, M. 1970. OGO-5 observations of quasi-trapped electromagnetic waves in the solar wind, *J. Geophys. Res.* **75**, 3735.

Scarf, F.L., Fredricks, R.W., Smith, E.J., Frandsen, A.M. & Serbu, G.P. 1971. OGO 5 observations of discrete whistlers and emissions during a large magnetic storm, *Space Sci. Lab. California, Doc.* 05402-6031-RO-OO.

Scarf, F.L., Fredricks, R.W., Green, I.M. & Russell, C.T. 1972. Plasma waves in the dayside polar cusp. 1. Magnetospheric observations, *J. Geophys. Res.* **77**, 2274.

Serbu, G.P. & Maier, J.R. 1970. Observations from OGO 5 of the thermal ion density and temperature within the magnetosphere, *J. Geophys. Res.* **75**, 6102.

Shabanskii, V.P. 1972. *Phenomena in the near-earth space*, Nauka, Moscow [In Russian].

Shaw, R.R. & Gurnett, D.A. 1972. Magnetospheric electron-density measurements from upper hybrid resonance noise observed by IMP 6, Res. Rep. 72-37, Dep. of Phys. and Astron., U. of lowa.

Shlyuger, I.S. 1974a. Self-modulation of a strong electromagnetic pulse reflected from the upper layer of the ionosphere, *ZhETF* **19**, 274 [In Russian].

Shlyuger, I.S. 1974b. Experimental investigations of nonlinear effects in the E and F regions of the ionosphere, *ZhETF* **20**, 722 [In Russian].

Silin, V.P. 1955. Investigation of a system of many particles with the help of the kinetic equation, *Trudy FIAN* **6**, 200 [In Russian].

Silin, V.P. & Rukhadze, A.A. 1961. *The electromagnetic properties of plasmalike media*, Gosatomizdat, Moscow [In Russian].

Silin, V.P. 1964. Nonlinear high-frequency conductivity of a plasma, *ZhETF* **47**, 2254 [In Russian].

Silin, V.P. 1965. Parametric resonance in a plasma, *ZhETF* **48**, 1679 [In Russian].

Silin, V.P. 1966. Kinetic plasma instability in a strong high-frequency field, *ZhETF* **51**, 1842 [In Russian].

Silin, V.P. 1973. *Parametric action of powerful radiation at a plasma*, Nauka, Moscow [In Russian].

Sipler, D.P. & Biondi, M.A. 1972. Measurements of $O(^1D)$ quenching rates in the F region, *J. Geophys. Res.* **77**, 6202.

Siscoe, G.L., Scarf, F.L., Green, I.M., Binsack, J.H. & Bridge, H.S. 1971. Very-low-frequency electric fields in the interplanetary medium: Pioneer 8, *J. Geophys. Res.* **76**, 828.

Sitenko, A.G. & Stepanov, K.N. 1956. On the oscillations of an electron plasma in a magnetic field, *ZhETF* **31**, 642 [In Russian].

Smith, P.H. & Hoffman, R.A. 1974. Direct observations in the dusk hours of the characteristics of the storm time ring current particles during the beginning of magnetic storms, *J. Geophys. Res.* **79**, 966.

Smith, R.L., Helliwell, R.A. & Yabroff, I.W. 1960. A theory of trapping of whistlers in field-aligned columns of enhanced ionization, *J. Geophys. Res.* **65**, 815.

Smith, R.L. 1961. Propagation characteristics of whistlers trapped in field-aligned columns of enhanced ionization, *J. Geophys. Res.* **66**, 3699.

Smith, R.L. & Brice, N.M. 1964. Propagation in multicomponent plasmas, *J. Geophys. Res.* **69**, 5029.

Stepanov, K.N., 1958. Kinetic theory of magnetohydrodynamic waves, *ZhETF* **34**, 1292 [In Russian].

Stepanov, K.N. 1959. On the generation of the electromagnetic field in a plasma, *ZhETF* **36**, 1457 [In Russian].

Stepanov, K.N. 1960. Cyclotron attenuation of electromagnetic waves in a plasma, *ZhETF* **38**, 265 [In Russian].

Stepanov, K.N. 1962. On the influence of the motion of electrons and ions on the transverse propagation of radio waves in a plasma. In: *Plasma Phys. and Thermonuclear Problems* No. 1, 52 [In Russian].

Stix, T.H. 1962. *The theory of plasma waves*, McGraw-Hill, New York.

Sturrock, P.A. 1965. Dipole resonances in a homogeneous plasma in a magnetic field, *Phys. Fluids* **8**, 88.

Suchy, K. 1972a. Attenuation of waves in plasmas, *Radio Science* **7**, 871.

Suchy, K. 1972b. The velocity of a wave packet in an anisotropic absorbing medium, J. Plasma Phys. **8**, 33.

Suchy, K. 1972c. Ray tracing in an anisotropic absorbing medium, *J. Plasma Phys.* **8**, 53.

Taylor, H.A., Brinton, H.C. & Smith, C.R. 1965. Positive-ion composition in the magnetosphere obtained from the OGO A satellite, *J. Geophys. Res.* **70**, 5769.

Taylor, H.A., Brinton, H.C. & Deshmukh, A.R. 1970. Observations of irregular structure in thermal-ion distributions in the duskside magnetosphere, *J. Geophys. Res.* **75**, 2481.

Taylor, H.A., Grebowsky, J.M. & Walsh, W.J. 1971. Structured variations of the plasmapause: evidence of a corotating plasma tail, *J. Geophys. Res.* **76**, 6806.

Terry P.D. 1971. Complex ray theory for ion cyclotron whistlers, *Nature*, **229**, 200.

Thome, G.D. & Blood, D.W. 1974. First observations of RF backscatter from field-aligned irregularities produced by ionospheric heating, *Radio Science* **9**, 917.

Tsytovich, V.N. 1967. *Nonlinear effects in plasma*, Nauka, Moscow [In Russian] [1970, Plenum Press, New York].

Utlaut, W.F. 1970. An ionospheric modification experiment using very high-power, high-frequency transmission, *J. Geophys. Res.* **75**, 6402.

Utlaut, W.F., Violette, E.J. & Paul, A.K., 1970. Some ionosonde observations of ionospheric modification by very high-power, high-frequency ground-based transmission, *J. Geophys. Res.* **75**, 6429.

Utlaut, W.F. & Cohen, R. 1971. Modifying the ionosphere with intense radio waves, *Science* **174**, 245.

Utlaut, W.F. & Violette, E.J. 1972. Further ionosonde observations of ionospheric modification by a high-powered HF transmitter, *J. Geophys. Res.* **77**, 6804.

Utlaut, W.F. & Violette, E.J. 1974. A summary of vertical incidence radio observations of ionospheric modification. *Radio Science* **9**, 895.

Utlaut, W.F. 1975. Ionospheric modification by high-power HF transmitters. A potential extended range VHF–UHF communications and plasma physics research, *Proc. IEEE* **63**, 1022.

Vas'kov, V.V. & Gurevich, A.V. 1973. Parametric excitation of Langmuir oscillations in the ionosphere in a strong radio-wave field, *Radiofizika* **16**, 188 [In Russian].

Vas'kov, V.V. & Gurevich, A.V. 1975a. Nonlinear resonance instability of plasma in the reflection region of an ordinary electromagnetic wave, *ZhETF* **69**, 176 [In Russian].

Vas'kov, V.V. & Gurevich, A.V. 1975b. Instability of plasmas in the reflection region of intense radio waves in the ionosphere, *Radiofizika* **18**, 1261; 1974. *ZhETF Lett.* **113**, 730 [In Russian]; (see also 1976, *Geomagnetizm i Aeronomiya* **16**, 50 [In Russian].

Vas'kov, V.V. & Gurevich, A.V. 1977. Resonance instability of small-scale plasma perturbations, *ZhETF* **73**, 923 [In Russian].

Vas'kov, V.V., Gurevich, A.V. & Smirnova, V.V. 1977. Disturbance of the upper ionosphere by a modulated strong radio wave, *Geomagnetizm i Aeronomiya* **17**, 832 [In Russian].

Walker, A.D.M. 1971. The propagation of very-low-frequency radio waves in ducts in the magnetosphere, *Proc. Roy. Soc. Lond.* **321A**, 69; 1972, *Proc. Roy. Soc. Lond.* **329A**, 219.

Walter, F. 1969. Nonducted VLF propagation in the magnetosphere, Tech. Rep. 3418-1, SU-SEL-69-061, Radioscience Lab., Stanford Elect. Lab., Palo Alto, Calif.

Whittaker, E.T. & Watson, G.N. 1927. *A course of modern analysis*, Cambridge University Press.

Williams, D.J., Fritz, T.A. & Konradi, A. 1973. Observations of proton spectra ($1.0 \leqslant E_q \leqslant 300$ keV) and fluxes at the plasmapause, *J. Geophys. Res.* **78**, 4751.

Williams, D.J., Barfield, T.A. & Fritz, T.A. 1974. Initial Explorer 45 substorm observations and electric-field considerations, *J. Geophys. Res.* **79**, 554.

Williams, D.J. & Lyons, L.R. 1974. The proton ring current and its interaction with the plasmapause: Storm recovery phase, *J. Geophys. Res.* **79**, 4195.

Yabroff, I. 1961. Computation of whistler ray paths, *J. Res. NBS* **65D**, 485.

Yeh, K.C. & Liu, C.H. 1972. *Theory of ionospheric waves*, Academic Press, New York.

Zabusky, N.J. & Kruskal, M.D. 1965. Interaction of solitons in a collisionless plasma and the recurrence of the initial state, *Phys. Rev. Lett.* **15**, 240.

Zabusky, N.J. 1967. *Proc. Symp. Nonlin. Part. Differ. Equations*, Academic Press, New York, p. 223.

Zakharov, V.E. & Karpman, V.I. 1962. On the nonlinear attenuation of plasma waves, *ZhETF*, **43**, 490 [In Russian].

Zeleznyakov, V.V. 1959. The magnetodamping emission and instability of charged particles in a plasma, *Radiofizika* **2**, 14 [In Russian].

Author index

Abrams, S. I 279
Ackerson, K.L. II 122, 229, 239
Adachi, S. I 217, 271
Aikyo, K. I 216, 238, 241, 243–5, 271 II 151, 224
Akasofu, S.I. II 242
Akhiezer, A.I. I 28, 37, 43, 61, 94, 101, 105, 109, 111, 118, 135, 207, 210, 271
Akhiezer, I.A. I 271
Alexander, J.K. II 100, 214, 224
Alexeff, I. I 204, 279
Aliev, Yu.M. I 193–5, 199, 271 II 224
Allen, E.M. I 271 II 224
Al'pert, Ya.L. I 22, 30, 37, 40, 46, 68, 71, 73, 76, 77, 79, 141, 143, 216, 217, 219, 221, 225, 227, 231, 233, 246, 249–51, 260, 264–7, 269, 271, 272 II 3, 4, 15, 38, 63, 65, 67–9, 75, 79, 80, 82, 89, 106–8, 160, 190, 224
Al'tshul', L.M. I 133, 135, 272
Anderson, R.R. II 128, 129, 132, 177, 196, 200, 218, 221, 225, 230, 231, 241
Andreev, N.E. I 200–4, 272
Angerami, J.J. I 15, 272 II 151, 152, 160, 161, 225, 241, 242
Appleton, E.V. I 49, 272
Ashour-Abdalla, M. I 78, 272 II 163, 189, 225
Astrelin, V.T. II 6, 26, 225
Aubry, M.P. I 217, 238, 272, 273 II 226

Bahnsen, A. II 228
Bailey, V.A. I 135, 273
Baker, B. II 225
Barfield, J.N. II 225
Barfield, T.A. I 109, 282
Barkhausen, H. II 160, 225
Barrett, P.J. II 6, 28, 225
Barrington, R.E. II 90, 111, 112, 138, 149, 150, 182, 225, 231, 235, 240
Bashilov, I.P. II 224
Baumback, M.M. II 234
Beghin, C. II 190, 225, 227, 231, 241
Belikovich, V.V. I 160, 161, 168, 169, 273
Bell, T.F. I 213, 273 II 232
Belrose, J.S. II 138, 149, 150, 225, 240
Benediktov, E.A. I 273
Benioff, H. II 103, 225
Berezin, Yu.A. I 135, 210, 211, 273
Bernstein, I.B. I 102, 133, 135, 273
Bernstein, W. II 191, 225
Binsack, J.H. I 280 II 240
Biondi, M.A. I 163, 273, 280
Bitoun, J. I 217, 238, 272, 273 II 186, 226
Blair, W.E. II 231
Bloch, J.J. II 227
Blood, D.W. I 169, 172, 281
Bloom, M.H. II 234
Bogashchenko, I.A. II 6, 25, 28, 29, 225, 226
Bohm, D. I 108, 273
Booker, H.G. I 27, 47, 171, 172, 174, 216, 217, 247, 273 II 226
Borisov, N.D. I 167, 273
Bossen, M. II 110, 217, 226
Bourdeau, R.E. II 5, 226
Bowen, P.J. II 5, 23, 34, 226
Bowhill, S.A. I 167, 273
Boyd, R.L. II 226
Brace, L.H. I 22, 274
Bremmer, H. I 216, 274
Brice, N.M. I 36, 277, 281 II 90, 111–13, 138, 163, 226, 230, 232, 233, 235, 240
Bridge, H.S. I 280 II 240
Brinton, H.C. I 281 II 227
Brown, L.W. I 23, 274 II 208, 212, 226, 236
Brown, P.E. II 236
Brundin, C.L. II 4, 226
Buchel'nikova, N.S. II 225, 230
Bud'ko, N.I. II 47–52, 55–57, 88, 90, 91, 226
Budden, K.G. I 47, 66, 217, 269, 274
Bullough, K. II 167, 226, 232, 234 241
Burns, T.B. II 117, 118, 230
Burtis, W.J. II 169–172, 226

Burton, R.K. II 171, 172, 226, 241

Cahill, L.J. II 241
Call, S.M. II 226
Callen, J.D. II 242
Calvert, W. II 180, 181, 183–6, 227, 228
Campbell, W.H. II 104, 227
Carlson, H.C. I 146, 147, 160, 181–3, 274, 276
Carpenter, D.L. I 15, 171, 177, 272, 274 II 149–51, 160, 161, 163, 166, 225, 227
Cartwright, D.G. II 97, 227, 233, 236
Cauffman, D.P. II 230
Cerisier, J.C. II 154, 227
Chan, K.W. II 135, 227
Chanteur, G. II 225
Chapman, S. I 39, 274
Chappell, C.R. I 15, 16, 17, 274 II 239
Cherepovskii, V.A. I 273
Chopra, K.P. II 3, 227
Christiansen, P. II 194, 227
Clark, D.H. II 17, 227
Clayden, W.A. II 6, 24, 25, 40, 228
Coffey, T.P. II 237
Cohen, H. II 225
Cohen, R. I 142, 160, 162, 167, 274, 281
Coleman, P.J. I 275 II 108–10, 217, 225, 228, 229, 236, 238–40
Conte, S.D. I 91, 275
Copal Rao, M.S. I 27, 217, 247, 274
Coronilleau, N. II 227
Coroniti, F.V. II 173, 190, 228, 233, 239
Cowling, T.G. I 39, 274
Cragin, B.L. I 167, 274
Crawford, F.W. II 186, 228
Crone, W.R. II 234
Crook, G.M. I 275 II 229, 239
Crystal, T.L. I 213, 277 II 163, 231, 232
Cummings, W.D. II 109, 217, 228

D'Angelo, N. II 122, 228
Daniell, G.J. I 217, 274
Davis, A.H. II 3, 228
Davis, L. II 240
Davis, T.N. II 225
Debrie, R. II 225, 241
Decreau, P. II 227
DeForest, S.E. I 21, 274
Dehmel, G. II 237
Denby, M. II 226
Deshmukh, A.R. I 281
Domrin, V.I. I 194–6, 274
Donley, J.L. II 5, 226
Dougherty, J.P. II 182, 186, 228
Dowden, R.L. II 151, 228
Doyle, P.H. I 93, 109, 275, 279
Dreicer, H. I 120, 141, 275
DuBois, D.F. I 135, 192, 275
Dubovoi, A.P. II 33, 34, 40, 46, 53, 54, 60, 61, 228
Dunckel, N. II 150, 151, 169, 170, 207, 208, 212, 222, 227, 228
Dungey, J.W. II 168, 228
Dwarkin, M.L. II 109, 217, 228
Dyce, R.B. I 47, 273
Dysthe, K.B. II 163, 228

Eckersley, T.L. II 160, 228
Edgar, B.C. I 216, 275 II 151, 228
Eidel'man, V, Ya. I 276
Eidel'man, Yu.I. II 226
Eidman, Yu.I II 225
Erukhimov, L.M. I 273
Estabrooks, M.F. II 186, 237
Etcheto, J. II 227

Fadeeva, V.N. I 91, 275
Fainberg, Ya.B. I 101, 109, 271
Fejer, J.A. I 135, 167, 274, 275 II 186, 228
Fialer, P.A. I 169, 170, 173, 275
Ficklin, B. II 228, 231
First, M. II 238
Fligel, D.S. I 272 II 106–8, 224
Forslund, D.W. II 130, 228
Försterling, K. I 54, 66, 275
Foster, H.G. I 277
Fournier, G. II 6, 228
Frandsen, A.M. I 280 II 239
Frank, L.A. I 18, 19, 20, 275, 277 II 122, 131, 136–8, 142, 144, 145, 156, 157, 195, 199, 210, 211, 218–21, 229–31, 239
Frankel, M.S. II 210, 229
Fredricks, R.W. I 14, 275, 280 II 99, 116, 129, 187–9, 197, 221, 228, 229, 233, 239
Fried, B.D. I 91, 109, 149, 275
Fritz, T.A. I 282
Frolov, V.L. I 273

Gaponov, A.V. I 275 II 229
Gardner, C.S. I 210, 275
Gendrin, R. II 97, 99, 104, 111, 160, 194, 227, 229
Georges, T.M. I 167, 275
Gershman, B.N. I 84, 90, 236, 275 II 147, 229
Getmantsev, G.G. I 160, 273, 276
Gey, F.C. II 233
Gibbons, W. II 226
Ginzburg, V.L. I 28, 39, 100, 276
Goe, G.B. II 181, 185, 227
Goldman, M.V. I 135, 192, 275
Goldstein, M.L. II 237
Gorbunov, L.M. I 203, 204, 276
Gordeev, G.V. I 120, 276

Gordon, W.E. I 146, 147, 160, 181, 183, 274, 276 II 176, 229
Gough, P. II 227
Graff, P. I 272, 273 II 226
Grard, R. II 4, 229
Grebowsky, J.M. I 281
Green, I.M. I 275, 280 II 229, 239, 240
Green, J.L. II 213–15, 229, 231
Greene, J.M. I 133, 273
Gross, E.P. I 103, 108, 273, 276
Gulel'mi, A.V. I 276 II 106, 229, 242
Gurevich, A.V. I 43, 120, 135, 137–41, 147, 150, 153–6, 167, 169, 179, 180, 184, 186–90, 192, 271–3, 276, 281, 282 II 3, 6, 21–5, 27, 28, 32–6, 38, 63–9, 75, 84, 92, 140, 224, 226, 229, 230
Gurnett, D.A. I 15, 37, 277, 280 II 100, 111–24, 128–32, 136–8, 142, 144–6, 150, 154–7, 159, 174–6, 186–9, 191–5, 198–201, 204–15, 217–22, 225, 229–31, 233, 234, 236, 238, 240–2
Gustafson, W.A. II 233
Guthart, H. II 124, 126, 218, 223, 231

Hagg, E.L. II 181–3, 231, 242
Hake, R.D. I 273
Hall, D.F. I 182, 277 II 6, 40, 231
Hallinan, T. II 225
Hamelin, M. II 231, 241
Hanson, W.B. II 240
Harp, R.S. II 228
Harris, I. II 228
Harris, K.K. I 274
Hartree, D.R. I 49, 277
Hartz, T.R. II 182, 231
Haselgrove, C.B. I 216, 238, 277
Haselgrove, J. I 216, 238, 277
Hashimoto, K. I 238, 277
Haslett, J.C. I 163, 277
Hayakawa, M. II 148, 154, 231
Helliwell, R.A. I 25, 26, 213, 273, 277, 279, 280 II 111, 151, 160–72, 226, 228, 232, 238, 240, 241
Henderson, C.L. II 5, 17, 23, 34, 226, 232
Hendrickson, R.A. II 233
Herring, R.N. I 217, 277
Hess, W.N. I 19, 277
Hester, S.D. II 6, 41, 42, 58, 232
Higgy, R.G. II 234
Higuchi, Y. II 105, 232
Hines, C.O. II 150, 232
Hoffman, J.H. II 232
Hoffman, R.A. I 19, 20, 280
Hoffman, W.C. II 151, 232
Holway, L.H. I 278
Holzer, R.E. II 135, 171, 172, 226, 227, 238, 240, 241
Howens, E.J. II 231
Huber, R.W. II 225
Hughes, A.R. II 226
Hultqvist, B. I 277 II 107, 232
Hurdle, C.V. II 6, 24, 25, 40, 228
Huxley, L.G.H. I 135, 277

Ignat'ev, Yu.A. I 273
Issledovaniya Kosmicheskogo Prostranstva I 19, 192, 277
Istomin, Ya.L. I 212, 277 II 163, 232, 233

Jacobs, J.A. II 104, 106, 232, 233
James, H.G. II 144, 147, 154, 227, 233
Jastrow, R. II 3, 233
Jew, H. II 4, 36–8, 45, 235, 238
Johnson, W.C. II 139, 140, 147–9, 168, 169, 234, 236
Johnston, T.W. II 6, 233
Jones, D. I 37, 277 II 194, 227, 229, 240
Jones, D.E. II 240
Joselyn, J.A. II 114, 233

Kaiser, M.L. II 100, 214, 224, 236
Kaiser, T.R. II 167, 226, 232, 233, 241
Kantor, I.J. I 181, 182, 277
Kapustina, O.V. II 224
Karpman, V.I. I 133, 135, 208, 210–12, 272, 273, 277, 278, 282 II 163, 232, 233
Kasha, M.A. II 6, 233, 237
Katsufrakis, J.P. II 163–6, 232, 238, 240
Kaufman, R.N. I 217, 246, 248–50, 256–8, 272, 278
Kawai, M. I 217, 278
Keeley, D.A. II 225
Kelley, M.C. II 233
Kellogg, P.J. II 97, 98, 227, 233, 236
Kemp, R.F. I 277 II 231
Kennel, C.F. I 117, 275, 278 II 150, 168, 171, 172, 188, 189, 225, 228, 229, 233, 235, 239, 241
Kenney, J.F. II 104, 105, 233
Kiel, R.E. II 4, 233
Kimura, I. I 216, 217, 238, 242, 243, 278 II 99, 150–2, 162, 163, 233, 235, 236
Kintner, P.M. II 114, 115, 124, 129, 217, 218, 233
Kirii, A.Yu. I 200, 272
Kitsenko, A.B. I 111, 115, 236, 278
Kivelson, M. II 239
Knaflich, H.B. II 104, 105, 233
Knecht, R.W. II 179, 234
Knight, D.J.E. II 233
Knyazuk, V.S. II 69–72, 234
Komrakov, G.P. I 273
Konradi, A. I 282
Korobkov, Yu.S. I 273

Kotik, Yu.S. I 273, 276
Kovner, M.S. I 111, 115, 278
Kraus, J.D. II 76, 234
Kraus, L. II 3, 55, 234
Kreppel, R. I 177, 178, 279
Kresin, V.Z. II 56, 238
Kruskal, M.D. I 133, 135, 210, 211, 273, 275, 282
Kugelman, P. I 279
Kumagai, H. I 277
Kurth, W.S. II 213, 214, 231, 234

Laaspere, T. II 139, 140, 147–9, 151, 168, 169, 234
Laird, M.J. I 217, 278
LaLonde, L.M. I 181, 276, II 176, 229
Landau, L.D. I 31, 34, 86, 278 II 76, 234
Lapshin, V.I. I 271
Lasch, S. II 147, 162, 227, 234
Lassen, H. I 49, 278
Laviola, M. I 279
Lederman, S. II 6, 234
Lefeuvre, F. II 167, 234
Leinbach, H. II 225
LeLevier, R.E. I 142–5, 159, 278
Lepping, R.P. II 231, 239
Liemohn, H.B. I 117, 278 II 104, 106, 161, 193, 233, 234
Lifshitz, E.M. II 76, 234
Liu, C.H. I 128, 282
Liu, V.C. II 4, 36, 38, 45, 46, 235
Lockwood, G.E. II 181, 182, 235, 237
Lucas, C. II 113, 235
L'vova, G.A. II 224
L'vovich, V.V. I 135, 278
Lyubarskii, T.Ya. I 135, 271
Lyons, L.R. I 19, 20, 282 II 114, 168, 233, 235

Maeda, K. I 216, 278 II 177, 225
Maier, E.J. II 238, 242
Maier, J.R. I 14, 280
Mainstone, J.S. II 104, 235
Mandel'shtam, L.I. I 193, 278
Mantei, T.D. II 228
Mariani, F. I 14, 278
Marsch, E. II 231
Martelli, G. II 227
Martin, A.R. II 4, 45, 235
Martin, D.F. I 273
Martz, J. II 225
Maslennikov, M.V. II 4, 235
Masson, F. II 228
Mather, W.E. II 225
Matsumoto, H. I 217, 278 II 98, 99, 163, 235, 236
Matthews, J.P. II 241
McAfee, J.R. II 180, 185, 186, 227, 235
McCleod, M.G. II 240
McCune, J.E. II 242
McEwen, D.J. II 90, 235
McGehee, J.H. II 233
McIlwain, C.E. I 15, 21, 274, 278
McKeown, P. II 13, 235
McKibbin, D.D. II 240
McNicol, R.W. II 104, 235
McPherron, R.L. II 108–10, 217, 226, 236, 238
Meckel, B.B. II 6, 236
Medved, D.B. II 9, 236, 242
Megill, L.R. I 163, 277
Melrose, D.B. II 100, 236, 242
Meltz, G. I 142–7, 159, 165–7, 191, 278
Mikhailova, G.A. II 224
Mikhailovskii, A.B. I 97, 98, 109, 111, 114, 278
Miller, G. II 4, 242
Miller, M.A. I 275, 278 II 229, 236
Miller, T.L. II 163, 166, 227
Mimura, I. I 277
Minkoff, J. I 169, 171–3, 176–8, 191, 279
Mironenko, L.F. I 276
Mityakov, N.A. I 161, 273, 276
Miura, R.M. I 275
Miyatake, S. II 97, 235, 236
Mlodnosky, R.F. I. 25, 279
Moiseyev, B.S. I 217, 222, 247, 260, 264–7, 269, 279
Monaghan, J.J. II 182, 186, 228
Monson, S.J. II 98, 233, 236
Montgomery, D. I 204, 279
Morgan, M.G. II 140, 234, 236
Moser, F.S. II 233
Mosier, S.R. II 118, 119, 123, 124, 142, 145, 146, 186, 219, 230, 236
Moskalenko, A.M. II 41, 69–74, 234, 236
Muldrew, D.B. II 183, 185, 186, 236, 237
Murukami, T. II 148, 237
Musman, G. II 237
Muzzio, J.L. II 120, 124, 237

Nagayama, M. I 279 II 224, 237
Nelms, G.L. II 181, 182, 225, 231, 237
Ness, N.F. I 14, 278
Neubauer, F.M. II 198, 231, 237
Neufeld, J. I 93, 109, 275, 279 II 239
Neugebauer, M. I 280
Newton, C.C. I 277
Nicolet, M. I 39, 279
Nishikawa, K. I 192, 279
Nishizaki, R. I 279 II 237
Nosachev, L.V. II 6, 39, 240
Nunn, D.A. I 217, 278 II 163, 237

Obayashi, T. II 106, 237
Oberman, C. I 279
O'Brien, B.J. II 120, 121, 159, 230
Ohtsu, J. II 231
Ondoh, T. I 15, 216, 238, 241, 243–5, 271, 279 II 148, 151, 224, 237
O'Neil, T. I 135, 279
Oran, W.A. II 6, 42, 43, 237, 241
Orr, D. II 233
Osborne, F.J.F. II 6, 233, 237
Ossakow, S.L. II 237
O'Sullivan, R.J. II 228
Ott, E. II 163, 241
Owens, H.D. I 19, 20, 275
Oya, H. II 181, 189, 237

Palmadesso, P. II 215, 237
Pan, Y.S. II 4, 237
Panchenko, Yu.M. II 47, 48, 51, 237
Papadopoulos, K. II 200, 237
Parady, B.K. II 241
Pariiskaya, L.V. II 230
Park, C.G. I 279 II 167, 227, 237
Patel, V.L. II 109, 217, 237
Paul, A.K. I 281
Pearse, C.A. II 3, 233
Pellat, R. II 163, 225, 228, 229, 233, 235, 238, 239, 241
Perkins, W.F. I 167, 182, 183, 279
Petschek, H.E. I 117, 278 II 168, 172, 233
Pfeiffer, G.W. II 230
Phillipp, W. II 231
Pigache, D. II 6, 229
Pitaevskii, L.P. I 214, 271, 272, 279 II 3, 28, 47, 51, 56, 75, 76, 79, 80, 82, 84, 224, 229, 230, 237, 238
Polovin, R.V. I 271
Pope, J.H. II 106, 238
Porter, D. I 279
Proceedings of the IEEE 1969 special issue II 179, 238

Radford, W.D. II 228
Radio Science (journal) 1974 special issue I 142, 279
Raghuram, R. II 232
Raitt, W.J. II 227
Rao, P.B. I 172–4, 271, 279 II 224
Rapoport, V.O. I 111, 273, 276, 279
Ratcliffe, J.A. I 47, 279
Rawer, K. I 47, 49, 279, 281
Razin, V.A. I 273
Reeve, C.D. II 151, 238
Rodriguez, P. II 111, 120, 129, 130, 156, 157, 174, 198, 218, 220, 230, 238
Roederer, J.G. I 19, 247, 279
Rorden, L. II 228
Rosenbauer, H. II 228, 231
Rosenberg, T.J. II 168, 238
Rostoker, N. II 125, 126, 238
Roux, A. II 163, 238
Rukhadze, A.A. I 28, 100, 276, 280
Russell, C.T. I 275, 280 II 103, 111, 116, 135, 154, 219, 226, 229, 236, 238–40
Russell, S. II 234
Rycroft, M.J. II 151, 238
Rytov, S.M. I 101, 216, 280

Sagdeev, R.Z. I 280 II 238
Saito, T. II 103, 238
Salimov, R.A. II 226, 230
Samir, U. II 5, 8, 9, 15–17, 23, 34, 36–8, 232, 237, 238, 241, 242
Sawchuck, W. II 22, 239
Sayasov, Yu.S. II 239
Sazonov, V.O. I 273
Sazonov, Yu.A. I 276
Scarabucci, R.R. I 217, 280
Scarf, F.L. I 14, 180, 275, 280 II 122, 125–7, 129–31, 141, 142, 154, 161, 176, 187–9, 193, 197, 198, 201–4, 218, 219, 221, 222, 228, 229, 231, 233, 234, 239, 240
Scherr, D.J. II 234
Schmitt, J.T.M. II 6, 239
Schwenn, R. II 231
Sellen, J.M. I 277 II 231
Serbu, G.P. I 14, 280, II 226, 239
Shabanskii, V.P. I 19, 280
Sharp, G.W. I 274 II 16, 240
Shavin, P.V. I 273
Shaw, R.R. I 15,280 II 186–9, 191–4, 204–7, 221, 222, 230, 240
Shawhan, S.D. I 277 II 99, 112, 128, 186, 229, 230, 240, 241
Shkarofsky, I.P. II 233
Shklyar, D.R. II 232, 233
Shlyuger, I.S. I 161, 187, 276, 280 II 140, 230
Showen, R.L. I 274
Sigov, Yu.S. II 4, 235
Silin, V.P. I 28, 135, 192–7, 199, 201, 203, 204, 271, 276, 280
Singer, S.F. II 4, 240
Sipler, D.P. I 273, 280
Siredey, C. II 190, 225
Siscoe, G.L. I 14, 280 II 108, 130, 217, 240
Sitenko, A.G. I 84, 90, 93, 94, 271, 280 II 240
Skvortsov, V.V. II 6, 39, 240
Smirnova, V.V. II 92, 230, 240
Smith, A.J. II 233
Smith, C.R. I 281
Smith, E.J. I 280 II 155–7, 171, 173, 174,

219, 227, 238–42
Smith, P.H. I 19, 20, 280
Smith, R.A. II 237
Smith, R.L. I 36, 217, 247, 277, 280, 281 II 90, 110, 138, 146, 150–3, 160, 220, 226, 230, 233, 241
Sonin, A.A. II 6, 41, 42, 58, 232
Stefant, R.J. II 104, 114, 229, 241
Stepanov, K.N. I 84, 90, 93, 94, 111, 115, 236, 271, 278, 280, 281 II 240
Stiles, G.S. II 163, 167, 241
Stilner, E.C. II 104, 227
Stix, T.H. I 28, 91, 96, 216, 281
Stone, K. II 227
Stone, N.H. II 6, 44, 237, 241
Storey, L.R.O. I 262–4, 268–70 II 160, 241
Stott, G.F. I 217, 269, 274
Sturrock, P.A. I 219, 281
Suchy, K. 47, 49, 217, 279, 281
Sudan, R.N. II 163, 241

Tanaka, Y. I 279 II 231, 237
Tartaglia, N.A. II 106, 241
Tatnall, A.R. II 167, 226, 241
Taylor, H.A. I 15, 281 II 139, 140, 227, 234, 236
Taylor, J.C. II 4, 241
Taylor, W.W. II 114, 116, 129, 186, 195, 196, 241
Tepley, L.R. II 103, 104, 241
Teplykh, A.I. I 273
Terent'ev, N.M. I 91, 275
Terry, P.D. I 217, 274, 281
Theis, R.F. I 22, 274
Thome, G.D. I 169, 172–4, 271, 279, 281
Thorne, R.M. II 133–5, 150, 154, 171, 235, 241
Tomchinskii, A.M. I 273
Tomljanovich, N.M. I 278
Trakhtengerts, V.Yu. I 273, 276, 279
Trimpi, M. II 232
Troitskaya, V.A. I 276 II 103, 106, 229, 242
Troy, B.E. II 5, 9, 15, 16, 238, 242
Tsurutani, B.T. II 157, 171, 173, 174, 240, 242
Tsytovich, V.N. I 135, 281

Utlaut, W.F. I 142, 160–2, 167, 168, 281

Vaglio-Laurin, R. II 4, 237, 242
Valeo, E.J. I 279
VanZandt, T.E. II 183, 184, 227, 234
Vas'kov, V.V. I 167, 169, 179, 180, 273, 281, 282 II 28, 48, 52, 56, 59, 75, 77, 83–5, 242
Violette, E.J. I 160, 167, 281
Virobik, P.F. II 239
Voots, G.R. II 213, 242

Walker, A.D.M. I 217, 282 II 160, 242
Walkup, J.F. II 227
Walsh, W.J. I 281
Walter, F. I 216, 238, 282 II 151, 242
Wang, C.Y. II 151, 234
Warren, E.S. II 181, 242
Watanabe, T. II 104, 232
Watson, G.N. I 3, 249, 282
Watson, K. II 55, 234
Weissman, I. I 279
Wentworth, R.C. II 104, 241
Whipple, E.C. II 16, 226, 242
Whitehead, J.D. I 167, 274
Whittaker, E.T. I 249, 282
Widhonf, V.C. II 234
Williams, D.J. I 19, 20, 282
Willmore, A.P. II 5, 23, 34, 226, 227, 235
Wilson, P.S. II 225
Winckler, J.R. II 97, 233, 242
Wrenn, G.L. II 5, 9, 15–17, 238

Yabroff, I.W. I 216, 238–40, 243, 280, 282, II 160, 240
Yaroslavtsev, A.A. II 34, 53, 54, 60, 61, 228
Yeh, K.C. I 128, 282
Young, T.S.T. II 189, 242
Yung, T.J. II 231

Zabusky, N.J. I 135, 210, 211, 282
Zakharov, V.E. I 135, 282
Zeleznyakov, V.V. I 282
Zhizhimov, L.A. II 239
Ziemke, R. II 225
Zmuda, A.J. II 228
Zuikov, N.A. I 273, 276
Zyunder, D. I 199, 271 II 224

Subject index

absorption coefficients I 81, 168, 180
absorption of particles II 12
accommodation of particles, partial and total II 11, 12
accumulation of particles near body II 63
Aerobee rocket II 149
AKR, *see* auroral kilometric radiation
Alfvén refractive index I 8, 53, 63
Alfvén velocity I 8, 43, 63, 203
Alfvén wave I 53, 58, 63, 64, 93, 94, 101, 102, 123, 124, 127, 203, 222, 237 II 104
Alouette satellites II 97, 149, 186
Alouette 1 II 90, 91, 110, 138, 151, 179
Alouette 2 I 23 II 112, 147, 148, 179–83, 193
ambipolar diffusion coefficient I 154
amplitude-modulation index I 186
angular dependence of electric field near body II 55, 56
angular dependence of particles near body II 19–25, 30, 33–41, 47–54, 57–61
angular dependence of temperature near body II 16, 17
anisotropic distribution of electron velocities II 178
anisotropic electron distribution II 130
anisotropic instability I 110ff
anisotropic Maxwellian distribution I 110, 117, 126
anisotropic pitch angle distribution II 100
anisotropic temperature distribution I 106, 107 II 100, 106
anisotropic velocity distribution I 106, 110ff, 116ff, 126 II 106
anomalous absorption I 167–9
anomalous Doppler effect I 111
Antarctica, *see* Siple, Eights station
Apollo model II 43, 44
Appleton–Hartree formula I 48, 49
Appleton–Lassen formula I 49
Ariel 1 satellite II 5, 17, 23, 34, 35
Ariel 3 satellite II 148, 167
Ariel 4 satellite II 167
artificial sporadic layer I 163, 164, 167 II 97
artificially created emissions II 194, 195
artificially stimulated emissions (ASE) II 162ff
ASE, *see* artificially stimulated emissions, stimulated emission
ATS I satellite II 109, 110
ATS 5 satellite I 21
attenuation coefficient I 70ff, 82ff, 147, 184
attenuation factor I 28, 38, 43, 45, 70, 71
aurorae II 213, 214
auroral activity II 136, 176
auroral hiss (AH) II 137, 142ff, 146, 147
auroral kilometric radiation (AKR) II 208ff, 222
auroral oval II 124, 144
auroral region I 21

backscatter I 163, 172, 191
band structure of chorus II 174
band structure of TNCR II 207
beam deformation I 190
beam instability I 31, 106, 107ff, 118ff II 100, 101, 178, 215
beam instability of longitudinal waves I 111ff
beam instability of transverse (e.m.) waves I 115ff, 123ff
beam of particles I 31, 107ff
beam of electromagnetic waves I 153, 184, 188
Bernstein modes I 102 II 185, 194
Bessel function (imaginary argument) I 102, 196 II 20, 61, 77
Bessel function (real argument) I 194, 261, 262 II 77
bi-Maxwellian function I 117
Boltzmann distribution function I 29, 30, 214
boundary conditions at body's surface I 7 II 10 (ch. 10)
boundary layer near body II 66

boundary problem I 246, 253
bow shock I 11, 14, 24, 206, 213 II 129, 130, 137, 170, 174, 175, 178, 196–9, 218, 220
branching spectra II 162
bremsstrahlung I 111, 115
broad-band emission II 133, 136–8, 142
broad-band noise II 207, 208
broad-band VLF emission II 154ff, 219
broad lines I 182
bunching of particles II 72
bursts of protons II 122
bursts of scattered radiation II 81, 82, 85
bursts of wave amplitude II 126, 128, 132, 133, 136, 142, 156, 175, 201, 207
bursts of X-rays II 168

caustic II 75, 84–6
centre scattering line I 175–8
Cerenkov absorption I 94
Cerenkov attenuation I 33, 34, 47, 94
Cerenkov damping I 89, 90, 92
Cerenkov excitation I 125 II 88, 89
Cerenkov growth rate I 110
Cerenkov radiation, coherent II 147
Cerenkov resonance I 82, 111, 112, 113, 125, 132 II 186
Cerenkov type emission II 144
Cerenkov–Vavilov effect I 33
Chapman layer I 143
characteristic plasma field I 137
characteristic time of nonlinear process I 133
chorus II 100, 110, 116ff, 119, 144, 148, 169ff, 207
chorus bands II 148
circular polarization I 51, 55
circular scanning antenna I 165, 166
classification of waves I 56
cloud-like inhomogeneities I 160ff
coherent radiation II 212
cold beam I 109
cold magnetoplasma I 35 (ch. 4)
cold plasma, cold beam I 113, 115, 120
cold plasma, hot beam I 111, 115, 122, 125
collisional plasma I 70ff, 128, 133
collision cross-section I 138
collision frequency I 36, 134, 138ff, 248 II 83
collision frequency, effective I 106
collision integral I 30, 37, 42 II 10
collisionless attenuation I 64, 169
collisionless plasma I 34, 35, 65ff, 128, 130, 193, 205 II 50
collisionless shock I 206, 213 II 198
collisions I 36, 222, 235 II 63, 75, 83
collisions, effect on wake of body II 27, 28
collisions of like particles I 42
combination waves I 192
complex frequency I 28, 107
complex mass I 37
complex refractive index I 26, 32, 73
complex tensor elements I 149
concentration regions near body II 31
condenser I 193
conductivity tensor I 32, 40
cone in electron density II 88, 89
cone in radiation polar diagram I 262–4
confluent hypergeometric function I 249
continuity equation I 208
continuum radiation II 137, 192, 204ff, 208ff, 222
Coulomb centre II 63, 64
Coulomb field II 33, 63, 64, 69
Coulomb interactions I 161
Coulomb logarithm I 36, 39
coupled equations I 54
coupling between harmonics I 205
critical collision frequency I 51
cross-modulation I 135, 151, 192
crossover frequency I 54, 66, 72, 78, 79, 81, 232–5, 270 II 112, 117
crossover region I 73
current-driven instability II 101, 138, 176, 177, 215
curtain spectrogram II 139
cutoff for guided waves I 247
cutoff for whistlers II 160
cutoff frequency I 50, 70, 255, 259, 268 II 117, 119, 138, 153
cutoff of ELF hiss II 117–9, 121, 122
cyclotron attenuation coefficient I 94 II 113, 160
cyclotron cutoff II 160
cyclotron damping I 89
cyclotron excitation I 125
cyclotron instability I 116
cyclotron resonance I 82, 111, 113, 123, 125
cyclotron resonance, electrons I 63
cyclotron waves I 104
cylinder in plasma stream II 26, 31ff, 42, 43, 47, 51, 52
cylindrical coordinates I 260

D region I 186, 192, 248 II 168
damping of ion whistlers II 113
Debye length I 2, 7, 86, 152, 179, 196, 207, 213 II 3, 6, 35, 42, 88, 126
Debye screening II 29, 33, 65, 66
decay instability I 160, 182, 203
decay line I 182
decay rate I 28
defocusing I 192

defocusing of beam I 184
demodulation I 187
dielectric moving body II 13, 15
differential scattering cross-section II 75–7, 81, 84
diffuse spectra of emission II 191–3
diffuse type resonances II 181
diffusion I 152 II 100
diffusion coefficients I 154, 179
dimensionless coordinate system I 260, 261
dipole field (magnetic) I 25 II 152
Dirac delta function I 261
disintegration of soliton I 210
disk in ion stream II 25, 32, 33
dispersion denominator II 88, 89
dispersion equation I 28, 32, 35, 53, 82, 194, 207, 219, 221, 252
dispersion length I 208
dispersive medium I 205, 206, 208
displaced Maxwellian velocity distribution I 107
distribution function (Boltzmann) I 29, 30, 214
Dodge satellite II 109, 217
Doppler broadening of resonance oscillations II 124
Doppler effect I 33, 34, 165, 166, 176
Doppler shift of excited waves II 129, 132
Doppler-shifted cyclotron resonance damping I 82
Doppler-shifted gyroresonance II 100, 124, 172, 196, 215
drift instability II 101, 129, 204
drift motion and instability I 127 II 100
ducts I 51, 235, 236, 243, 251, 260 II 160, 161, 166
dusk bulge (of plasmapause) I 15–17

E layer I 162
E region I 186, 192
Earth-ionosphere wave guide II 168
Earth's bow shock I 11, 14, 24, 206, 213 II 129, 130, 137, 170, 174, 175, 178, 196–9, 218, 220
Earth's magnetic field I 7, 12
Earth's rotational energy II 99
effective collision frequency I 106
effective length of wake, for scattering II 85
effective mass (complex) I 37, 71, 77
effective mass of ion mixture I 69
Eights station, Antarctica II 168
elastic diffuse reflection of particles II 11
electric field, angular distribution near body II 55, 56
electric field, effect on particles near body II 29ff
electron-acoustic waves I 96ff II 198
electron-cyclotron wave I 102ff
electron density I 7, 10, 16, 18
electron density, altitude dependence I 22, 23 II 152, 187
electron–electron collisions I 42, 235
electron flux at body surface II 14
electron gyrofrequency I 25, 26, 27, 56 II 136, 144, 155, 160, 180, 191ff, 214
electron gyroresonance I 49, 140 II 96, 100
electron–ion collisions I 185
electron–neutral collisions I 186
electron oscillations II 98
electron resonances II 91
electron ring current II 148
electron run-away condition I 119
electron sound I 97
electron temperature I 2, 6, 10, 14, 138ff, 146 II 9, 16ff, 124, 194
electron velocity distribution II 189
electron viscosity I 43
electron-whistler I 56, 64, 90, 95, 115, 116, 126, 224, 247, 251 II 104, 110, 120, 159
electrons injected into plasma II 97, 191
electrostatic gyroresonance instability II 196
electrostatic instability II 144, 215
electrostatic noise II 122, 124, 136, 137, 191–3
electrostatic resonance oscillations II 209
electrostatic waves I 98 II 96, 125, 128, 129, 132, 175, 176, 187, 188, 198, 200, 212, 216–221
ELF, *see* extremely low frequency
ELF hiss II 144
ELF pulsations II 108ff
ELF transverse waves II 108
ellipsoidal body II 20ff
elongated irregularities I 1152
energy density of waves II 125, 126, 131, 216 (ch. 20)
energy density, time average I 218
energy flux I 216, 218 II 216
energy sources for waves II 99
energy spectra of electrons II 154, 172
enhancement factor of trapped energy I 257–9
enhancement of dipole field in plasma I 266, 267
enhancement of particle concentration II 31, 72
enhancement of power grid harmonics II 167
enhancement of whistler II 165
equations of motion of charged particles I 40

equator, geomagnetic II 135, 136, 155, 161, 169, 173, 174, 188, 220
erosion by particles II 12
error function II 28, 36, 64
ESRO la satellite II 17
Euler–Lagrange system equations I 238
evanescent I 268
evaporation of particles from surface II 10, 12, 13
experiments, bodies in plasma II 5
experiments, near Sun II 133
experiments, waves and oscillations II 96f, 191f
Explorer satellites II 97
Explorer 2 satellite II 214
Explorer 8 satellite II 5
Explorer 12 satellite II 109, 217
Explorer 20 satellite II 179, 184
Explorer 31 satellite II 5, 8, 15–17, 35–7, 92
Explorer 45 satellite I 19, 20 II 114, 115, 128, 129, 176, 195, 196, 218
extraordinary fast wave I 56, 92, 230, 231 II 207, 214
extraordinary slow wave I 56, 92, 227, 255
extraordinary wave I 48, 53, 74, 140, 144, 161, 178, 179 II 110, 115
extremely low frequency (ELF) I 48, 52, 62, 65, 70, 93, 118, 222ff, 230, 233, 248, 270 II 98, 103 (ch. 16), 217, 218

F region I 161 II 75
F-1 layer I 161, 162
F-2 layer I 128
F-2 region I 5, 22
falling tone (faller) II 161, 165, 166, 168, 173, 174
far zone of body in plasma II 4, 8, 9, 22, 23, 29, 31, 34, 41, 47ff, 66, 68, 88
fast ion-acoustic wave II 125
fast magnetoacoustic wave I 63, 64, 66, 90, 93, 95, 102, 123, 224, 225
fast-moving body II 19 (ch. 11), 75 (ch. 13), 87 (ch. 14)
feedback mechanism for ASE II 164
Fermat's principle I 238
field-aligned ducts I 260 II 160, 161, 166, 203
fine structure of HF emissions II 190
five branches of dispersion curves I 56, 57
floating spike II 182, 183
flow instability II 100
flow of plasma II 3 (chs. 9–14), 36, 38
focusing of charged particles II 7, 31, 38, 40, 41, 43, 46, 47, 53, 61, 72, 73, 88
focusing of waves I 153, 167, 184ff, 241 II 84
$f^{(0)}$F2 I 143, 158, 161, 166, 169, 190, 191
force line, magnetic, curvature I 27
force line, magnetic, length I 26
formations in trough I 17, 18
Fourier–Bessel transform I 261
Fourier component of electron perturbation II 76, 88
Fourier transform I 261
FR 1 satellite II 154
fractional hop whistler II 111, 149
free energy II 100, 101
free particles near body II 63–5
frequency dispersion I 32
frequency modulation II 173
Fresnel zone in wake of body II 83, 86
frictional forces of colliding particles I 43
fringe pattern ionograms II 184, 185
fundamental equations I 28 (ch. 3)

galactic noise I 208, 209, 212
gas flow I 1
Gaussian beam I 154, 189
GBR radio transmitter II 168, 169
Gemini/Agena satellite II 5, 15–17, 92
geomagnetic tail I 24, 25 II 96, 100, 101, 108, 137, 138, 174–6, 187ff, 195, 208, 209, 214, 220–2
geometrical optics I 216, 247
GEOS 1 satellite II 194
GEOS 2 satellite II 122
gravity waves I 128, 164
Gross gaps I 103
group delay time II 106–8, 112
group propagation time II 104
group refractive index I 219 II 152
group velocity I 60, 216 (ch. 8), 218ff, 260ff II 90, 106, 112, 121
group velocity direction I 218, 219, 220, 233ff
group velocity opposite to wave velocity I 236
growing lines I 181
growth rate I 28, 106 (ch. 6), 127–9, 131, 198, 199, 202–4 II 92
guiding of waves I 27, 51, 228, 239, 247 II 103, 109, 110, 160, 166
gyrofrequency, electron I 7, 25 II 180, 195, 214
gyrofrequency, ion I 7, 56, 71ff II 77
gyrofrequency multiples I 89 II 180
gyroresonance I 95, 138 II 96, 164
gyroresonance instability II 100, 106, 148
gyroresonance, odd half-integral II 97, 187ff
gyroresonant interaction II 168, 204
gyrosynchrotron emission II 210

half-integral gyroresonances II 97, 187ff, 194, 212
half-integral resonances II 182
half-plane in plasma stream II 26
Hamilton's principle I 238
harmonics of carrier frequency I 184, 205, 208, 209
harmonics of electron gyrofrequency II 180, 184, 193, 195, 204
harmonics of Langmuir frequency II 97, 98, 181, 199, 200
harmonics of power frequency II 166, 167
harmonics of proton gyrofrequency II 123, 124
harmonics of upper-hybrid frequency II 97, 180, 181, 186, 215
Hawkeye 1 satellite II 114, 115, 122,129, 136, 154, 195, 210, 211, 214, 217–21, 231
heat conduction I 143, 154
heat-flux instability II 130
heating of ionosphere I 142ff II 97
heating type instability I 128, 141, 153, 167 II 100
heating type nonlinearity I 135, 136ff II 98
helicon I 101
Helios 1 satellite II 131, 200, 201, 218, 221
Helios 2 satellite II 131, 132, 198, 201, 222
helium ion density I 15
helium ions I 65, 230
helium whistler II 111, 112
HF, *see* high frequency
high energy electrons and protons I 19
high energy particles I 19
high frequency (HF) I 93 II 98, 178 (ch. 19), 221, 222
high pass noise II 213
hiss II 100, 116ff, 119, 128, 129, 135, 138, 169ff, 220
hole in bulge I 190, 191
hook II 161, 162
horizontal gradients in ionosphere II 151
hot ions I 126
hot plasma, cold beam I 114, 124
hybrid frequencies I 67
hybrid resonance I 68 II 89, 101
hydrodynamic approximation I 127, 128 II 30
hydromagnetic approximation I 37
hydromagnetic whistlers I 22, 24, 93, 251 II 100, 103ff
hysteresis of electron temperature I 140

image decay lines I 182
IMP 6 satellite I 16, 18 II 122, 129–31, 136, 155, 174, 186, 188, 191–3, 198, 201 204–6, 208, 212–4, 218, 221, 222
IMP 7 satellite II 188
IMP 8 satellite II 131, 175, 176, 188, 195, 198, 199, 201, 208–10, 212, 214, 220–2
impedance (wave) of free space II 120
inclination of Earth's magnetic field I 27
incoherent scattering I 175, 179
inelastic reflection of particles II 11
infinities of refractive index I 48, 49, 65
inhomogeneity caused by heating I 152, 153–8, 160ff
inhomogeneity in height I 143
inhomogeneous plasma I 142, 246
inhomogeneous structure, resonance of I 246
injection of electrons into plasma II 97, 191
Injun 3 satellite II 110, 111, 116–8, 121
Injun 5 satellite II 111, 118, 119, 121, 122, 124, 141–50, 156, 157, 219, 220
Injun 6 satellite II 220
instability I 21, 106 (ch. 6), 127
instability near body II 4, 87 (ch. 14)
intensification of emission II 116ff, 162ff
intermediate zone of body in plasma II 4, 7, 21, 31ff, 58
intermixing, turbulence I 22
internal gravity waves I 128, 164,
interplanetary medium II 187ff
interplanetary plasma I 10ff II 130
interplanetary shock II 222
interplanetary space I 5, 7 II 217, 218, 222
intersection frequency I 54
inverted-V electrons II 137, 215
ion-acoustic oscillations I 180
ion-acoustic waves I 98ff, 118–20, 124, 181, 201, 204, 207–9, 212, 236 II 30, 88ff, 96, 123ff, 198, 218, 223
ion composition I 14, 15 (Table 2.3) II 141
ion-cyclotron damping II 112
ion-cyclotron instability II 176, 215
ion-cyclotron resonance I 63
ion-cyclotron wave I 9, 102ff II 100, 104, 110ff, 220
ion-cyclotron whistlers II 110ff
ion flux at body surface II 14
ion gyrofrequency I 56, 71ff
ion gyroresonance I 48
ion–ion collisions I 42, 46
ion–ion hybrid frequency I 69–71, 231–4, 270
ion-Langmuir frequency I 8 II 124, 128
ion-Langmuir waves I 236, 237
ion mass, effective I 14
ion oscillations II 98
ion temperature I 2, 6, 10, 14
ion viscosity I 43
ion whistler I 56, 64, 93, 94, 222 II 110ff

ion whistler O^{++} or He_2^+ II 112
ionic sound I 197
ionization time II 12
ionized clouds I 2, 160, 163
ionogram I 161–3, 166–8
ionogram, fixed frequency II 184, 185
ionogram, topside II 179ff
ionosonde I 161–3
ionosonde, rocket-borne II 179
ionosonde, satellite borne II 179ff
ionosphere I 128, 184, 217, 218 II 36, 75, 77, 81, 82, 84, 117, 125, 149, 159, 168, 169
irregularities I 2
ISIS 1 satellite I 22 II 97, 122, 179
ISIS 2 satellite II 148, 179
isolated packets of waves I 135
isotropic plasma I 88, 96, 98, 107, 130, 139, 148, 184, 199, 200, 204, 207, 214 II 36, 88

Javelin 8 rocket II 123, 124
Javelin 8–45 rocket II 128

K_ε and K_σ functions I 149, 150
K_p, index I 17 II 135
kinetic attenuation I 94 II 113
kinetic correction to refractive index II 160
kinetic energy density II 125, 126
kinetic equation I 29 II 3, 19, 44, 87
kinetic instability I 106
kinetic refractive index II 101
kinetic theory I 2, 147ff, 236, 247
knee in outer ionosphere II 160
Korteweg–de Vries equation I 210
Kramp function I 47, 90, 91 II 34, 48, 58, 59, 78, 88
K 9M 26 rocket II 154

L value I 15, 16, 20, 24, 117
laboratory measurements, body in plasma II 24, 39, 42
laboratory plasma experiments II 191
Landau damping (Landau attenuation) I 34, 83, 86, 89, 118, 133, 196 II 123, 133
Landau excitation I 82
Langmuir electron oscillations I 9
Langmuir frequency, electrons I 13, 57, 58, 175, 199, 207, 230, 255 II 76, 178, 180, 184, 185, 191ff, 203, 204, 208, 214
Langmuir frequency, ions I 8, 13, 65 II 99, 128, 142
Langmuir ion oscillations I 98
Langmuir oscillations I 114, 118, 192, 197, 212 II 101, 191, 196, 198, 199, 201
Langmuir resonance I 50, 228
Langmuir–Tonks waves I 98,
Langmuir waves I 84–6, 96, 132, 175, 193, 204, 236 II 96, 202
large body I 7 II 46, 47, 52, 57, 66ff
Larmor radius, electrons I 2, 7, 36 II 59
Larmor radius, ions I 2, 7, 36 II 3, 6, 53
Larmor resonance I 224
lateral deviation I 143, 146
LF, *see* low frequency
lightning II 102, 110, 159, 167, 168
linear polarization I 51, 54, 55, 60, 66, 73
linearization II 3
lion's roar LF waves II 154ff, 220
lobes in scattering function II 78, 86
lobes in wake structure II 30, 50, 51, 55, 75, 89
longitudinal Langmuir waves I 107, 110 II 201
longitudinal VLF waves II 136
longitudinal waves I 85, 198ff II 96, 185
longitudinal waves transformed to transverse II 203
Lorentz force I 29
loss cone II 106
low frequency (LF) I 93 II 98, 141, 142, 148, 159 (ch. 18), 219, 220
lower-hybrid frequency I 64, 66, 67, 69, 93, 103, 136, 194, 199, 224, 225, 241, 243 II 96, 134, 135, 138ff
lower-hybrid resonance I 9, 48, 50, 56, 61
luminosity (airglow) I 163
Luxemburg–Gorki effect I 135

Mach cone II 30, 88
magnetic activity I 15, 17
magnetic antenna II 120–2, 124–8, 141, 142, 155, 157, 169, 198, 207
magnetic conjugate point on Earth II 97, 104, 111, 159, 163–5, 169
magnetic field, effect on particles near body II 26ff
magnetic field of Earth I 24 II 99
magnetic pressure I 94, 143
magnetic storm I 19, 20 II 108–10, 114, 116, 128, 129, 142, 148, 173, 174, 189, 195, 203, 211, 212, 218, 221
magnetoacoustic transverse wave I 237
magnetoacoustic wave I 56, 62, 63, 127, 236 II 96
magnetobraking I 33, 82
magnetohydrodynamic waves I 94, 101ff
magnetopause I 5, 10, 11, 24, 213 II 106, 169, 197, 204
magnetosheath I 5, 11, 24 II 122, 155, 156, 158, 197, 212, 217, 219, 220
magnetosphere I 5, 11, 217 II 102, 108, 149, 167, 187ff, 217, 221, 222
magnetospherically reflected (MR)

whistlers II 150, 152
magnetotail II 205, 208, 209, *see also* tail, geomagnetic
Mandel'shtam–Brillouin combination scattering I 204
Mariner 4 satellite II 108, 217
mass, complex I 37, 71, 77
mass, effective, of ion mixture I 69
mass spectrometer I 17 II 140
maximum of refractive index I 72, 78
Maxwell–Boltzmann distribution II 3
Maxwellian distribution function I 30, 31 II 92, 99
Maxwell's equations I 29, 261
mean free path I 2 II 19, 75, 91
mean free time (electron) I 130
meteors II 12
micropulsations, magnetic II 101
midfrequency noise II 213
minimum of refractive index I 72, 78
mixing type instabilities II 100
modified Alfvén wave I 53, 58, 63, 64, 93, 124, 127, 228, 237 II 104
modified ion-acoustic oscillations I 196
modifying the ionosphere I 142 II 97
modulation index I 212
momentum conservation in collisions I 38
Morse code dashes II 162, 164, 165, 168
MR whistlers, *see* magnetospherically reflected whistlers
multicomponent plasma I 37, 39, 46, 65, 73 II 5

NAA radio transmitter II 162, 167
narrow-band hiss II 122, 141, 143
narrow-band plasma waves II 198
narrow-band spectra of emissions II 191–3, 195, 200
narrow-band wave packets II 140, 141, 160
NCR, *see* nonthermal continuum radiation
near-Earth plasma I 10ff, 11 (Fig. 2.1), 134
near zone of body in plasma II 5, 7, 26, 30, 31ff, 38, 66, 68
neutral approximation near body II 4, 19ff, 30, 31, 35, 36, 38, 51
neutral particles I 39, 138 II 20, 21
neutral particles near body, experimental measurement II 23, 24
neutral sheet I 24, 25 II 174–6, 189, 195
noise bands I 177 II 186, 195, 206
non-ducted waves II 149ff
non-Maxwellian distribution I 246 II 193
nonisothermal plasma I 200, 204, 207, 237 II 23, 30–2, 38, 39, 46, 50, 88, 91
nonisothermal sound I 98, 197, 207
nonisothermal velocity of sound I 8, 101, 118 II 6
nonisothermality factor I 137
nonlinear differential equations I 143
nonlinear effects I 130 (ch. 7) II 5, 98, 101, 102, 178, 179
nonlinear instabilities I 128
nonlinear Landau damping I 132
nonlinear refractive index I 215
nonlinear waves I 204ff
nonlinearity parameter I 208, 209
nonthermal continuum radiation (TNCR) II 204ff, 208ff
NPG radio transmitter II 162
NSS radio transmitter II 168
nu whistlers (ν whistlers) II 150, 153
numerical solution for particle motions II 31

oblique scattering I 180ff
OGO 1 satellite II 89, 152–5, 169, 170, 172, 207, 208, 212, 219, 222
OGO 2 satellite II 120, 124, 140, 141, 218, 223
OGO 3 satellite I 1,20 II 135, 154, 161, 169, 219
OGO 4 satellite I 22 II 120, 140, 141
OGO 5 satellite I 16 II 109, 110, 116, 122, 129, 130, 133, 134, 141, 142, 145, 154, 156, 157, 171, 173, 187–90, 197, 202, 203, 219, 221, 222
OGO 6 satellite II 139, 140, 147–9, 168
Omega (radio navigation system) II 162, 167, 168
opacity, regions of I 70
opaqueness frequency bands I 235
opposite directions of wave and group velocity I 236
orbits of particles II 63
ordered velocities I 40
ordinary fast wave I 56, 230, 231
ordinary ion wave I 62
ordinary wave I 48, 53, 73, 143, 161, 167, 168, 175, 178, II 110, 115
outer ionosphere I 67, 70, 73, 130, 142, 155, 156, 161, 239 II 107, 112, 117, 123 (ch. 17), 160, 179ff, 217–20
OVO 3 satellite II 141
OV3 3 satellite II 126, 127
oxygen ions, atomic I 65, 230 II 36

P 11 satellite II 124–6, 218
packets of waves II 90, 142, 179
parabolic beam I 190
parametric decay instability I 182, 192 II 98, 179
parametric excitation I 198
parametric instability I 128, 160, 182, 193, 198 II 185

parametric resonance I 193, 199
parametric type nonlinearity I 135, 192ff
partial reflection I 163
particle distribution near body II 62ff, 73, 92
particle flux I 106, II 137, 138
PC magnetic pulsations II 109, 217, 223
PC-1 II 109, 110, 217
PC-3, PC-4, PC-5 II 109ff
pearl-type micropulsations II 103, 206
pencil beam I 169
pendulum I 193
periodic structure near body in plasma II 27, 28
permittivity of plasma I 134, 189, 205 II 75
permittivity tensor I 28, 35, 37ff
perturbation method I 253 II 76
perturbation of electron density I 157–9 II 75, 88
PH, *see* plasmaspheric hiss
phase-modulation, phase modulation index I 187
phase perturbation I 185
phase velocity I 28
photoemission II 12, 92
Pioneer 5 satellite II 108, 217
Pioneer 8 satellite II 130, 131, 201, 202, 218, 222
Pioneer 9 satellite II 130, 218
pitch-angle diffusion II 167
pitch-angle distribution, electrons II 172
pitch-angle scattering II 100, 168
plane inhomogeneity I 252
plane-stratified plasma I 247, 251ff
plasma diagnostics II 62
plasma dispersion function, *see* Kramp function
plasma flow I 29, 31
plasma flow velocity II 16
plasma frequency I 8, 18, 50
plasma instability I 21
plasma instability near body II 4, 87 (ch. 14)
plasma lines I 175–9, 181, 183
plasma oscillations I 2, 8, 21, 33, 60, 97, 106, 182 II 10, 99ff
plasma parameters, near-Earth and interplanetary I 12, 13 II 6
plasma temperature I 134, 135
plasmapause I 5, 10, 11, 14–17, 22, 24, 239, 241 II 100, 106, 114, 128, 129, 133–5, 148, 160, 162, 163, 166, 169, 176, 186, 187, 204, 208, 212, 218
plasmasphere I 5, 11, 15ff, 22, 24, 243 II 133, 135, 136, 153, 154, 162, 163, 166, 167, 191, 193, 217, 219
plasmaspheric hiss (PH) II 133ff, 219
plasmatrough I 18 II 169, 172, 187ff, 196, 204ff, 208ff, 214, 221, 222
plate in plasma stream II 26, 27
point body I 7
point source I 260ff
Poisson equation I 29, 31, 208, 214 II 3, 31, 44, 88
polar cap I 22 II 101, 109, 122, 195, 217, 220
polar cusp II 121, 122, 142, 195, 203, 204, 214, 219, 221
polar HF radiation II 100
polar hiss II 137, 139, 140, 148, 176
polar ionosphere II 136, 139, 147, 156
polar magnetosphere II 138
polar zone observations of hiss II 121, 122, 124, 156–8
polar zone (polar region) I 22 II 137, 140, 143, 148, 213–15, 217, 219, 220, 223
polarization coefficient I 53, 73
polarization of wave I 53ff II 135, 214
polarization, random II 135
polarization reversal I 54, 55, 60 II 117
postmidnight chorus II 173, 174
potential barrier I 213
potential distribution near body II 44–6, 63, 68, 73
potential of body in plasma II 6, 10, 11, 13ff, 62
potential of body in plasma, experimental measurement II 15, 16
potential well I 133
power grid harmonics II 166
power spectrum of hydromagnetic whistler II 105, 106
Poynting vector I 192, 216, 247ff, 252ff II 119
Poynting vector, time average I 218
precipitation of electrons II 122, 168, 172, 215
pressure and nonlinearities I 135, 205
pressure, gas kinetic I 94, 143
pressure, magnetic I 94, 143
pressure on electrons I 213ff
probe measurements of plasma density II 203
proton bursts II 122
proton density I 15, 16, 18
proton flux I 20, 21 II 176
proton gyrofrequency II 108, 110, 117, 123, 124, 155
proton gyrofrequency harmonics II 123, 124
proton gyroresonance II 110, 123
proton plasma I 268 II 36
proton ring current II 129

proton whistler II 110–2
protonosphere II 149–51
protons I 65, 230
protons, suprathermal streams II 130
pulsations of magnetic field II 108ff, 217, 223
pulse transmission II 178–80
pulse type emission II 132, 144

quasi-electrostatic waves II 137
quasi-equilibrium plasma I 207
quasi-isotropic propagation I 228
quasi-longitudinal propagation I 51, 222, 224, 248ff, 253
quasi-monochromatic packet I 216
quasi-monochromatic waves II 102, 162
quasi-neutral approximation II 45, 76
quasi-neutral plasma II 35
quasi-periodic chorus II 174
quasi-periodic ELF waves II 109
quasi-periodic structure of wake II 91
quasi-periodic transverse wave II 110
quasi-perpendicular propagation I 51
quasi-plane waves I 216
quasi-static electric fields II 124
quasi-stationary body II 62 (ch. 12)
quasi-transverse propagation I 46, 51, 230, 255, 256

radiation belts I 19 II 100, 135, 167, 210
radio signals as trigger II 140
radio wave scattering II 75 (ch. 13)
radiotelegraph signals II 162, 168
rarefaction in wake II 7, 8, 29, 30, 31, 36, 39, 72, 73, 88
rarefaction regions near body II 31
ray paths II 135
ray tracing I 238
recombination coefficient I 154
reflection of ELF waves II 118, 119, 121
reflection of particles I 10, 11 II 68, 69
reflection of whistlers II 120, 150
reflectrix I 192
refractive index I 28, 35 (ch. 4), 70ff, 82ff, 147 II 115, 216
refractive indexes, equal I 51, 54, 55
relaxation time for scattered intensity I 178
relaxation time for temperature I 137, 154
remote resonance II 183
resonance I 32, 87
resonance absorption I 169
resonance attenuation coefficient I 94, 95
resonance branches I 194ff
resonance cone I 262–4, 267, 269 II 146
resonance excitation I 178 II 90
resonance frequencies I 68, 71, 103
resonance oscillations of plasma II 96
resonances (spikes) in topside sounders II 179ff
resonant electrons I 133
resonant interaction of waves and wake II 90
reversal of sense of polarization I 66, 81
reversal points of whistler ray I 243
reversed Storey cone I 269, 270
ring current in magnetosphere II 114, 128, 129, 148, 155
ring electron distribution II 189
rising tone (riser) II 161, 162, 165, 166, 168, 173
Roberval station, Quebec II 163, 166
rockets II 97, 123
rough surface of body II 81
run-away electrons I 141

S 3–3 satellite II 124–6, 217
satellite I 134 II 83, 89, 123
saturation of whistler enhancement II 165
saucer-shaped emissions (SSE) II 142ff
scattering cross-section I 172–4, 178 II 75–9, 84
scattering function II 77–9
scattering of electrons II 91
scattering of particles II 11
scattering of radio waves II 10, 75 (ch. 13)
second-harmonic gyroresonance I 85, 88, 180
self-action I 151, 167, 184ff
self-action factor I 185
self-focusing instability I 167
self-focusing of beam I 184, 188, 190
self-modulation I 151, 184
shadow in particles near body II 73
shadow zone for waves II 84, 85
sheath near body surface II 66–8
shock, collisionless I 206, 213 II 198
shock front II 101, 142, *see also* Earth's bow shock
shock wave I 206, 213 II 30, 50
shot effect II 208
side scattering I 175
Siple station, Antarctica II 163–7
sky maps of scattering I 164–6
slow electron-acoustic wave II 159
slow ELF waves II 127, 128
slow magnetoacoustic wave I 102
slowly moving body II 69ff
small body I 7 II 41, 47, 55–60, 62ff
smoothly-varying plasma I 238
solar-flare II 201, 203
solar wind I 5, 7, 9–11, 14, 213 II 96, 99, 130–2, 187ff, 205, 218, 221, 222
solitions I 135, 206, 209–12

sound velocity I 8, 210
sound waves I 197
source size (AKR) II 215
space probes I 134
spatial dispersion I 32, 61, 63, 82ff, 103, 222, 236
spatial inertia I 32
spatial inhomogeneity of electrons I 131, 132
spatial nonuniformity I 107, 205
spectra of ELF waves II 114, 115
spectra of scattered waves I 175ff, 180, 181
spectrogram (proton flux) I 21
spectrogram (swept frequency radar) I 170, 171
spectrogram of waves II 91, 104, 105, 111, 112, 116–21, 123, 124, 128, 132, 138–56, 159, 161, 162, 165–77, 188–90, 195, 196, 206, 208, 209
spectrograms in colour II 119, 144–6, 150, 151
spectrum analyser, high time resolution II 173
specular reflection of particles II 11, 20
specular reflection of waves II 78
specular scattering I 175, 182
spherical body II 20ff, 25, 28, 36, 39–56, 68–72, 77, 81–4
spherical coordinates I 238, 266
spherical waves II 83, 84, 86
spikes II 180ff
sporadic layer I 161–4 II 97
square plates in plasma II 45, 46
SSE, *see* saucer-shaped emissions
stabilization time for temperature I 137, 160
standing waves I 184
stationary body, large II 66ff
stationary body, small II 62ff
steady-state process I 207
steepening of wave front I 206, 211
stimulated emission (ASE) II 161ff
Storey cone I 262–4, 269
streams of electrons II 197, 210
streams of particles II 6
striction type nonlinearities I 136, 215
strong pump waves I 128, 135, 192
strongly ionized trapping layer I 255
structure of Earth's magnetic field I 24
structure of plasma near Earth I 11
subprotonic whistlers II 150
superposition, principle of I 134
supersonic plasma flow II 6, 19
supersonic speed I 209
suprathermal electron beams II 147
suprathermal proton stream II 130
surface materials II 13
tail, geomagnetic I 24, 25 II 96, 100, 101, 108, 137, 138, 174–6, 187ff, 195, 208, 209, 214, 220–2
tail of velocity distribution I 86 II 100
temperature distribution near body II 16, 17, 92
temperature instability II 106
temperature of plasma I 134, 135
temporal decay rate I 196
temporal inertia I 32
temporal variation of electron temperature I 146ff
tensor elements (permittivity) I 81, 83ff, 214, 215, 221, 253, 261
terrestrial kilometric radiation (TKR) II 208ff
thermal energy in plasma regions II 216ff
thermal motion I 83ff, 236
thermal velocity of electrons, ions I 6, 39 II 160
third branch of refractive index I 84
three (ion) component plasma I 221ff, 230ff, 241, 269
threshold field I 200–3
TKR, *see* terrestrial kilometric radiation
TNCR trapped nonthermal continuum radiation *see* nonthermal continuum radiation, continuum radiation
topside sounding II 179ff
total internal reflection I 253–5, 258–60
trains of signals II 103, 104, 106
trajectories I 216 (ch. 8), 238
transition frequencies I 268, 269
transport of particles I 152
transverse electromagnetic waves I 109 II 203ff
transverse propagation II 151, 152
transverse wave I 193, 202ff II 96, 120, 200
transverse whistlers II 150, 174–6
trapped particles II 63
trapped whistlers I 243
trapping boundary II 144
trapping cone I 223, 224, 234
trapping of particles by waves II 102
trapping of waves I 62, 251ff II 96, 149ff, 203ff, 212
trapping of waves by magnetic field I 243ff II 157, 160
triggered emission II 140, 162ff
trough region of magnetosphere, plasmatrough I 17, 18 II 204ff
turbulence I 22
turning of Poynting vector I 247ff
two component plasma, electrons and one ion I 56ff
two (ion) component plasma I 76ff
two-stream instability II 101, 130, 198, 199

ULF (ultra low frequency) II 99
ultrashort radio waves I 163
unguided rays, waves I 241ff II 154
units I 3 II 216, 218
universal function for ion density perturbation II 49–53, 55, 57, 58
unstable plasma I 133 II 99
upper hybrid frequency I 179, 194, 198, 228 II 96, 180, 181, 193, 203, 204, 209, 215
upper hybrid resonance I 49, 50, 56, 61, 103, 180, 236 II 180, 181, 186

V, inverted-V electrons II 137, 215
V-shaped emissions II 142ff
Vanguard 3 satellite II 159
velocity (thermal) of electrons, ions I 6
very low frequency (VLF) I 48, 52, 62, 65, 70, 93, 118 II 87, 98, 123 (ch. 17)
very low frequency resonance oscillations I 62
very small irregularities I 169, 179
virtual height I 161
virtual range II 180
viscosity I 43
VLF, *see* very low frequency
Voyager 1 and 2 satellites II 201, 218, 222

wake of body in plasma II 7–9, 16, 17, 19, 20, 26, 27, 30, 32, 33, 38, 42, 43, 46, 47ff, 75 (ch. 13), 87 (ch. 14)
walking-trace whistlers II 151
warm plasma I 82 (ch. 5)
wave guidance I 27, 51
wave normal, measurement of direction II 172
wave packet II 90, 156, 164, 174
wave–particle interaction I 33 II 102, 135, 164, 178, 179
wave shape I 209
wave surface I 221
wave–wave interaction II 102, 178, 189, 200
weakly ionized trapping layer I 252
whistler I 56, 63, 92, 93, 239ff II 150
whistler mode I 63, 66, 93, 224, 225, 239, 247, 263, 266 II 97, 136, 144, 157, 159 (ch. 19), 219, 220
whistler precursors II 151
whistler ray trajectories I 239–46 II 135, 150, 151, 153
whistling atmospheric I 22, 24 II 90, 91, 103, 111, 119, 138–40, 150, 159ff

X-ray bursts II 168

Z mode I 268, 269
zeros of refractive index I 50, 70
zones around body in plasma II 4ff, 21ff, 31ff, 47ff, 57
zones of altitude I 6, 12, 13